LASER MATERIALS
PROCESSING

MATERIALS PROCESSING – THEORY AND PRACTICES

VOLUME 3

Series editor

F.F.Y. WANG

NORTH-HOLLAND PUBLISHING COMPANY
AMSTERDAM · NEW YORK · OXFORD

LASER MATERIALS PROCESSING

Edited by

MICHAEL BASS
Center for Laser Studies
University of Southern California
Los Angeles, California, USA

1983

NORTH-HOLLAND PUBLISHING COMPANY
AMSTERDAM · NEW YORK · OXFORD

©NORTH-HOLLAND PUBLISHING COMPANY, 1983

ISBN 0 444 86396 6

Published by:
NORTH-HOLLAND PUBLISHING COMPANY
AMSTERDAM · NEW YORK · OXFORD

Sole distributors for the USA and Canada:
ELSEVIER SCIENCE PUBLISHING COMPANY, INC.
52 VANDERBILT AVENUE
NEW YORK, N.Y. 10017

Library of Congress Cataloging in Publication Data
Main entry under title:

Laser materials processing.

  (Materials processing, theory and practices; v. 3)
  Includes index.
  1. Lasers—Industrial applications. I. Bass,
Michael. II. Series.
TA1677.L365 1983   621.36'6   82-8209
ISBN 0-444-86396-6

Printed in The Netherlands

INTRODUCTION TO THE SERIES

Modern technological advances place demanding requirements for the designs and applications of materials. In many instances, the processing of materials becomes the critical step in the manufacturing processes. However, within the vast realm of technical literature, materials processing has not received its proper attention. It is primarily due to the lack of a proper communication forum. Since the materials processing is intimately concerned with specific products, those who are experts have no need to communicate. On the other hand, those who are involved with a different product will develop, in time, the technology of materials processing when required.

It is the objective of this series, Materials Processing – Theory and Practices, to promote the dissemination of technical information about the materials processing. It provides a broad prospective about the theory and practices concerning a particular process of material processing. A material process, intended for one technology, may have an applicability in another. This series provides a bridge between the materials engineering community and the processing engineering community. It is a proper forum of dialogues between the academic and the industrial communities.

Materials processing is a fast-moving field. Within the constraints of time and printed spaces, this series does not aim to be encyclopedic, and all-inclusive. Rather, it supplies an examination of material processes by the active workers. The view will be, by necessity, subjective. But the view will include both near-term and long-term prospectives. It is the fondest hope of this general editor that the volumes in this series can serve as first reference books in the field of materials processing.

Franklin F.Y. WANG
Stony Brook, New York

PREFACE TO VOLUME 3

When the laser was demonstrated in 1960, it caused a major flurry of speculation about its potential uses. Such speculation far outstripped the capabilities of the then existing lasers and it was not long before lasers became known as "solutions in search of problems". Given time to accomplish the necessary basic research, to engineer more effective and reliable lasers and to explore applications suited to laser technology this image changed dramatically. In so far as materials processing was concerned, 1965 was the turning point. During that year researchers at the Bell Telephone Laboratories invented the Nd:YAG and CO_2 lasers. These two devices went on the become the mainstays of materials processing applications because they can produce high average and peak powers and because they were soon engineered into reliable devices suitable for manufacturing environments. It is not clear whether the materials processing community's interest in these lasers spurred their development or vice versa. What is clear is that today Nd:YAG and CO_2 lasers are routinely used to process a bewildering variety of materials in many interesting ways.

This book is designed for both the experienced laser user and the engineer or scientist as yet unfamiliar with the subject. I have written a very brief description of Nd:YAG and CO_2 lasers and laser systems to introduce the newcomer to laser terminology and to provide enough background material on lasers to reduce one's hesitation to employ these devices. The bulk of the book is devoted to discussions of some of the most important materials processing activities in current use or development. Concern for generality and the limitations of space caused me to select articles about processes in general and not specific treatments. Thus the articles are descriptive of the general mechanisms and principals responsible for the process achieved. Articles about specific laser materials processing tasks such as writing on video discs or cutting cloth, while very important, are in the literature where interested parties can find them. Since most laser materials processing articles have been concerned with the interactions of CO_2 lasers with metals

I have included one entire article concerning Nd:YAG laser applications. In addition I have made sure to include in this book an article about the very important process of laser annealing of semiconductors. While most of the material treated in this book is devoted to the laser processing of metals, there is in the chapters on cutting, welding, machining and drilling some mention of processing nonmetals such as rubber, plastics, ceramics, paper and wood. Chapter 9 entitled "Considerations for Lasers in Manufacturing" is essential to such a book as this since after one has read the article or articles which cover his area of interest and has decided to consider laser materials processing, he must know what steps should be followed in order to successfully implement his decision.

Laser materials processing takes advantage of the high power and good focusability of laser beams to deposit a great amount of heat into a selected region of a target. The irradiated material heats due to optical absorption and the heated material may then cool mostly by conduction into the bulk, and, in small part, by radiation and convection into the surrounding atmosphere. When the material is heated past the point of making a phase transition, one obtains a laser induced materials process. This may be hardening or annealing of metals, vaporization to remove material (i.e., to cut, drill or shape), or melting and rapid solidification to form new phases of alloys.

The absorbtivity and thermal properties of materials are strongly dependent upon temperature and so the thermodynamic problem of laser heating is nonlinear. In addition the problem, in its most general form, is three dimensional and therefore extremely difficult to solve in closed form. Analyses of laser heating are presented in most of the chapters of this book. In each case, the analysis is simplified to suit the particular process described. For example when discussing laser hardening of metals the analysis does not have to include consideration of melting or vaporization.

As might be expected, optical absorption is critical to any laser materials processing endeavor. Concern for this material's property has resulted in some debate over which laser to use in any given case. For example, metals are in general more absorbtive at 1.06 μm, the Nd:YAG laser wavelength, than they are at 10.6 μm where the CO_2 laser operates. Ignoring all other considerations one would then choose a Nd:YAG laser to process metals. However, the total power that can be derived from a CO_2 laser far exceeds that of a Nd:YAG laser and many simply, cheaply and reliably applied coatings are available to enhance a metal's 10.6 μm absorptivity. Thus, into the choice of laser, one must fold such other processing considerations as desired processing rate, metal thickness and the acceptability of requiring an absorbtive coating.

Materials with high thermal conductivity are often hard to process with CW lasers, Instead, short duration laser pulses are used to achieve the desired heating before too much heat is lost to the bulk of the sample. This question is of major concern when processing such metals as Cu and Al and requires a careful examination of the irradiation temporal waveform.

Nd:YAG and CO_2 lasers have established clearly defined roles in a great majority of laser materials processing applications. They will retain their respective positions because of the quality of their engineering and the large body of supportive data to back up their use. However, certain less traditional applications such as semiconductor annealing, selective chemistry and laser induced deposition may require different sources. Thus, there will be roles for more recently developed lasers in the future of laser materials processing.

It is clear that in materials processing lasers are no longer searching for problems. Today, problems are seeking out the laser as in the case of semiconductor annealing. This book provides the information necessary to evaluate laser materials processing and to make valid decisions concerning the implementation of this advantageous technology.

Michael BASS
Los Angeles, California
March, 1982

CONTENTS

CHAPTER 1

LASERS FOR LASER MATERIALS PROCESSING

MICHAEL BASS

University of Southern California
Los Angeles, California 90007, USA

Laser Materials Processing, edited by M. Bass
© *North-Holland Publishing Company, 1983*

Contents

1. Introduction

This chapter is a very basic introduction to lasers for laser materials processing for a scientist or engineer who is comtemplating using this new technology. As such it does not include a number of details about the lasers and the techniques for their proper utilization. Instead the chapter attempts to provide the reader with enough information so that he may meaningfully discuss his laser requirements with potential suppliers. The reader familiar with lasers in general and Nd:YAG and CO_2 lasers in particular may wish to proceed to other parts of this book.

A laser is a source of optical frequency radiation that is controllable and intense. These properties make it an appropriate source to use in producing desired material surface modifications. The three principal components of a laser are sketched in fig. 1.1 and are, 1, the amplifying medium, 2, the means of exciting the medium to its amplifying state, and 3, an optical resonator. The amplifying medium determines the laser wavelength and the type of excitation that is required. A medium is potentially useful as an optical frequency amplifier if the population of its energy levels can be altered so that there are more emissions than absorbtions per unit time. When so altered the population is said to be "inverted". The means for creating the inverted population can be irradiation with incoherent light which is optically absorbed in order to create excited atoms or the passage of an electrical discharge through gaseous media to produce collisionally excited atoms or molecules. The former is the process used for exciting an Nd:YAG crystal to lase at 1.06 μm and the latter is used to excite a CO_2 gas laser at 10.6 μm. These two are the most commonly used materials processing lasers and will be discussed in some detail in this chapter.

The light emitted by an amplifying medium is not by itself monochromatic and unidirectional as it is from a laser. These properties are imposed by placing the medium in an optical resonator. Such a device causes light emitted parallel to its axis to be reflected back and forth through the amplifier. (This light is said to oscillate in the resonator.) When the amplifier gain equals the round trip losses in the resonator the combination of amplifier and resonator is at the threshold for lasing. Upon reaching or exceeding threshold the light in the excited resonator traveling parallel to its axis is amplified many times. The fraction removed on each pass by transmission through the output coupler results in a laser beam. Such multiple use of the amplifier makes possible highly energetic laser outputs. In addition only those wavelengths and spatial distributions that are exactly reproduced in a round trip through the resonator fully satisfy the conditions for oscillation. Since the process of amplification is stimulated emission in the gain medium, these are the only optical fields present in the output. As a result the laser beam is unidirectional, virtually monochromatic and is distributed in a spatial pattern that is determined by the design of the resonator. In summary then a laser is a source of light which:

(1) is nearly monochromatic;
(2) is nearly unidirectional;
(3) has a spatial distribution imposed by the optical resonator; and
(4) can be very energetic.

A collimated beam of light with any particular spatial distribution can be delivered to almost any position. This is accomplished by propagating the laser's output to the target in a beam handling system (a combination of reflective and transmissive optics) designed for the task. In addition to employing optics that can handle intense laser beams without failure, the beam handling system must contain sufficient degrees of freedom in posi-

1. A means to *amplify light.*

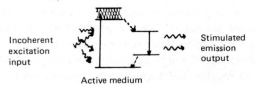

2. A means to *excite* the amplifying state.

3. A means to provide *optical feedback.*

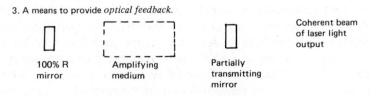

Fig. 1.1. Schematic of the principal parts of a laser.

tioning and aiming the beam to allow all the conceivable target area to be irradiated.

The effect that one desires to achieve on the target determines the distribution of the beam that is finally delivered. For example, if one wants to drill holes, cut or weld materials then one wants the most tightly focused beam possible. This requires a TEM_{00} or lowest order gaussian beam distribution and low f/number optics. On the other hand the most effective surface heat treatments (i.e.: hardening, alloying, cladding, etc.) require a more uniformly irradiated area. In this case a higher order mode or a multimode laser and high f/number optics are needed.

Though a variety of beam delivery systems have been designed and are in operation, all but a few, very simple systems are designed for a specific set of user requirements. These are often dictated by the workpiece configuration, the workpiece mass and the area to be treated. For example, optical simplicity and reliability are much easier to design and maintain if the optical components remain fixed and the workpiece is translated past the beam. This is feasible if the workpiece is small and lightweight and the area to be irradiated is readily accessible. Examples of this combination can be found in the laser welding of circuit encapsulations or in the scribing of Si wafers. In each case the workpiece can be translated under the beam at rates which result in economically feasible processing. When the workpiece is large and complex as in the case of welding turbine blades into turbine discs, it is more appropriate to translate the focused light beam over the target. Sometimes, as in laser assisted machining and laser machining, it is preferable to move both the workpiece and the light beam in order to cover the area required.

In order to quantify a laser-materials interaction process one must determine several properties of the laser beam. The laser wavelength determines both the optics used to deliver the light and the diagnostic devices used to measure it. Power meters capable of accurately measuring several kW of laser power are available. These devices are either air or water cooled calorimeters with a thermocouple or other temperature sensor attached to measure the laser induced heating of a calibrated absorber.

The laser's temporal waveform and its temporal stability must be determined using a proper fast detector. At 10.6 μm this detector is either a cryogenically cooled Au doped Ge crystal, a pyroelectric or a photon drag detector. These are mentioned in order of increasing speed. In general, a commercially available Au doped Ge detector will be sufficient for determining the temporal properties of a high power CW CO_2 laser. These detectors are quite sensitive and with response times $< 10^{-8}$ s are certainly fast enough to determine temporal fluctuations that might affect most processes.

Measurement of high power laser beam spatial distributions has generally been accomplished by burning a target (i.e. wood, plastic, thermosensitive paper, etc.) and inspecting the burn. While this is quite adequate for most casual requirements a better measurement is necessary for adequate experimental research and process parameter measurements. Thus various means of scanning the intense beam with small apertures or with IR sensitive TV systems have been developed.

The beam spatial distribution and divergence (obtained from measurements of the beam diameter at different places along the beam path) determine the size of the area to which a given optical system can focus that beam. This then determines the irradiation intensity and the volume of the material directly laser treated.

During the development of a laser materials interactions process the measurements mentioned above must be carried out. However, when a process is established it is only necessary to monitor the laser power and temporal stability continuously. Occasional checks on the beam spatial distribution will be adequate if the laser and beam handling optics are mechanically stable.

The interaction of laser light and matter is often affected by the ambient gas at the workpiece surface. For example, in laser cutting steels and titanium a jet of oxygen gas delivered coaxially with the laser beam results in much faster cutting. This is due to the exothermic reaction of the heated metal with the gas and the mechanical process of blowing away the molten material. However, when welding titanium or titanium containing alloys it is essential to keep oxygen away from heated material. Control of the ambient atmosphere in laser materials processing is obtained through judicious use of gas jets and shields. In some processes, the atmosphere itself provides the materials which react with the surface layer. In others, the ambient may be selected to suppress undesired reactions or to direct the reactants to the irradiated area. It also is often desirable to use a flow of gas to protect the final optical element from damage due to backstreaming ejecta from the workpiece.

The combination of intense laser light and an adequate delivery system is useful as a directed energy source. It can deliver energy to a target with no mechanical contact and no source related workpiece contamination. Laser energy can be highly localized so that only selected sites are processed. The only requirements that laser processing places on the ambient atmospheres are that it should not absorb the laser light and not contaminate the irradiated material. In addition, laser irradiation can produce extremely rapid temperature changes which can result in valuable new properties of materials.

2. The Nd:YAG laser

2.1. Properties of the amplifying medium

The Nd:YAG laser is one of the most versatile and important lasers yet developed. Its excellent laser potential derives from a combination of trivalent neodymium ions (Nd^{3+}) with the excellent optical and thermal properties of the host crystal yttrium aluminum garnet (YAG). When excited by light from an electric discharge lamp Nd^{3+} ions in YAG in a proper optical resonator can cause lasing at a number of wavelengths. The most efficient laser operation is obtained when the cavity is designed for 1.0641 μm and so this is the wavelength used in Nd:YAG lasers for materials interactions.

Pure YAG is a transparent colorless crystal. When doped with ~1% Nd^{3+} ions and when it is free of contaminants and color centers it has a clear light blue color. The visible absorption bands of the Nd^{3+} ions result in this coloration. However, it is the unseen absorbtion bands near 0.81 and 0.89 μm that are primarily responsible for exciting the $4F_{3/2}$ level of Nd^{3+} so that 1.0641 μm lasing can occur. The lower level of the laser transition, the $^4I_{11/2}$ level, is 2000 cm^{-1} above the ground state and so at room temperature is nearly empty. This means that the Nd^{3+} laser operates on a so called "four level" system and that the threshold for lasing should be low. In fact the lasing threshold will be determined by the cavity losses and internal losses in the laser rod due to impurities and scattering. In certain Nd:YAG lasers designed for use as range finders the threshold can be at less than 1 J of input to the flashlamp.

2.2. Means of excitation

The Nd:YAG laser can be efficiently excited by optical radiation from xenon or krypton filled discharge lamps. In general Kr is more useful for low current density discharges (i.e. CW) than is Xe. In pulsed operation Nd:YAG lasers are usually pumped by Xe filled flashlamps. This allows one to take advantage of better spectral matching between the high current density Xe discharge light and the Nd absorption lines in YAG.

The most efficient optical pumping cavity for a solid state, optically pumped laser is the close coupled cylindrical ellipse. This type of pump cavity is generally designed to minimize the distance between the axes of the lamp and the laser rod. As a result, the direct coupling between the lamp and the rod is maximized. The light radiated in directions that are not directly intercepted by the laser rod is reflected by the gold plated elliptical cavity walls and focused into the rod. By this combination of direct and indirect pumping the close coupled ellipse cavity can produce both low lasing threshold and acceptable output efficiency. As a result the ellipse is in general use in both pulse pumped and CW Nd:YAG lasers.

In some high power CW lasers the manufacturers employ a double ellipse pump cavity. The double ellipse employs two ellipses with a common focus. The laser rod is placed along this focus and one lamp along the other focus of each ellipse. In this manner higher total pump power can be delivered to the laser rod in order to obtain higher total outputs.

2.3. Cooling requirements

The laser rod, pump lamp and pump cavity must be cooled to carry off the input energy which is not used to produce laser light. In fact Nd:YAG lasers convert at most 3% of the electrical input to the pump lamp into useful laser energy. Thus a 100 W CW Nd:YAG laser requires ≥ 3.3 kW of input. The cooling system must carry off the unused input energy while retaining good lamp life and optical quality in the pump and laser cavity components.

The pump cavity can be flooded with water which on flowing through cools the cavity, the rod, and the lamp. This technique is the simplest cooling scheme but has two significant drawbacks. One is the loss of efficiency due to absorption of pump light in the water and the other is reduced quality imaging of the lamp in the rod due to turbulence in the water. To overcome these problems the cavity is often cooled by a flow of water around the outside surface while at the same time the rod and lamp are cooled by flowing cooling water through an annular region between the rod (lamp) and a transparent cooling jacket. The thickness of the annulus is generally made as small as is compatible with good cooling and the desired optical properties of the pump cavity. It is essential in this to cool the lamp electrodes and the electrode holders in order to prevent lamp failure due to severe thermal loading in these regions. Thus the cooling water must be deionized to prevent the power supply from shorting out through the water.

Most manufacturers employ a closed cycle cooling water flow and a heat exchanger to cool the cooling water. These aspects of the laser systems must be examined and evaluated when considering a purchase.

It is important to note that the cooling of the laser rod results in a temperature gradient across its radius. It is hottest on axis and coolest on its circumference where it is in contact with the coolant. The thermal gradient results in mechanical stresses which along with the temperature coefficient of the rod's index of refraction contribute to a net positive lensing of the rod and to an induced birefringence. Both of these effects must be considered in designing an Nd:YAG laser for a specific mode and beam divergence. In fact, the pump light induced lensing is the only reason that an optically excited solid laser can oscillate in gaussian modes while using two flat cavity mirrors.

2.4. Manners of operation

Nd:YAG lasers have been operated in a variety of ways in order to obtain desired energies or powers and specific temporal waveforms. The principal means of obtaining a desired waveform is by selection of the pump waveform and the properties of selected intracavity switches. In one case where the laser need only produce as much energy as possible in a pulse of several hundred microseconds one uses a long pulse pump lamp and allows the laser to run free. In this manner of operation as much as 2–6 J may be obtained from a single rod with a waveform made up of many ≈ 1 μs long relaxation oscillation spikes and total duration ≈ 1 ms. With proper cooling such a long pulse laser can be operated at rates as high as several hundred pulses per second. This type of laser is useful for welding, cutting and certain hole drilling applications.

More precise work may be accomplished by tailoring the pump pulse to produce a maximum of stored energy in a time of the order of the radiative lifetime of the excited Nd^{3+} ions. If the cavity is shuttered during the time of building this population inversion and then, as the inversion reaches its maximum, is opened one can obtain up to 1 J of output energy in a pulse as short as ≈ 10 ns. The procedure for this type of operation is called Q switching since it involves switching the cavity Q from very low to very high to obtain an output. These short intense pulses are useful for hole drilling and ranging.

When run CW the Nd:YAG laser is extremely useful for a variety of materials processing requirements. However, certain others such as marking and cutting require high peak powers. Thus the CW laser often operated in a repetitively Q-switched manner. Pulses of 10–200 ns are obtained in this manner at repetition frequencies as high as 5 kHz. The average power is retained in repetitively Q switched operation while the peak power can be increased by as much as 10^3 times.

In all manners of operation one must select the desired spatial mode characteristics. Often these can be relaxed to allow for more modes and therefore more total energy. Precision processes such as cutting and trimming of electronics components may require that the laser be restricted to its lowest mode. In any case since it can be shown that the maximum heating for a given laser power occurs if the beam is circular one should require this symmetry of the laser's output. Also the beam handling system must not distort the final focused beam distribution.

2.5. Specific user concerns

The Nd:YAG laser user must be specifically concerned with selection of the proper manner of operation for his application. The chapters which follow will explain how that choice should be made. While using any laser system one must check on the condition of the optical components and, in particular, any coated surface in the optical train. Coatings for 1.06 μm of excellent quality and resistance to laser induced failure can be obtained. Nevertheless, these components are the most likely to be damaged during laser use. At 1.06 μm optical components can be made from such excellent materials as fused quartz and so are far more rugged than transmissive optics for use at 10.6 μm.

3. The CO_2 laser

3.1. Properties of the amplifying medium

The CO_2 laser operating at 10.6 μm was reported by Patel in 1964. He obtained 1 mW output in the original unit which used pure CO_2 gas excited by an electric discharge. Very soon after the laser was announced, it was pointed out that a much more efficient system of excitation could be obtained in a mixture of CO_2 and N_2. The process of collisional transfer of energy between nearly resonant vibrational states of the molecules of N_2 and CO_2 was proposed and found to increase the overall laser efficiency from by a factor of 10^3. Studies of the energy levels of the CO_2 molecule then showed that the lower level of lasing transition did not decay rapidly to the ground state. Thus, during laser operation this state's population would grow and reduce the population inversion nearly to the point of quenching lasing. To remedy this problem He was added to the discharge to collide with the CO_2 molecules and relax those in the lower laser level into the ground state. With this change the output of CO_2 laser had been increased by 10^5 in a little more than a year. Today's CO_2 lasers all employ a mixture of CO_2, N_2 and He as their active media. The variety of these devices is obtained from the type of discharges used, the manner of gas flow and the choice of the optical cavity design.

While the CO_2 laser efficiency can be quite high (theoretically it could be 40%) there is the need to cool the components of this device as well as there is in the case of the Nd:YAG laser. Most of the cooling of the active medium is accomplished outside the optical resonator. The gas is flowed through the discharge region and either exhausted or it is recycled after processing to remove contaminants. The electrodes and the discharge confinement walls must be cooled by a coolant flow and, if the laser is operated at high power, so must the mirrors.

3.2. Major CO_2 laser devices

There are three major classes of CW CO_2 lasers in general use in laser materials processing. As indicated in fig. 1.2, they are the sealed-off axial discharge laser, the flowing gas axial discharge laser and the transverse flow laser. The means of flowing the laser gas determines how rapidly the hot and therefore partially absorbtive gas can be removed from the laser resonator. Clearly faster flows can lead to higher upper limits on the power output.

The sealed-off laser has the advantage of not needing a supply of laser gas or a gas handling system. However, its output power is generally no more than 50 W and its lifetime is limited by the dissociation of the CO_2 into CO and O_2. Most materials processing requirements call for more power and longer lifetimes and so the flowing gas axial discharge laser was developed. In this device the discharge is operated parallel to the cavity axis and the gas flow is in the same direction. This is the most conventional high power CO_2 laser design. By properly extending the length of the discharge, controlling the gas flow and designing the electrodes these devices can produce CW laser powers as high as 4 kW. In fact one such laser was assembled with 280 m of discharge length and produced a 8.8 kW. The output of this laser and most other conventional lasers which exceed 1 kW total power is generally made up of two or more spatial modes. Flowing gas CO_2 lasers with outputs less than 1 kW can be prepared with stable optical resonators that can sustain efficient TEM_{00} mode oscillation.

A more rapid removal of hot laser gases from the laser resonator is achieved by flowing the gas perpendicular to both the cavity axis and the discharge. Once removed the gas can be cooled, the contaminants and dissociation products can be compensated, and the gas can be reused. These laser systems can achieve very high output powers from smaller mechanical packages than the flowing axial discharge device. However, the transverse discharge is an electron beam sustained process and therefore the transverse flow laser is electronically more complex than the conventional laser. Transverse flow lasers are most useful in applications requiring in excess of 2 kW of laser power.

Both types of discharges can be pulsed. The pulsed conventional discharge can produce pulses of ≤ 1 ms duration while a pulsed transversly

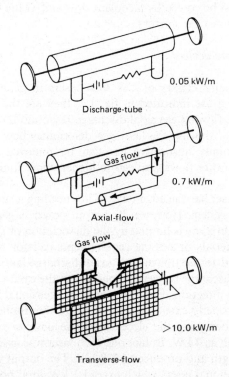

Fig. 1.2. Types of electric gas lasers showing practical limits of laser output power per meter length of amplifying medium. (From AVCO Circular #7705.)

excited atmospheric pressure (TEA) laser can produce pulses that are typically ≤ 200 ns in duration. In the latter the output consists of an ~ 100 ns gain switched pulse and if N_2 is present in the laser gas there will be an ~ 2 μs tail containing laser output due to collisional excitation. This combination of a short leading pulse and a lower intensity trailing component may be a useful waveform for certain applications. It should be noted that the TEA laser output waveform is composed of trains of very short pulses called "mode-locked spikes" which result from nonlinear interactions between CO_2 molecules and CO_2 laser light. These pulses can be detrimental to some research projects but in general are not a significant concern in laser materials processing phenomena.

3.3. 10.6 μm laser optics

Since CO_2 lasers operate at high powers at 10.6 μm the optical materials which can be used for components are severely limited. Transmissive 10.6 μm optics are made of germanium, zinc selenide, sodium chloride or potassium chloride. It is common to find polished copper or molybdenum metal mirrors used for reflective optics. Sometimes these mirrors are overcoated with gold or dielectric materials to enhance their reflectivity or to protect the surface from corrosion. Germanium and zinc selenide with appropriate dielectric coatings are often used for transmissive mirror substrates. The former is not transparent to the red (0.632 μm) light from HeNe lasers that are often used to align the cavity optics while the latter is. While this is only a convenience in aligning laser cavities, it is an important criterion in the selection of the material to be used for lenses and other transmissive optics. Very high power CO_2 laser systems employ water cooled, metal, reflective optics exclusively.

The alkali-halides (i.e. NaCl and KCl) are relatively inexpensive 10.6 μm transmissive materials for use as windows, beam splitters and lenses. However, they are hydroscopic, hard to optically polish, and require special care in handling. In short they are not well suited to other than laboratory use.

In considering beam handling optics for both Nd:YAG and CO_2 laser systems one often thinks only in terms of combinations of discrete optical components. At 1.06 μm it is feasible to use single fiber optics to deliver 100 W of power for extended periods. Such a system is in use in the treatment of gastrointestinal bleeding by laser induced photocoagulation. At 10.6 μm solid core fibers are being researched but are not as yet readily available. Flexible metal walled waveguides have been developed which can deliver over 1 kW of continuous 10.6 μm power.

3.4. *Specific user concerns*

In the selection of a CO_2 laser for materials processing one must be concerned with: (1) the power that will be required; (2) the gas handling requirements and gas usage; (3) the cooling provisions for the cavity optics, discharge chamber and electrodes; and (4) the choice of beam handling optics. These concerns involve the choice of which type of gas flow geometry one will use and the selection of either transmissive or reflective beam optics.

CHAPTER 2

LASER CUTTING

W.M. STEEN AND J.N. KAMALU

Imperial College of Science and Technology
London SW7 2BP, UK

Laser Materials Processing, edited by M. Bass
© North-Holland Publishing Company, 1983

Contents

1. Introduction

The general characteristics of the laser cutting process are sufficiently attractive and unique to explain the great breadth of applications which have or are being developed. They also suggest that many new cutting applications will develop in the near future far beyond those currently adopted. Consider, for example, the following characteristics of the laser for cutting:

(1) The cut can have a very narrow kerf giving a saving in material.

(2) The cut can have a very narrow heat affected zone (HAZ) and therefore give low thermal distortion.

(3) The cut edges can be smoother than a band saw cut edge.

(4) The cut has square edges.

(5) There is no mechanical stress on the edge other than that from the HAZ, i.e. no edge burr as from a guilotine, or stress relaxation problems as with cutting rubber etc.

(6) The workpiece does not need holding against tool drag. Workholding demands are reduced to merely a locating role.

(7) It is possible to cut through friable, brittle, soft or hard material or composites with almost equal ease.

(8) There is no tool wear (contact free energy transfer).

(9) The cut can be made equally easily in all directions, and it can be started anywhere.

(10) Blind cuts can be made.

(11) Cutting speed is fast compared with a band saw.

(12) Easily automated by numerical control.

(13) Capable of multistation operation.

(14) Low operating noise.

(15) Possible to weld cut edge directly.

(16) Possible to stack-cut plastics, fabric, etc.

(17) Tooling changes are mainly "soft".

(18) No chips, therefore saving of material and less dust problem with asbestos, fibre glass etc.

(19) The light beam is inertialess therefore high tool speeds are possible.

In fact the laser can be a high quality, fast cutting tool. Its major weaknesses are the limited cutting depth which is a function of power and the capital cost which is also a function of power. This second point, although true, has tended to be greatly exaggerated recently particularly since laser prices have been steady for several years while the price of most other machine tools has risen faster with inflation.

In this Chapter the laser cutting process is discussed by first describing the physical mechanism whereby a laser is able to cut, then illustrating how it is currently used in industry and some of the current research developments which are being considered. This is followed by a discussion of how the process can be modelled theoretically. There then follows a section on the practical performance of CW lasers in cutting with tables and graphs of the relationships between the principal laser operating parameters and the cut speed and quality. The operating parameters considered are: laser power, beam diameter, mode structure, F number of focusing optics, nozzle design, gas flow and composition, arc augmentation. Pulsed lasers are treated separately together with the specialist cutting processes such as hole drilling, surgical uses of lasers, etching, scribing and resistance trimming.

2. Laser–material interaction: a brief outline

When a laser beam strikes an opaque surface, heat transfer occurs by photon/electron interactions with free or bound electrons (the reverse Bremsstrahlung effect). These interactions raise the energy state of the electrons in the conduction band and the mechanism is the same as for classical thermal conduction. The excited electrons collide with lattice phonons and with other electrons or give back their radiation by spontaneous emission. It is worth noting that the photon energy, E, from a laser is almost too low to generate X-rays, e.g. for 10.6 μm radiation $E = hc/\lambda = 0.12$ eV, where h and c are Planck's constant and the speed of light, respectively, and λ is the radiation wavelength ($= 10.6$ μm). Therefore, X-rays are not generated in laser processing. It is calculated by both Duley (1976) and Ready (1971) that thermal equilibrium is established very quickly since the mean free time of electrons in a conductor is 10^{-13} s. It is only in the case of new powerful picosecond pulsed lasers (10^{-12} s pulses) that the classical heat transfer mechanism may not apply. The applications of lasers, in particular, the CO_2 laser, are limited by:

(a) the surface reflectivity;

(b) the need to use low F number optics to produce a small spot size;

(c) at high power densities gas breakdown occurs leading to beam absorption and reflection above the workpiece.

The reflectivity of a surface is a function of the material, nature of the surface, the level of oxidation, temperature, wavelength and power density of incident radiation. It is difficult to isolate the relative importance of the different parameters in determining the ultimate surface reflectivity but all workers agree (e.g., Duley 1976, 1979, Ready 1978) that the reflectivity falls dramatically with increase in incident power density, time of exposure, and temperature. If the temperature rises sufficiently evaporation occurs and a dynamically stable void called a keyhole is created. The keyhole acts as a black body as far as radiant absorption is concerned.

2.1. Five different ways in which a laser can be used to cut different materials

(i) *Vapourisation*. The beam energy heats the substrate to above its boiling point and material leaves as vapour and ejecta – requires around 10 times the energy of (ii).

(ii) *Melting and Blowing*. The beam energy melts the substrate and a jet of inert gas blows the melt out of the cut region – requires around twice the laser energy of (iii).

(iii) *Burning in reactive gas*. The beam energy heats the material to the kindling temperature which then burns in a reactive gas jet; as in (ii) the jet also clears the dross away – requires around 10 times the energy of (iv) for some materials.

(iv) *Thermal stress cracking or controlled fracturing*. The beam energy sets up a thermal field in a brittle material e.g. glass such that it can guide a crack in any direction.

(v) *Scribing*. A variant on (i) whereby a blind cut is used as a stress raiser allowing mechanical snapping along the scribed lines.

What distinguishes these five mechanisms is how the heat is subsequently coupled, and its effects on the workpiece.

2.1.1. Vapourisation cutting

In vapourisation cutting the material is removed from the cut slot or kerf as vapour or ejecta. For this purpose very high power densities are required ($\sim 10^8$ W/cm^2). This is the usual cutting mechanism for pulsed lasers or in the cutting of materials which do not melt, for example wood, carbon, and some plastics.

It follows from the use of such high power densities that very steep temperature gradients are created in the workpiece which cause high localised thermal stresses. Steverding (1970, 1971) has calculated that a 10^9 W/cm^2 10 ns pulse of 10.6 μm radiation could cause a maximum stress of 1–200 kbar. A further surface stress arises from recoil pressure due to the evaporating surface atoms. A simple one dimensional heat balance shows that the velocity of the vapour front into the work piece is given by the expression

$$V = F_0 / [\rho L + C\rho(T_v - T_0)] \text{ m/s}, \tag{2.1}$$

where: F_0 = absorbed power density (W/m^2); ρ = density (kg/m^3); L = latent heat of fusion and vapourisation (J/kg); C = heat capacity (J/kg°C); T_v = vapourisation temperature (°C); and T_0 = ambient temperature (°C).

For tungsten, the vapourisation velocity at 10^8 W/cm^2 will be around 10^3 cm/s ($F_0 = 10^{12}$ W/m^2, $\rho = 19\ 300$ kg/m^3, $L = -94$ J/kg*, $C = 194$ J/kg°C**).

This gas ejection drags molten debris with it, so scouring the hole and establishing the molten walls of the keyhole (Andrews and Atthey 1976). The density of the vapour is around $1/1000$ that of the solid and hence exit gas velocities of the order of 10^6 cm/s would be expected if there were no shock waves. The pressure required to cause this flow of vapour can be estimated from Bernouilli's equation:

$$\Delta p = \rho V^2 / 2 = 4 \times 10^8 \text{ N/m}^2.$$

A further pressure wave is generated from the rapidly changing thermal stress field.

It must be remembered that the recoil pressures are built up in the time, t_v, which is the time taken to initiate the full rate of vapourisation. This time can be simply calculated from the same heat balance, assuming the heat penetrates only by conduction, the workpiece is opaque, and the laser pulse is a step function in time and is spread uniformly over the surface with no heat losses by radiation and convection. Then (Ready 1971)

$$T(0, t) = (2F_0 / K)(\alpha t_v)^{1/2} / \pi, \tag{2.2}$$

where α is the thermal diffusivity and K is the thermal conductivity of the material. From eq. (2.2),

$$t_v = \frac{\pi}{4} \left(\frac{T_v - T_0}{F_0} \right)^2 KC\rho.$$

The resulting impulse, $\Delta P t_v$, represents a localised compressive "knock" which could reach high enough values to cause damage. Withers and Wilshaw (1973) have shown that this compressive knock can be reflected

*Hultgren (1963).
**Smithells (1976).

within the workpiece causing tensile surface stresses. This gives rise to chipping in brittle materials. They also show that the shape of the laser pulse is important. Particularly important is the rate of fall off of stress and not the peak stress.

Thus the mechanics of vapourisation cutting can be summarised as follows.

(a) The incident beam strikes the surface. It is partly reflected and the remainder is absorbed. As the surface heats up the reflectivity falls due to:

    (i)   increased temperature affecting the electronic structure of the workpiece (Ready 1978);

    (ii)   surface shape changes due to strain from thermomechanical stresses and later evaporation producing a blackbody hole;

    (iii) vapour or gas breakdown on the workpiece surface causing a superficial plasma;

    (iv) surface chemical changes such as oxidation affecting the workpiece electronic structure (Duley 1976).

All these events proceed extremely quickly. Herziger et al. (1971) calculates that the absorption coefficient μ increases in 10^{-7} s to 10^3 cm^{-1}, and its value can be calculated from $\mu = A \exp(-B/t)$ where A and B are experimental constants and t is time.

(b) The surface temperature rises to its boiling point rapidly enough to avoid considerable melting by thermal conduction.

(c) Vapour comes away from the surface at near sonic speeds. The acceleration force required to remove the vapour fast enough creates a stress wave within the material which upon reflection within the workpiece may cause chipping in brittle materials for powers $> 10^9$ W/cm^2. It also raises the pressure at the vapourising front which may raise the vapourisation temperature (Anisimov 1967).

(d) The vapour takes with it molten particles, and erosion fragments to create a hole. Gagliano and Paek (1971) estimated by counting particles on a high speed film of a laser drilling that 60% of the material was removed as molten droplets.

(e) The thermal field around the hole, has a steep thermal gradient leading to high mechanical stresses for power densities $> 10^8$ W/cm^2 producing a stress field similar to a point load. This will also be reflected within the material as a stress wave.

(f) Superheating has also been observed in the case of very rapid pulsed drilling. The hot vapour coming from the "keyhole" may, if hot enough, reflect and absorb the incident laser beam due to its high electron density. This leads to the concept of optimum power density. For stainless steel this was found to be 5×10^8 W/cm^2 by Hamilton and James (1976) who measured it by maximising the function hole depth/total incident energy. Above this power the vapour absorption blocks the increased power, and the absorption wave begins to move away from the workpiece surface towards the laser.

(g) If the material is partially transparent heat will be absorbed internally, ahead of the vapourising front which can lead to internal boiling occurring in the form of sub-surface explosions – a theory first put forward by Gagliano and Paek (1971) and supported by the experimental observations of several other workers e.g. Haller and Winogradoff (1971) who noted the explosive removal of K_2O from a SiO_2PbOK_2O glass.

Control of the temporal shape of a laser pulse is important for effective cutting. In fact, it can be argued that there is an ideal pulse shape for a given material and application (Herziger et al. 1971). Modern YAG or Nd glass lasers are capable of producing pulses whose spatial and temporal modes can be carefully controlled (Herziger et al. 1974). A typical laser pulse consists of a series of very short spikes following an overall envelope (fig. 2.1, Battista and Shiner, 1976).

The spiky nature of the glass laser output is an important advantage in drilling. The higher peaks cause great vapourisation and erosion and thus leave less recast material. For example, according to Battista and Shiner (1976), a 10 J pulse lasting 900 μs from a glass laser showed 80% more penetration than the same power from a YAG laser. (Comparative data like these are not necessarily proof since the exact laser rod quality is crucial. Kato and Yamaguchi (1968) compared two apparently identical ruby rods which differed in drilling ability.) The YAG laser, however, is easier to cool and can hence be pulsed faster.

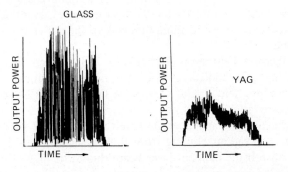

Fig. 2.1. Output characteristics of Nd lasers (Battista and Shiner 1976).

The sharp initial spike which is usually present in the envelope of a single pulse (fig. 2.2) serves not only to rapidly reduce the reflectivity but also to effect the bulk of the vapourisation. In doing so it generates a plasma which blocks further vapourisation. When the plasma has decayed towards the tail of the pulse vapourisation continues.

The shape of the resulting hole is usually tapered. The sides are so arranged as to allow sufficient absorption to maintain a liquid wall. Thus position of focus, depth of focus and reflection channelling are important. Duley and Young (1973) recorded (by high-speed photography) the growth of a hole in fused quartz using a 300 W CW CO_2 laser. The cutting/drilling sequence was:

(a) rapid initial growth, fig. 2.3; hole profile is dependent on radial intensity;

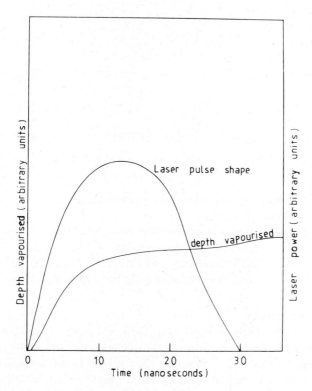

Fig. 2.2 Schematic representation of the history of vapourisation by a Q-switched laser (Ready 1965).

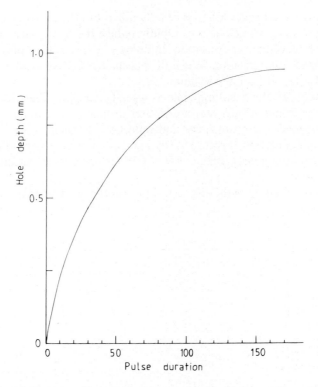

Fig. 2.3. Variation of hole depth with pulse duration (Duley and Young 1973).

(b) sides steepen to allow sufficient absorption to maintain a liquid wall, with vapourisation at the surface;

(c) when power intensity at the surface is no longer able to sustain a vapourisation front, liquid forms at the tip of the hole and oscillates to minimise surface tension causing the tip to oscillate off centre; molten product flows up and down a hole causing the depth to oscillate as well.

2.1.2. Fusion cutting
In fusion cutting the laser beam is of sufficient power to evaporate a hole (known as a "keyhole") into the workpiece, as noted before for vapourisation cutting, but instead of removing the material as vapour and ejecta it is here removed as a molten product by an auxiliary jet, usually mounted coaxially with the laser beam. The result is a clean cut made with around one tenth of the power density required for vapourisation cutting.

The physical mechanism involved in fusion cutting can be summarised as follows.

(a) The laser beam strikes the workpiece surface, some of its energy is reflected while the rest is absorbed, evaporating a small hole.

(b) The small hole (keyhole) acts as a blackbody and absorbs all the laser energy. The hole is surrounded by molten walls which are held in place by the fast flow of vapour ejecta.

(c) The melting isotherm penetrates the workpiece, and pressure from the auxiliary gas jet blows the molten material away.

(d) The workpiece moves and so the "hole" is traversed leaving a slot. The laser beam now strikes only the leading edge of this slot from which a steady or pulsing flow of molten material is blown.

There is little vapour interference with the laser beam since it is blown clear. At very slow cutting speeds (of thin material), most of the laser beam will pass straight through the kerf (Duley and Gonsalves 1972). Increasing the speed causes more of the beam to strike the material, thus automatically increasing the coupled power to the material and explaining the fairly wide operating region for good quality cuts obtainable with the laser. Arata et al.

Fig. 2.4. Typical striation pattern on laser-cut surface of mild steel.

(1979) measured the temperature of the cutting face and showed that it increased with cutting speed, a natural result of this mechanism. In thicker materials the laser evaporating action is not fast enough or the molten product does not move fast enough and so the beam may be reflected within the kerf on the cutting face. A cut will then be achieved if the molten product can escape before it is frozen by the cold gas jet. All laser cut edges show a striation pattern since:

(i) The cutting process is initiated at one power level with the resultant oxygen burning being stopped at a lower power level.

(ii) The slope of the cutting face is so steep the power density on it cannot sustain the melting process so a step in the cutting face forms, causing intermittent advance of the cutting face.

(iii) As cutting proceeds absorbing or reflecting plasma or smoke may cause an intermittent action.

2.1.3. Reactive fusion cutting

If oxygen or a reactive gas is used in place of the inert gas just discussed then another heat source is created due to the ignition of the substrate. The mechanism becomes far more complex. The keyhole as before has molten walls. The gas blowing through moves these walls and heat and mass transfer occur. The rate of burning depends on the rate of mass transfer to the molten dross and diffusion through the dross to the ignition front. The higher the oxygen flow rate the faster the chemical reaction and removal of material, this is accompanied by a faster cooling of the reacted oxide at the cut exit. Outside the region where the kindling temperature is achieved the oxygen jet acts as a coolant reducing the heat affected zone.

By comparing the cutting rate with argon and that with oxygen (see figs. 2.17 and 2.18), it is apparent that about 60% of the cutting energy is supplied by the chemical reaction with oxygen during the cutting of steel. This percentage is considerably more in the cutting of a reactive material such as titanium (see fig. 2.18).

Having essentially two energy sources there are two cutting regimes as discussed in section 6.1.6. There is a region where the oxygen burning rate is faster than the laser traverse speed typified by a wide rough kerf and another region where the laser travels faster than the oxygen burning rate typified by a narrow, relatively smooth kerf. The transition between the two can be quite abrupt (see fig. 2.41).

2.1.4. Controlled fracture

Brittle material which is vulnerable to thermal fracture can be quickly and controllably severed by laser spot heating. The laser heats a small zone causing thermal gradients and consequent severe mechanical strain which in some brittle materials will result in crack formation. Since the area of maximum mechanical strain is that which has been heated by the laser, the laser is able to lead the crack in almost any direction provided balanced thermal gradients can be maintained. However this is not possible around a sharp corner, nor for a cut approaching an edge at an angle. It is also risky for cutting out closed shapes. The process as analyzed by Withers and Wilshaw (1973) is fast and requires very little laser power. In fact, the power input should not be so high as to melt the workpiece surface because this will spoil the cut edge. Lumley (1969) gives some experimental results of the thermal fracturing of several materials.

The controlling parameters for controlled fracture cutting are the laser power and spot size together with the workpiece thickness, width and length, as shown in figs. 2.5 and 2.6 and table 2.1.

2.1.5. Scribing
This is a process in which a laser makes a groove or series of perforations on the surface of the workpiece. The quality of the groove is usually considered best if there is a very small heat affected zone and little surface debris. These two conditions are satisfied by the use of low energy, high power

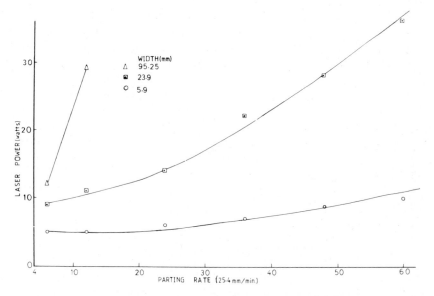

Fig. 2.5. Laser power versus parting rate for high alumina (> 99%) ceramic 0.7 mm thick (Lumley 1969).

density pulses which operate by removing material principally as vapour (see section 2.1.1).

3. Industrial application of laser cutting

There are different types of laser systems that are designed to meet specific demands. These systems are usually either continuous power (usually referred to as continuous wave or CW lasers), e.g. the CW CO_2 laser, or pulsed systems such as neodymium glass or neodymium yttrium aluminium garnet (YAG) lasers. (There are also argon or ultraviolet lasers which can be operated in either pulsed or continuous modes to give a special wavelength for some specific purpose, e.g. eye surgery). The general set-up for a laser cutting operation is usually as shown in fig. 2.7. In this arrangement, cutting can be carried out in any of the following ways (Atkey 1978, Forbes 1976): a stationary laser and a moving workpiece; stationary laser, stationary workpiece, and moving optics; or moving optics and moving workpiece.

The automation of these movements is usually achieved in either of two ways:

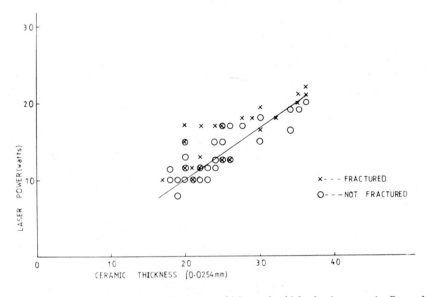

Fig. 2.6. Power required for separation versus thickness for high alumina ceramic. Rate of separation was 305 mm/min (Lumley 1969).

Table 2.1
Scribing parameters for various materials (Lumley 1969)

Material	Width (mm)	Length (mm)	Thickness (mm)	Tech. used	Spot dia. (mm)	Laser powers (W)	Separation rate (mm/s)
	23.8	114.3	0.7	spot focus	~ 0.250	8	304
99% + Al$_2$O$_3$	36.5	15.2	1		~ 0.380	16	76.2
	23.8	114.3	0.7		~ 0.380	7	304
Micro-crystalline	25.4	76.2	0.43		~ 0.25	5	304
alumina	25.4	76.2	0.63		~ 0.25	7	304
	25.4	76.2	0.89		~ 0.25	11	304
Glass	25.4	76.2	1.2		0.38	3	304
Corning 0211	25.4	76.2	1.6		0.50	9	304
Soda lime	25.4	76.2	1.0		0.50	3	304
glass	25.4	76.2	1.0	line focus	0.5 × 12.7	10	304
Sapphire	25.4	25.4	1.2	spot focus	0.38	12	76
Crystalline quartz	9.5	25.4	0.8		0.38	3	609
Ferrite sheet	50.8	35.0	0.2		0.38	2.5	1220

(1) use of a pantograph system where one limb of the pantograph is a photoelectric line-following device and the other limb is the laser or laser optics;

(2) Numerical control: the required movements, distances, or speeds are digitised and the signals are fed to a stepping motor or encoded engine. Traversing corners presents problems in optical-following machines due to

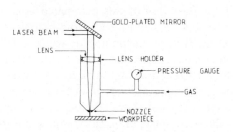

Fig. 2.7. Gas-jet laser cutting arrangement (schematic).

restrictions in speed imposed by an upper limit in optical-following speed. In such cases, the processing speed has to be slowed down at the corners and, in certain applications, the power must be reduced accordingly if overheating is to be avoided (Boehme and Herbrich 1978). A 30% automatic power control is possible. This speed restriction does not exist in numerically-controlled machines.

The setting-up and alignment of an optical guidance system requires time and manual positioning. But, in a numerical machine, these two operations can be performed automatically by the incorporation of auxilliary functions.

A laser cut has high precision and reproducibility (as determined by kerf width and heat affected zone measurements). In fact the precision of a laser cut has been quoted (Atkey 1978) as ± 0.05 mm in 20 mm gauge stainless steel. The commonly accepted precision of an optical following device is ± 0.3 mm while for a numerically-controlled table (Anderson 1978, Berry 1978) it is ± 0.1 mm.

The cutting of contoured material requires accurate, fast response systems with automatic height adjustment. Such systems usually incorporate devices that are based on (a) capacity effect, (b) feeler electrodes (Boehme and Herbrich 1978), and (c) inductance probes.

There are several industrial processes in which laser cutting is currently being used for the large scale production of different materials. It is not possible to list all these processes but we shall, in this section, review some of the most important applications. Those areas which utilise pulsed lasers will be discussed separately from CW applications.

3.1. CW laser applications

(a) Dieboard cutting. This was the first industrial cutting application of lasers, utilising a BOC Falcon 200 W machine installed at William Thyne Ltd in the UK in 1971. The dieboards which are cut by this fully automated machine are used in the manufacture of cartons and the current aim is to make full use of the flexibility of this automated laser cutting system by digitising the carton plans (Serchuk 1978). Roto-graphics have come up with a system known as the British "carton line" in which a line-shared computer is programmed to draw all standard-shaped cartons and also provide a digital tape (of the particular drawing) to direct a laser die-board cutter. The entire design process is simple and very brief because the carton dimensions flap sizes and caliper of material are entered on a computer terminal and the computer produces complete drawings within about three minutes via a plotter. If the drawings are acceptable, a digitised tape for a dieboard cutter is produced. The use of laser cutting has cut down the production time for these dieboards to about one tenth of the time it would normally have taken with the use of a motorised jig-saw (Lunau 1970) or by sticking blocks together.

(b) Quartz tubes. Quartz tubes are used in the manufacture of halogen lamps and traditional cutting methods of nicking and cracking usually give unsatisfactory results by wasting a lot of material. Diamond saw cutting generates dust, a wide kerf and some breakages. Laser cutting of quartz tubes for the lamp industry reduces dust and also results in the sealing and fire-polishing of cut edges (Webster 1976, Megaw and Spalding 1976). According to Fletcher (1977), laser cutting of quartz tubes with outer diameters between 8 and 13 mm at Thorn Lighting (fig. 2.8) gives a kerf width of about 0.5 mm, compared with 1.5 mm from mechanical cutters.

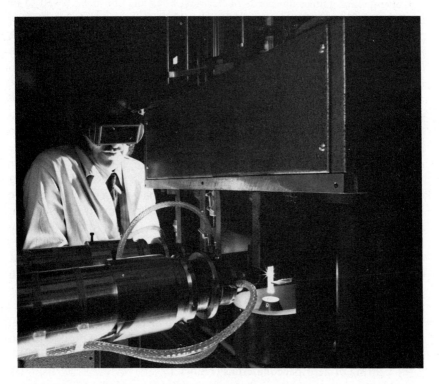

Fig. 2.8. Quartz tube for quartz halogen lamps being cut by a 500 W laser at Thorn Electric Ltd (Courtesy of Thorn Electric Ltd).

This represents an economic saving of 4 m/h with a production capacity of 4000 pieces/h.

(c) Profile cutting of thin metallic or non-metallic sheets (job shop cutting). The flexibility of an automated laser cutting machine results in considerable economic advantage when cutting batches of complicated sheet type parts of up to 5000 pieces (Atkey 1978). This is due solely to the savings in tooling cost. Cutting tolerances claimed by Laser Cutting Ltd in 20 mm gauge stainless steel is ±0.05 mm which is more accurate than die stamping and cheaper for batches of limited size.

The cutting of letters, motifs and designs for the sign industry (Berry 1978) on a one-off basis is as easy as a long run. Slot cutting in 4 mm thick automotive hubs at 15 mm/s with a 1200 W laser gives almost double the productivity due to tool changes being only the software on a numerically controlled laser (Anderson 1978).

Cutting filter gauzes at speeds of between 6–12 m/min has proved to be similarly profitable (Boehme and Herbrich 1978).

Other examples of profile cutting are: standard rings, discs, orifice plates, articles in spring steel and valve plates (Berry 1978). According to Forbes (1976), a 30% saving in the cost of manufacturing complicated valve plates in small batches of up to 500 is possible if laser cutting is used.

(d) Cloth cutting. Single layers of cloth can be cut at speeds of up to 30 m/min using sophisticated beam steering devices based on a long focal length large diameter oscillating mirror. The cloth is also stack-cut up to thicknesses of over 5 cm provided certain plastic materials, such as PVC, are not present (Tandler 1971).

The main advantages in using a laser to cut cloth is the lack of fraying and dust. Also, some synthetic fabrics form thermally sealed self edges. The cloth does not need clamping and does not suffer from stress relaxation after cutting so that a more accurate cut results. Furthermore, the auto-mated laser allows a quick response to fashion changes (Eleccion 1972).

Nylon/polyester strips were usually cut with hot knives producing six rough-edged strips every $4\frac{1}{2}$ s. A 250 W laser did the same job in 2 s with a smooth fused edge. Since this manufacturer makes 350 000 cuts/week the laser route saves 24 production hours/week (Evans 1976).

Rubber fabric can be cut accurately since there is neither mechanical strain nor vulcanisation adjacent to the cut (Berry 1978). The process is also fast and flexible.

(e) Cutting of aerospace materials (Huber 1977).

(i) Hard brittle ceramics such as silicon nitride, used in turbine engines, can be cut ten times faster with a laser than with a diamond wheel, there are no tooling costs and any shape can be cut (Yessick and Schmatz 1975).

(ii) Titanium alloy components (Diels 1976) can also be cut to advantage by a laser. The titanium splice plate for the F14A horizontal stabiliser was cut by Grumman Aircraft Corporation using a laser. Previously, these parts were chemically milled, chiselled or ground from a 150 cm × 340 cm × 9 mm plate. The laser was able to handle larger plates as well as cut faster; the cut required less finishing and the result was a saving of 17.6 man hours/aircraft and $1350 in material/aircraft in 1976. Wing stringers in titanium used to take 2 h to prepare by bandsaw and routing. The laser took only 20 min. Curved nacelle skins of 2.3 mm titanium with a minimum radius of 125 cm were cut with a 58% saving in time compared to chemical etching.

(iii) Aluminium alloys (Wessling 1977, Kahaner 1977). Due to the high reflectivity only high powered lasers (> 3.5 kW) can be effectively used for the processing of aluminium. Using 6 kW of laser power Wessling (1977) reports savings in set-up time, and material representing an overall saving of 60%–70%, compared to the conventional methods of blanking and router cutting.

The fatigue properties of laser cut edges were similar to those from the more costly milling or routing operations. However, these properties were lower than those of the parent metal due to surface roughness, as opposed to metallurgical damage. These results are currently being studied by the major aircraft manufacturers.

(iv) Boron–epoxy (used as the skin on the tail of the F15 aircraft). McDonnell Douglas Corporation cut sheets at 4 min/sheet whereas they previously required 8 h to cut (Financial Times, 24th March 1976).

(v) Ti-coated aluminium honeycomb plate consisting of 30 mm of aluminium honeycomb with 0.2 mm Ti on both sides was previously sawn at 100 mm/min with some stress damage. A 350 W laser was able to cut the same material at 5000 mm/min with no stress damage (Hoffman 1974).

(f) Cutting of electrical panel doors (Anderson 1978). A uniform spacing is required between the door and door frame of an electrical panel box. Previously, the box was made by stamping. Since stamping requires substantial excess material for the blank holder of the trim die, the door had to be stamped from one sheet while the frame was made from another. The laser does not have to grip any edge and the only material lost is from the 0.15 mm kerf width. Thus a 1000 W laser cutting the 1.2 mm thick, low carbon steel at 15 mm/s was able to cut the door out, then cut around again to give the correct spacing, all from one sheet – a straight saving of at least 50% in material.

(g) Cutting fibreglass. The advantages of cutting fibreglass by laser are in the reduction of dust, no cracking as is usual with drilling or sawing, a sealed cut edge and considerably reduced tool wear. Diamond or tungsten carbide tools last only a few minutes when cutting holes in "Kevlar" glass polymer composite ducting (Anderson 1978). Cutting speeds are also high with a laser. Anderson (1978) reports 53 mm/s with a 1500 W laser while Yessik and Schmatz (1975) reports 25 mm/s with a 400 W laser cutting 3.3 mm thick glass-filled polyester.

(h) Skiving electric cable. An interesting application of blind cutting is given by Webster (1976) in which a laser cuts the outer insulation on a copper cable but does not cut the copper cable due to its higher reflectivity and thermal conductivity. Mechanical skiving gives variable results and some loss of metal. The reason for wanting to remove the plastic covering is either for making electrical connections or for salvaging the copper (Webster 1976).

(i) Prototype car production (Hoffmann 1974, Ford 1974). In prototype car production small batches of parts are required in various sheet material such as gaskets, deep-drawn sheet, or lead-coated steel sheet. This is an ideal application for lasers. The Ford Company in Cologne installed a 400 W laser in 1974 and report that by previous methods of shears, nibblers, and grinding machines operated by hand, 10 components in St.37-2 steel were produced per hour, while 120 components were produced per hour by using a laser. Greater flexibility in the design of dash panels and accuracy of production was also achieved with a laser (Yessick and Schmatz 1975).

(j) Cutting alumina (Longfellow 1973) and dielectric boards (Berry 1978). The alumina substrate for a microelectronic device such as a microwave circuit must be accurately cut both for its location in a housing and its effect on the circuit. A 250 W laser has been able to do this, although some microcracks are formed at the beginning and end of cuts or at cuts of partial penetration. This is possibly due to stresses incurred by resolidifying material. The competing processes are:

(i) rough cutting and grinding which is costly, time consuming and subject to breakages;

(ii) cutting in the unfired state ("cookie cutting") and then final grinding which is subject to thermal warping;

(iii) forming an alumina log of the correct cross section which is subsequently fired, sliced and surface polished.

Alumina sheets are also scribed and drilled by pulsed lasers. Dielectric boards are also cut with no tearing of particles and no subsequent treatment being required (Berry 1978).

(k) Manufacture of air conditioning ducts (Berry 1978, Hoffman 1974). The galvanised or plastic-coated steel sheet (St. 37-2, 0.5–2 mm thick) for air conditioning ducting was previously cut by hand after a time-consuming marking-out operation; subsequent machining caused some distortion.

Using a 350 W laser and photoelectric guidance, a clean profile cut was produced which saved time and material through avoiding marking-out and tool changing as well as introducing a fast, single operation cutting process. The cost of the installation was almost paid for in one year. The company claimed a saving of $50 000 in the first year (Forbes 1976). The low heat affected zone in laser cutting kept the zinc or plastic coating intact close to the cut (Berry 1978). The cutting speed on the machine was 50 mm/s (Forbes 1976), and the machine utilisation has been reckoned at 75%.

(l) Heavy duty regenerator manufacture (American Machinist, Vol. 122, No. 2, 1978). Heavy duty regenerators (4.6 m high) for large gas turbines are manufactured by the Garrett Corporation Airesearch Manufacturing Co., California. They were previously made by stamping out parts on a blanking press. The blanking die for such large parts can cost up to $100 000 (1978 price). They installed a 2.5 kW laser with a numerically controlled contouring system which cuts out the tube sheets, side plates and fins. The laser is competitive on speed of production and saves the total cost of the blanking dies and their upkeep. It is also responsive to design changes.

(m) Helicopter blade manufacture. The Westland Lynx helicopter blade, which is 3 m long, was previously milled and surplus material removed – an operation that took 35 min with 10 special cutters being scrapped per blade. The laser does the same job in 100 s using only a few pence of electricity. The cutting rate was 66 mm/s through 2 mm thick stainless steel sheet (Forbes 1976). It is later formed into a D section for the rotor leading edge. It has been calculated that this installation paid for itself in 10 months (Forbes 1976).

(n) Cutting thin-walled stainless tube (Hoffmann 1974). 0.2–2 mm wall thickness tube can be cut with no mechanical deformation, burr, or tool costs.

(o) Furniture industries. The laser can cut through even the hardest wood with ease up to depths of 3.8 cm (Peters and Banas 1977). Because of the possibilities of profile cutting it is not surprising to find that many furniture businesses use lasers as a normal woodworking tool. The mahogany leg shown in fig. 2.9 was cut at 17 mm/s with 1700 W power using a line following device. This represents a saving of between 50–80% in processing time, and an even greater saving of material since no side cuts are required.

An interesting development is the use of lasers in marquetry where the narrow kerf width allows stack cutting on all but high quality marquetry,

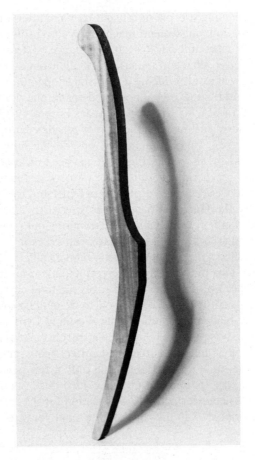

Fig. 2.9. 3 cm thick mahogany table leg cut by a 1.7 kW laser at 17 mm/s (courtesy of Control Laser Ltd).

while pattern burning as in fig. 2.10 forms a new development in wood working.

(p) Perforating irrigation pipes (Webster 1976, Evans 1976). *In situ* perforated pipes are used for drip irrigation of such crops as sugar cane. The tubing is usually polyethylene extruded at 90 m/min and perforated with 0.5 mm holes at regular intervals. The Reed Irrigation System, California uses four 500 W CO_2 lasers 24 h/day and 7 days/week to drill four holes every second. The beams optically track the pipe as it is extruded. Alternative methods left burns or debris which subsequently cause blockages.

Fig. 2.10. Laser on wood engraving using a 500 W CO_2 laser (courtesy of Pyroform Ltd).

(q) Perforating cigarette paper (Webster 1976). Cigarette paper is perforated to lower the nicotine and tar levels (which smokers inhale) by diluting the smoke. Traditionally, these holes were produced mechanically. Coherent Ltd have made a machine with four beams from one CO_2 laser head firing into the fast moving paper strip through a mask. A 300 m length of bobbin can be perforated in 6 min with no flaps to close the hole again and precise holes which are debris free are obtained (figs. 2.11 and 2.12). Cutting paper has been studied (Ainsworth 1978). The laser cut edge is found to be of a high quality, and stronger than a sheared edge which tends to pull the fibres apart. Proper control of the laser cutting process can avoid discolouration due to charring.

(r) Flexograph print rolls (Webster 1976). Rubber flexograph print roles are used for printing wallpaper, textiles and packaging materials. After conventional manufacturing, rubbers show thermal or stress-induced relaxation. The laser being a non-contact high speed process causes none of these disfigurements, allowing an order of magnitude improvement in accuracy with tolerance of ±0.0025 mm.

Fig. 2.11. A laser machine for perforating filter tip paper for cigarettes (Courtesy of Coherent Ltd).

There are currently two methods of machining a contoured pattern onto the surface of a rubber roll as in fig. 2.13.

(1) Cover the roll with a copper mask to protect the image areas. The copper image is made by etching a layer of copper held on a plastic sheet. The laser is then run over the entire cylinder, it is reflected by the copper but removes 0.5–3 mm depth of rubber elsewhere.

(2) Modulation of the inscribing laser. The artwork is scanned photoelectrically and the reflected signal modulates the machining laser.

The advantages of using a laser for this process are: (i) no mechanical forces and therefore no stress relaxation; (ii) no thermal relaxation if power and speed are correctly adjusted; (iii) it is a direct method and therefore there is no distortion in mounting a flat rubber engraving onto a cylindrical roll; (iv) the resulting quality has brought this laser process into competition with high quality processes such as photogravure.

(s) Cutting radioactive material (UKAE Annual Report 1975/6). In order to avoid having heavy machinery, as well as machinery that will need regular and frequent maintenance in a radioactively hot area the laser is

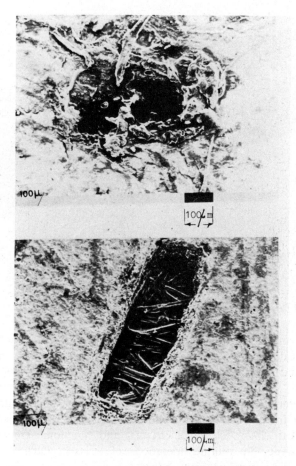

Fig. 2.12. The differences between a hole made with a pin and that from the laser machine shown in fig. 2.11 (Courtesy of Coherent Ltd).

being used by the UKAE for machining in hazardous environments. The laser itself is outside while only the beam is passed into the hot area. The beam can be used over considerable distances and is, of course, unaffected by radioactivity. Simple lightweight jigging and mechanical manipulators are all that is required within the hot zone. At present the laser is used to cut the radioactive wrappers off fuel rods in the reprocessing of fuel, but many more uses for the laser are being considered since it is a nearly ideal power source for these hostile environments.

Fig. 2.13. Flexographic print roll machined by laser (Courtesy of Laser Techniques Ltd).

3.2. Pulsed laser applications

Pulsed systems are used principally as evaporative drilling machines. Their main application areas are in evaporating a blind groove as in resistance trimming and scribing or in evaporating a hole. The following are typical examples:

(a) Resistance trimming – see section 6.3 on resistance trimming. The active trimming of thick and thin film resistance for hybrid microelectronic circuitry (Karr 1975, telephone circuits) and functional trimming of integrated circuits are unique operations of a laser which are becoming essential to the microelectronics industry.

(b) Drilling diamonds (Sharp 1978). Diamonds are used as fine dies for the manufacture of thin wire. The central hole in the diamond has to have a precise double bell mouth shape for the wire extrusion process. Mechanical drilling was a slow process and necessitated many changes of the tiny drills used. A Nd/YAG laser drilling from both sides into a rotating diamond produces an axisymmetric hole with a double bell mouth, which requires only a small amount of final shaping. It saves both time and materials. The comparative costing is given by Sharp (1978) in which he shows that a die made by laser costs between $\frac{1}{10}$ to $\frac{1}{4}$ that made mechanically depending on the hole size. The fabrication time alone is reduced from around 20 h to 1 h.

Drilling of diamonds is also done by laser to remove impurities, such as voids, twins, or inclusions. The tiny laser hole is subsequently filled by a material with a suitable refractive index and is invisible (Prifti 1971).

(c) Rotor balancing (Design News 11–20–78/W). The laser's ability to remove material without contacting the workpiece is well illustrated by dynamic balancing applications. The small gyroscope or other high speed rotor is mounted in a balancing machine which notes the out-of-balance and signals a laser to evaporate certain parts of the rotor. The balancing can be done at normal running speeds (~ 6300 rpm). Removal rates are 0.1 to 10 mg/pulse, preference being for short pulses. (Eleccion 1972).

A similar application is the balancing of balance wheels in clocks which is practiced by the Bulova Watch Company. Using a Nd/YAG laser precision was increased by a factor of 10 and the adjustment time was reduced to $\frac{1}{20}$.

(d) Miscellaneous hole drilling. Drilling small diameter holes is mechanically difficult. It is found that the laser is cost effective for hole diameters between 0.01 mm to 1.5 mm (Kocher 1975) and by using a trepanning action larger holes up to 1 mm deep have been made with commercial gains. There are also advantages in the flexibility of the hole size produced. The examples of ways in which these small holes are used are too numerous to list. The trade literature for pulsed lasers is full of examples. Here it will suffice to enumerate some examples with some references.

(i) Holes in bubblers in gas/liquid absorption (Young 1978).

(ii) A 250 μm hole drilled 2.5 mm up the axis of an 850 μm diameter surgical needle, using a ruby laser. The suture cord is then threaded into the end of the needle, and the end crimped. The inconvenience of the double threaded approach is thus avoided.

(iii) Holes in teats for babies (Webster 1976), with no resealing as is the case for a punched hole.

(iv) Aerosol valve components (Webster 1976). Drilling of 0.33 mm diameter vapour tap holes in aerosol valve housing as practised by Newman Green Inc., Ill. (Evans 1976) saves $10 000/y at 1976 prices in mould repair costs compared with core moulding.

(v) Pump spray nozzles drilled with 0.04 mm holes by laser should save Canyon Manufacturing $20–30 000/y at 1976 prices (Evans 1976).

(vi) Holes in turbine power generator combusters and transition ducts made of hastalloy X (a Ni–Cr–Mo–Fe alloy) (Engel 1978). This alloy is tough and gummy, necessitating frequent tool changes and yet some 500 000 holes/y at 0.76–2.5 mm diameter through 0.165–2.7 mm thick alloy are required.

Mechanical drilling was slow, taking 60 s/hole and causing extrusions at both the exit and entrance requiring a post drilling operation for removal.

Mechanical punching was fast but the smallest hole that could be produced this way was 3.2 mm and the resulting hole had excessive taper in the lower half.

Electrochemical machining (ECM) was too slow at 180 s/hole.

Electric discharge machining (EDM) was the same as mechanical drilling at 58 s/hole but more expensive.

Electron beam machining was fast at 0.125 s/hole but suffered from a large capital cost compared with a YAG laser.

The laser took 4 s/hole with good accuracy. The hole was made by beam defocusing for holes up to 0.3–0.45 mm or focusing and trepanning above that. The accuracy in hole diameter was ±0.02 mm.

(vii) For drilling a fine hole in a sealed component case to allow the introduction of a required atmosphere and the use of the same laser to reseal by spot welding (Young 1978).

(viii) Holes in watch bearing jewels or Al_2O_3 (Kocher 1975, Shkarofsky 1975).

(ix) Optical apertures: Holes with 0.2 μm diameter have been obtained with a frequency-quadrupled YAG laser (Kocher 1975).

(x) Drilling of holes in thin films of tellurium, bismuth or indium for optical recording systems. This is a new and fast developing field, requiring the rapid production of precisely located (< 1 μm) diameter holes.

(e) Scribing (Eleccion 1972). For cutting glass, silicon, or ceramic. Typical scribing speeds 5–15 cm/s with a YAG laser Q switched to a few kHz (Kocher 1975).

4. Current research developments of the cutting process

4.1. Continuous power cutting

From an understanding of the physical mechanisms involved in cutting the potential areas for development are listed below.

To increase:
(1) the total energy input by use of
   – higher power lasers,
   – additional energy sources,
   – improved energy coupling;
(2) the power density by
   – decreasing the spot size;
(3) the ease of removal of molten products by
   – increased drag;
   – increased fluidity.

These will now be considered in turn.

4.1.1. Increase of total energy input

(a) *Higher power lasers.* Large lasers are being developed and currently a 100 kW research laser is being operated in the United States. The Avco Everett Co. currently market of 20 kW machine. As industrial machines these large lasers have three drawbacks: firstly, the very large capital outlay; secondly, the lower reliability compared with less powerful machines; thirdly, the lower process efficiency due to the higher plasma temperatures produced at the laser/workpiece interface. This last point is illustrated by noting that theoretically a plasma at around 30 000°C would be totally reflecting due to its high electron density.

The advantages of high powered lasers are principally found in their ability to treat high reflectivity materials such as aluminium, copper, and gold, where powers greater than 3.5 kW have been found necessary (Wessling 1977) in order to start a keyhole against the initial high reflectivity. A secondary advantage of high powered lasers is the faster processing which they allow. This not only results in greater productivity but also a reduction in the heat affected zone and thermal distortion.

It is possible to get high powers from a low power rated laser by either Q-switching or discharge pulsing. The high power is achieved only momentarily during each pulse. For example a 30 W CW Nd:YAG laser has peak powers when Q-switched of 50 kW. While a 2 kW axial flow CO_2 laser with a pulsed discharge tube would be expected to have a peak pulse power of 5 kW. This is a relatively cheap method of achieving some of the advantages of high powered lasers.

According to Saunders (1977), up to 20% improvement in penetration is obtainable by pulsed welding while Hitachi (1974) puts this figure at about 50%. Similar improvements would be expected in the cutting situation.

(b) *Additional energy sources.* Higher energy inputs are attractive provided they can be achieved at little additional expense. The addition of a coaxial or off axis oxygen jet, one of the first major developments of laser cutting, was first reported in by Houldcroft (1968) and has been discussed in greater detail by Adams (1970). Oxygen assisted laser cutting of materials which react exothermically with oxygen deposits additional energy into those areas which are heated above their kindling temperatures. It has the additional advantages of cooling those areas not above their kindling temperature thus reducing the heat affected zone and increasing the drag on any molten product. In addition it blows any absorbing and reflecting plasma or smoke away from the laser beam while protecting the laser optics.

Various improvements or variations on this theme are currently being studied. These include:

(i) Use of faster jets, in particular supersonic. Results here have been disappointing due to excessive cooling of the molten product and shock waves interfering with the fine focus on the laser beam.

(ii) Use of multiple jets: concentric jets can increase the jet potential core and so give a higher stagnation pressure above the cut slot, thus improving slot velocities. Grumman (1976) uses an oxygen jet with a surrounding helium flow (Diels 1976). Other variations include the simultaneous use of jets above and below, cross blowing beneath to remove the underbead, and crossblowing over a venturi to introduce suction from below (Steen, 1977).

(iii) Changes in gas composition: chlorine can react exothermically with most metals. It is also more dense than oxygen and thus has more drag on the molten product, with many metals the resulting halides are volatile at the high temperatures involved in laser cutting or at least have a lower melting point. With these advantages it is a pity that chlorine gas is lethal. The use of gases such as SF_6 would generate a very hot plasma the radiation from which could be kept within the cut kerf by appropriate gas velocities. This could have advantages in fire-polishing the cut edges of plastic or glassy materials. However among its other disadvantages SF_6 interferes seriously with the lasing action of a CO_2 laser if any gets in the cavity. Gas mixtures which are only capable of burning at high temperatures would give enhanced energy not only at the laser interaction points, but also later on where the molten product might be solidifying and preventing cutting.

An alternative method of introducing more energy is by adding an electric arc to the laser interaction zone (Clarke and Steen 1978). The arc will preferentially root to the hottest areas and thus cooperates with the laser. This is discussed in more detail later.

(c) *Improved energy coupling.* There are two different coupling mechanisms to consider. There is the initial coupling when the focused beam strikes the flat surface of the workpiece and there is the later coupling of the laser power within the kerf. The first is a problem in understanding reflectivities (Duley 1976), initial plasma formation (Schawlow 1977) and surface profiles under thermal stress. The second is a highly complex situation involving multiple reflections off a reacting rippled molten surface. If a fundamental understanding of these processes is obtained, some practical advances would be expected. At present this is not the case. Some experiments showing the different coupling with plane of polarisation and angle of incidence have opened up this subject (Olsen 1982); while others have shown improvements in cutting by using antireflectance coatings such as DAG graphite or pulsing with a high leading power spike.

A pulsed laser is usually most effective in cutting by evaporation, thus taking advantage of the very high peak powers. Cutting by evaporation has some advantages in that the heat affected zone is almost eliminated. However, the thermal effects that will accumulate from the presence of molten residue could be serious enough to crack such a brittle material like alumina (Longfellow 1973). In cutting and welding with pulsed lasers, (either CO_2 or solid state), considerable attention is now being directed towards the variation in power with time during each pulse. It is felt that in pulsed cutting, a high leading edge spike is needed to break down any surface reflection problems, thus allowing the rest of the pulse to be absorbed, (Chun and Rose 1970). Plasma effects are also reduced. Numerous examples are cited by Saunders (1977) in which he illustrates that pulsed cutting may have advantages over CW with heat sensitive materials since the evaporative mode of cutting leaves less heat in the workpiece.

4.1.2. Increased power density

Cutting is a process which is dependent on the power density of the energy source, either as power/unit area or power/unit length. In either case the spot size of the laser beam on the workpiece surface is as important as the total laser power. The spot size expected depends on the incident beam diameter falling on the focusing optics together with the power distribution within the incident beam, known as the mode structure. Improved optics design and laser cavity optics are thus areas of current research interest. However, it is by no means established experimentally that the smaller the focused spot size the better the cut. In fact, on the contrary the optimum spot size is a function of material thickness to allow adequate material removal from the kerf.

4.1.3. Increased removal rate of the molten product

(a) *Increased drag.* This can be achieved by suitable gas jets as discussed in section 4.1.1.

(b) *Increased fluidity.* Some promising results have been obtained by blowing slagging agents such as lime (calcium oxide) into the kerf with a view to lowering the dross melting point and reducing its viscosity. However, this process also increases the thermal load. An exothermically reacting slagging agent would be expected to be more effective – but probably too costly.

Also ultrasonic vibration has been shown to reduce the resolidified thickness in hole drilling due to increased fluidity of the molten product (Mori and Kumehara 1976). These experiments also reported an increased hole penetration of around 10% with a vibration amplitude of 25 μm.

5. Theoretical models of the laser heating process

The effects of material processing with a laser depend upon many variables, some of which are not easy to control or monitor. For example the beam diameter is probably functionally more significant than the incident power but it is not easy to measure – or even define (Courtney and Steen 1978). Similarly, the effects of such parameters as mode structure, surface reflectivity, convection and radiation losses are difficult to establish quantitatively. It will, therefore, be very useful to be able to theoretically predict experimental trends in order to establish our understanding of the physics of the process and also be able to determine the approximate value of important parameters such as the thermal cycle at any location in the workpiece, its maximum temperature, cooling rates and thermal stresses, together with the basic operating parameters of cutting or welding rates for full penetration.

The approaches made so far fall into three classes: (1) analytic solutions; (2) numerical solutions; and (3) semiquantitative solutions.

5.1. Analytic solutions

If the heating time is much longer than the free time between collisions ($\sim 10^{-11}$–10^{-13} s) (Bar-Isaac and Korn, 1974) for the electrons within the workpiece it can be argued that classical heat conduction equations will apply to the laser heating process. In this case the differential equation governing heat flow within the workpiece will be:

$$\frac{u}{\alpha}\nabla T + \nabla^2 T(x, y, z, t) - \frac{1}{\alpha}\frac{\partial T}{\partial t}(x, y, z, t) = -A(x, y, z, t)/K,$$

where: T = temperature at a location (x, y, z) at time t; α = thermal diffusivity; K = thermal conductivity; u = velocity of movement of workpiece relative to the heated spot; and A = source term – heat production per unit volume/unit time.

From this equation it is apparent that a set of solutions is possible in which the source term A varies in spatial distribution and duration. There are also solutions for a stationary workpiece ($u = 0$) or a moving workpiece.

In all analytical solutions it is necessary to neglect the nonlinear term due to radiant heat losses. In most continuous heating processes this is not justified but some pulsed processes would suffer negligible radiant loss in the time considered. Convection losses are usually, but not necessarily neglected, although their effect is usually less than 1% of the input power unless there is some very unusual jet system used. Most analytic solutions make no allowance for material lost during cutting. They have been extensively reviewed by Shkarofsky (1975). However, the most useful solutions are summarised here:

(1) overall heat balance;
(2) extended surface source – one-dimensional heat flow;
(3) point heat source;
(4) disc sources; and
(5) line sources.

5.1.1. *Overall heat balance*

If all the incident energy was used in melting the material which is to be removed then $P(1 - r) = Vkt\rho(C\Delta T + L_m)$, where the variables are defined in section 6.1. Allowing for some evaporation as well as melting,

$$P(1 - r_f) = Vkt\rho(C\Delta T + L_m + fL_F),$$

where f = fraction of material melted which is subsequently evaporated.

Models of this sort can be used to indicate trends, except for very narrow cuts – see table 2.2 (Moss and Sheward 1970).

Table 2.2
Cutting speed predictions from various models (Bunting and Cornfield 1975)

Material	Thickness (cm)	Width, for plasma torch (cm)	Simple heat balance (Moss and Sherward 1970)	Moving line (Rosenthal 1946)	Moving band (Arata and Miyamoto 1972)	Moving cylindrical (Bunting and Cornfield 1975)	Experimental (cm/s)
Copper	0.0071	0.038	49.9	12.2	14.7	12.5	20.3
	0.0071	0.024	103	32.6	38.4	5.51	5.49
	0.124	0.57	25.9	8.1	6.4	10.9	8.7
	0.124	0.33	41.6	13.5	10.8	17.7	11.7
	3.18	0.64	5.45	2.08	2.33	2.12	0.847
Iron	0.0127	0.069	30.8	10.2	10.3	4.9	5.5
	0.02	0.075	7.48	2.99	5.63	3.08	1.58
	2.5	0.516	3.17	1.16	1.30	1.44	1.7

Some semiquantitative data on drilling rates have also been derived this way (Masters 1956).

Dabby and Paek (1972) have used this model to predict subsurface explosions in material where the penetration of the radiation before absorption is significant (i.e. partially transparent materials). Their predictions have been experimentally verified by Slivinsky and Ogle (1977).

5.1.2. Extended surface source – one-dimensional heat flow

Assuming the workpiece is opaque and absorption occurs in the surface layer, the heat flow equation can be solved. Ready (1971) gives the one-dimensional solution for the temperature distribution as:

$$T(z,t) = \left[2F_0(\alpha t)^{1/2}/K\right]\mathrm{ierfc}\left[z/2(\alpha t)^{1/2}\right] \tag{2.3}$$

and in particular the surface temperature is

$$T(0,t) = (2F_0/K)(\alpha t/\pi)^{1/2}, \tag{2.4}$$

where $F_0 = P_{tot}(1 - r_f)$ and r_f = reflectivity, P_{tot} = total incident power, K = thermal conductivity, α = thermal diffusivity, and t = time.

Other one-dimensional solutions exist for partially transparent workpieces and also for opaque workpieces subject to power pulses with regular shape (Ready 1971). Both of these solutions are so difficult to handle that the advantages of a one-dimensional assumption are lost.

5.1.3. Point heat source

The solution of the heat conduction equation which assumes that there are no radiant or convective losses and that the source is an instantaneous point source located at x', y', z' is (Carslaw and Jaeger 1959)

$$T_{\mathrm{inst.pt.}}(x, y, z, t)$$

$$= \frac{Q(x', y', z')}{8(\pi \alpha t)^{3/2}} \exp\left(-\frac{(x - x')^2 + (y - y')^2 + (z - z')^2}{4\alpha t}\right), \tag{2.5}$$

where Q = quantity of heat liberated by the instantaneous point source. For axial symmetry in a thermally thin sheet, in fact for two-dimensional heat flow (a form of line source)

$$T_{\mathrm{inst.pt.}(r,t)} = \frac{Q}{4\pi K d t} \exp\left(-\frac{r^2}{4\alpha t}\right), \tag{2.6}$$

where: d is the thickness of the thin sheet, $r = (x^2 + y^2 + z^2)^{1/2}$, and $d^2/4\alpha t \ll 1$.

Point source solutions (either instantaneous, pulsed, or continuous) always give an unrealistically high central temperature. A consideration of heat flow through a small control volume surrounding the heat source shows that all the incident energy must escape down a thermal gradient from an infinitesimal point and, therefore, infinite temperatures are required. At some distance – well beyond the actual beam diameter – the solution should compare with reality if it were not for radiant effects becoming significant as heat flow spreads over a large area.

The interest in instantaneous point sources lies in the fact that they allow rapid estimates to be made. They are also a starting point for disc source solutions and solutions over periods of time.

For a moving continuous point source in a thin material (Carslaw and Jaeger 1959, Duley and Gonsalves 1972)

$$\exp(vx/2\alpha) = 2\pi k\, DT \left[\varepsilon PfK_0(rV/2\alpha) \right]^{-1},$$

where: r = radial location; P = incident power; ε = emittance at 10.6 μm; f = fraction of beam striking sheet; K = thermal conductivity; D = sheet thickness; K_0 = Bessel function of the second kind and zero order; v = cutting speed; and α = thermal diffusivity.

This model has been successfully used by Duley and Gonsalves (1972) to predict the performance of a laser for cutting stainless steel. They also included a term, f, the fraction of the beam striking the surface as opposed to that passing through the kerf.

5.1.4. Disc sources

By integrating a set of instantaneous point sources of varying intensity, a disc source solution of any defined power distribution can be obtained. The method is described by Carslaw and Jaegar (1959).

The solution for an instantaneous gaussian disc source in which the power distribution is $E(r) = E_0 \exp(-r^2/R_B)$, where R_B is defined as the beam radius measured to the point where the power intensity falls to $1/e$ of the central value (sometimes R_B is defined at the $1/e^2$ position) is

$$T_{\text{inst.gauss.}}(r, z, t) = \frac{E_0 R_b^2}{\rho c (\pi K t)^{1/2}(4\alpha t + R_b^2)} \exp\left(-\frac{z^2}{4\alpha t} - \frac{r^2}{4\alpha t + R_b^2} \right).$$

$$(2.7)$$

This may be integrated with respect to time to give the continuous gaussian disc source solution.

$$T_{\text{pulsed gauss.}}(r, z, t)$$

$$= \frac{F_0}{\pi K} \left(\frac{\alpha}{\pi}\right)^{1/2} \int_0^t \frac{p(t - t')\,\mathrm{d}t'}{t'^{1/2}(4\alpha t' + R_b^2)} \exp\left(-\frac{z^2}{4\alpha t'} - \frac{r^2}{4\alpha t' + R_b^2}\right). \tag{2.8}$$

The solution of this equation for a normalised pulse shape has been given by Ready (1971) in the form of dimensionless plots.

If the power input does not vary with time then eq. (2.8) reduces to:

$$T_{\text{cont.gauss.}}(r, z, t)$$

$$= \frac{F_0}{\pi K} \left(\frac{\alpha}{\pi}\right)^{1/2} \int_0^t \frac{\mathrm{d}t'}{t'^{1/2}(4\alpha t' + R_b^2)} \exp\left(-\frac{z^2}{4\alpha t'} - \frac{r^2}{4\alpha t' + R_b^2}\right)$$

$$\tag{2.9}$$

which, for the central point, gives

$$T_{\text{cont.gauss.}}(0, 0, t) = \frac{F_0}{\pi K} \left(\frac{\alpha}{\pi}\right)^{1/2} \frac{2}{(4\alpha R_b^2)^{1/2}} \tan^{-1}\left(\frac{4\alpha t}{R_b^2}\right)^{1/2}$$

$$= \frac{F_0}{\pi^{3/2} K R_b} \tan^{-1}\left(\frac{4\alpha t}{R_b^2}\right)^{1/2}, \tag{2.10}$$

and for steady state situations

$$T_{\text{cont.gauss.}}(0, 0, \infty) = \frac{F_0}{2(\pi)^{1/2} K R_b},$$

where F_0 = total incident power. As a result

$$\left(\frac{T\pi K D_b}{P_r(1 - r_f)}\right) = \pi^{1/2} \simeq 1.77. \tag{2.11}$$

In the case of thin sheets, $(R_0^2 \ll 4\alpha t)$, the solution becomes that of two-dimensional heat flow (Ready 1971).

For most laser applications the gaussian power distribution is approximately realised in practice but with special rastering and beam spreading devices for surface heat treatment a uniform disc source is of interest.

The solution for a disc source moving in the direction of the beam axis, i.e. drilling, is discussed by Gagliano and Paek (1971) together with some stress analysis.

5.1.5. Line sources

Swifthook and Gick (1973) established for a line source that a particular isotherm is described by:

$$Y = 4Ur\{1 - [K_0^2(Ur)/K_0'^2(Ur)]\}^{1/2}, \qquad (2.13)$$

and

$$X = 2\pi\exp[UrK_0(Ur)/K_0'(Ur)]/K_0(Ur), \qquad (2.14)$$

where:

$Y = Vb/\alpha = 2Ub = $ normalised isotherm width;
$X = W/aS = $ normalised power per unit depth;
$U = V/2\alpha$;
$V = $ traverse speed of the beam;
$\alpha = -K/C\rho$;
$K = $ thermal conductivity of the material;
$C = $ thermal capacity of the material;
$\rho = $ density of the material;
$b = $ total melt width;
$r = $ radial distance from the source;
$W = $ total absorbed power;
$S = \int_0^T K dT$;
$T = $ rise in temperature above ambient;
$a = $ depth of penetration;
$K_0 = $ Bessel function of the first kind and zero order; and
$K_0' = $ differential of K_0 with respect to Ur.

Eqs. (2.13) and (2.14) can be solved numerically. However, Swifthook and Gick (1973) used the asymptotic expansions of the modified Bessel function to obtain approximate values of Y and X as: at high speeds

$$X \simeq (8\pi eUr)^{1/2} \qquad Y \simeq 4(Ur)^{1/2};$$

or

$$Y \simeq (2/\pi e)^{1/2} X \simeq 0.483X;$$

at slow speeds:

$$X \simeq 2\pi/\ln(2e^{-\gamma}/Ur); \qquad Y \simeq 4Ur; \qquad (2.15)$$

where γ is Euler's constant $= 0.577$. Therefore, $6.3/X \simeq \ln(4.5/Y)$ or $Y \simeq \exp(\ln 8 - \gamma - 2\pi/X)$. This, written in full, becomes

$$Habv/w = (2/e\pi)^{1/2} \simeq 48\%. \qquad (2.16)$$

This is the limiting process efficiency or melting ratio for penetration welding. A similar analysis for a surface heated pool leads to a value of 34%. A line source is not capable of calculating the depth of penetration, nor is it able to show how the isotherms are effected by variations in the beam diameter. To show this variation with beam diameter a cylindrical source is better, though this too will not calculate the depth of penetration. Bunting and Cornfield (1975) describe such a model for plasma cutting. It is also relevant for laser cutting. They give tables comparing different model predictions of cutting speeds to experiments, as in table 2.2.

This table illustrates how line sources and simple heat balances tend to exaggerate the possible cutting speeds. This trend is expected since they have not always allowed for all the major heat losses in the system such as: (a) removal of molten material (only the simple heat balance allows for this); (b) radiation losses; (c) convection losses; (d) conduction losses (only the simple heat balance does not allow for this).

With very thin cuts the loss of material is less significant and overall heat balances become inaccurate; but so do line sources if the kerf width is equal to or less than the diameter of the experimentally heated zone.

5.2 Numerical solutions

Finite difference models (e.g. Steen 1977, Mazumder and Steen 1980). Finite difference or finite element analysis starts with the same differential equations which are based on the conservation of energy over an element. These equations are then expressed in finite difference form and solved numerically using a computer.

Their advantage over analytical solutions is that nonlinear terms such as radiation losses are easily included, they are also adaptable to simulating variable reflectivity, thermal conductivity etc. However, allowance for too many varying terms leads to stability problems when iterating for a solution. Their disadvantage is that they only supply a solution in a numerical form for a given specific situation. It is thus laborious and costly in computing time to establish functional graphs between process parameters.

5.3. Semiquantitative methods

In understanding highly complex physical phenomena the semiquantitative model can be extremely useful. In this approach the situation is simplified to a one-dimensional case or the like and so solved.

For example, Klemens (1976) reduced the heat flow in the plasma-filled keyhole to a tractable form by arguing that the heat flow will be essentially at right angles to the keyhole and that it is approximately correct to assume that a linear heat flow exists at the leading edge of the keyhole and at orthogonal positions to this.

Dutta (1974) treated the melting and oxidising front of a material being cut in a similar fashion.

6. Practical performance results

6.1. Cutting

6.1.1. Effect of laser power

Consider the simplest theoretical model [eq. (2.17)] in which we assume that all the absorbed energy is used in melting or evaporating the material and none is lost by radiation, convection or conduction. We find that:

$$P = \eta V w t \rho \left(C_p \Delta T + L_f + m' L_v \right), \tag{2.17}$$

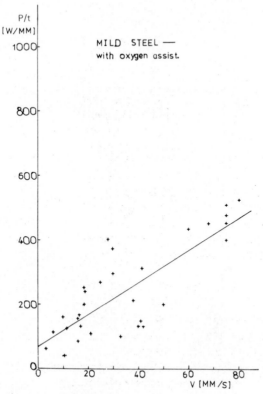

Fig. 2.14. Graph of P/t versus V for mild steel. See also fig. 2.15.

where: P = absorbed energy = $P_i(1 - r_f)$, P_i = incident power;
r_f = reflectivity;
η = reciprocal of process efficiency;
V = cutting speed;
w = kerf width;
t = workpiece thickness;
ρ = workpiece density;
C_p = workpiece specific heat;
ΔT = temperature rise to melting point T_m;
L_f = latent heat of fusion;
m' = mass fraction melted which subsequently evaporates; and
L_v = latent heat of evaporation.

This equation can be rearranged so that the main process variables are on one side: thus

$$P/Vt = \eta w \rho \left(C_p \Delta T + L_f + m' L_v\right). \tag{2.18}$$

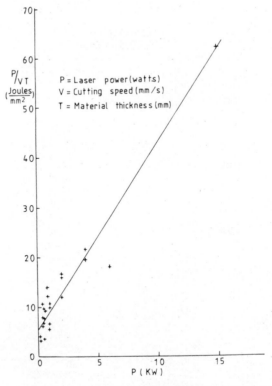

Fig. 2.15. Graph of P/Vt versus P for mild steel taken from literature data (see table 2.3).

At the maximum cutting speed (see section 6.1.5 for the definition) it is reasonable to assume that all the terms on the right hand side of eq. (2.18) are approximately constant including the process efficiency, η, and the kerf width w. Thus, as a first approximation it is possible that P/Vt is a constant for a given material. Plotting P/t versus V for most of the published data on mild steel (fig. 2.14) shows that the broad sweep of the results fits this concept to ± 100 W/mm of cut depth. Thus for mild steel the energy/unit area severed, P/Vt, is (6.0 ± 3) J/mm^2 (from the average shape of fig. 2.14) but for individual results the figures vary from 3–60 J/mm^2 as shown in fig. 2.15.

The scatter is due to a variety of factors:

(i) The definition of the maximum cutting speed used by the data source. The maximum severing speed, V_c (see fig. 2.16), is about 1.2 times the maximum speed for a high quality cut, V_b. Both speeds are used in papers and trade literature as the maximum cutting speed and are rarely defined. This alone accounts for a large scatter. The early results of Adams (1970) and Lunau (1970) were far more efficient than the results of those who followed. By good fortune the Welding Institute laser used was built along a wall necessitating cutting towards the laser, which unknown to them was optimal for their beam's plane of polarisation.

(ii) The other major causes of variation are the operating conditions. The effect of nozzle design, optics used etc., together with interference from plasma effects.

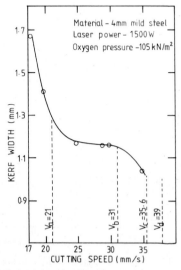

Fig. 2.16. Variation of kerf width of top surface with cutting speed.

The trend shown in fig. 2.15 strongly hints at a reduction in process efficiency with increased power. This may be due to increased spot size and kerf width, increased plasma absorption or that the cutting speed quoted by Locke and Hella (1974) for a 15 kW laser, was not the maximum cutting speed but just a successful cutting speed, in which case the figure is merely misleading on the performance capabilities of a 15 kW laser.

This form of analysis can be extended to other materials as shown in figs. 2.14, 17–19, while table 2.3 shows values of the P/Vt for a wide variety of materials.

Using these values of P/Vt, an estimate can be made of the cutting speed expected for a given laser power and material thickness. If the material to be cut is not shown in the table a material in the table with similar thermal capacity, density and latent heat would be expected to show a similarity in cutting behaviour.

The major parameters defining cut rate and thickness are therefore laser

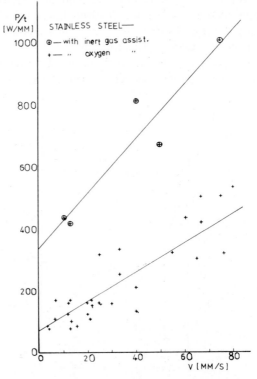

Fig. 2.17. Graph of P/t versus V for stainless steel.

power and material properties (assuming that the laser beam is being
focused to around 200–500 μm).

Secondary parameters are:

   (a) the optical system;

   (b) surface reflectivity; and

   (c) nozzle design and gas supply.

These are now discussed in turn.

6.1.2. Optical system

There are three parts to be considered.

(a) The laser cavity design – which produces a certain power distribution
across the beam (the mode structure) and a certain exit beam diameter
which affects the F number of all subsequent optics.

(b) The focusing optics – depth of focus, spot size and position of focus
relative to workpiece surface.

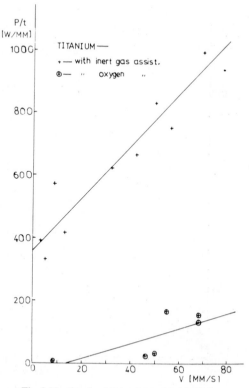

Fig. 2.18. Graph of P/t versus V for titanium.

(c) The beam coupling with the workpiece – reflectance.

As has been discussed in section 4.1, the aim of the optical system should be to get an optimum power density absorbed in the workpiece.

(a) *Cavity design.* A gaussian or near gaussian power profile would be preferred to a higher order mode since it can be focused to a finer spot, thus giving a higher power density. However, Forbes (1975) noted that for optimum cut quality the laser spot size had to be increased from 0.1 mm to 0.2 mm when changing thickness from 2.5 mm to 7.6 mm steel using a 500 W laser. So the optimum mode structure is not quite as clear as may be thought at first.

(b) *Focusing optics.* The focused spot diameter, D, can be calculated from diffraction theory to be $D = 2.4F\lambda = 25.4F$ μm for 10.6 μm radiation, where D is the diameter where the power intensity has fallen to $1/e^2$ of the central value, and F is the F number of the optics used. For a biconvex lens

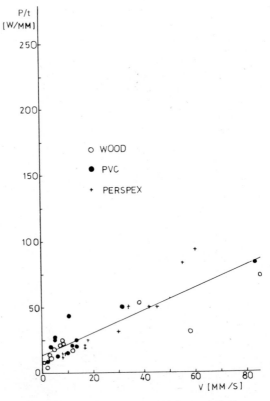

Fig. 2.19. Graph of P/t versus V for non-metals.

Table 2.3
Cutting performance results from the literature

Power, p (W)	Thickness, t (mm)	Cutting speed, V (mm/s)	P/Vt (J/mm)	Ref.
Stainless Steel				
850	7.87	6.35	17.0	Adams (1970)
"	5.08	12.7	13.17	"
"	3.175	12.7	21.1	"
1000	0.78	271	4.73	a
"	1.57	178	3.58	"
"	3.175	76	4.14	"
500	"	30	5.25	"
"	1.57	76	4.2	"
250	3.175	4.2	18.74	"
"	1.57	25	6.36	"
"	0.78	55	5.83	"
1000	1	133.33	7.5	b
"	2	66.67	7.5	"
"	6	13.33	12.5	"
300	1	65	4.6	"
"	2	21.67	6.9	"
"	3	13.33	7.5	"
250	0.5	75	6.7	c
"	1.0	33.33	7.5	"
"	1.5	21.67	7.7	"
"	2	11.67	10.7	"
"	3	3.33	25	"
2000	4.8	66.7	6.2	Clarke and Steen (1978)
"	6.3	25	12.7	"
"	2.8	20	7.1	"
800	5	11.67	13.7	Russel (1975)
2000	6	33.3	10.1	"
"	3	50	13.3	"
2500	2.5	75	13.3	a
"	3.1	40.5	19.9	"
"	6.0	13.0	32.05	"
1300	0.254	169	30.3	"
1200	0.147	254	32.1	"
1000	0.254	148	26.6	"
300	0.147	33	51.5	"
250	0.147	33	51.5	"
2000	12	6.67	25	Russel (1975)
800	3.0	11.67	23	"
400	1.5	32	8.3	"

Table 2.3
Continued

Power, p (W)	Thickness, t (mm)	Cutting speed, V (mm/s)	P/Vt (J/mm)	Ref.
Mild Steel				
600	54.0	6.17	18.0	d
4000	16.75	19.1	12.5	"
400	3	16.67	8.0	Russel (1975)
800	2	28.3	14.13	"
2000	2	83.3	14.13	"
400	1.5	25	10.67	"
4000	2.54	60.5	26.03	a
"	"	73	21.57	"
3000	"	40.5	29.2	"
2000	"	35	22.5	"
100	1.9	80	6.58	"
"	2.2	68	6.68	"
"	4.7	38	5.6	"
200	4.75	10.6	4.0	Lunau and Paine (1969)
320	16.2	76.2	0.26	Harry and Lunau (1972)
1000	1.0	150	6.67	b
"	2.0	75	6.67	"
"	6.0	16.67	10.0	"
"	8.0	11.67	10.7	"
300	1.5	50.0	3.0	"
"	3.0	33.33	3.45	"
500	3.2	16	9.76	Forbes (1978)
"	1.6	41.5	7.53	"
"	0.9	75	6.74	"
850	22.86	29.63	1.25	Adams (1970)
Titanium				
240	4.32	2582	0.02	Harry and Lunau (1972)
250	3	216.67	0.38	c
"	8	50	0.63	"
"	10	46.67	0.54	"
"	40	8.33	0.75	"
850	5.08	55	3.04	Adams (1970)
400	1.0	50	8.0	Russel (1975)
800	0.5	50	32.0	"
1500	1.0	166.67	9.0	"
2500	6	13	32.05	a
"	4	32	19.5	"
"	3	50	16.7	"
"	2.5	70	14.3	"
1000	1.5	42	15.9	"
"	3.1	5	64.5	"
200	0.51	3.39	115.6	Lunau and Paine (1969)

Table 2.3
Continued

Power, p (W)	Thickness, t (mm)	Cutting speed, V (mm/s)	P/Vt (J/mm)	Ref.
Perspex				
250	20	5.83	2.14	c
"	12	11.67	1.78	"
"	10	13.33	1.88	"
"	5	31.67	1.6	"
"	3	83.33	1.0	"
850	31.7	5.1	5.25	Adams (1970)
2000	9	200	1.11	Clarke and Steen (1978)
"	6	292	1.14	"
"	3	500	1.33	"
200	25	1.66	4.8	Lunau (1978)
"	10	3.33	6.0	"
"	4.6	10.6	4.1	"
800	32	5.0	5	Russel (1975)
2000	9	166.7	1.33	"
850	1.25	200	3.4	Adams (1970)
PVC				
250	3	55	1.52	c
"	5	34.17	1.46	"
"	10	18.17	1.38	"
"	16	9.17	1.7	"
"	5	45.83	1.1	"
"	13	16.67	1.15	"
"	5	41.67	1.2	"
"	8	30.0	1.0	"
"	12	16.67	1.25	"
"	15	12.17	1.37	"
"	16	9.17	1.7	"
"	17	8.33	1.76	"
"	20	8.08	1.55	"
300	3.2	60	1.56	Harry and Lunau (1972)
250	1.143	228.6	0.96	Evans (1976)

Table 2.3
Continued

Power, p (W)	Thickness, t (mm)	Cutting speed, V (mm/s)	P/Vt (J/mm)	Ref.
Wood				
1000	9	125	0.9	b
"	12	56.67	1.47	"
"	19	38.33	1.37	"
300	19	11.67	1.35	"
250	18	3.33	4.17	"
"	8	58.33	0.54	"
"	12	6.67	3.1	"
2000	15	15	3.12	Clarke and Steen (1978)
"	25	33.33	2.4	"
"	25	16.66	4.8	"
"	25	25	3.2	"
"	15	58.33	2.3	"
"	13.5	18.33	8.07	"
"	19	43.33	2.42	"
850	13.95	25.4	2.4	Adams (1970)
200	50	1.67	2.4	Lunau (1978)
"	18	3.33	3.34	"
"	25	1.25	6.4	"
350	4.8	88.33	0.83	Harry and Lunau (1972)
300	16	4.65	4.0	"
"	3.8	10.10	7.8	"
200	50	2.08	1.92	"

References for table 2.3

[a] Sinar Laser Systeme, Hamburg, FRG.
[b] La Soudre Autogene Francaise (SAF).
[c] Messer Griesheim, Reprints 6/75e and 32/76e.
[d] United Technologies Research Center.
Forbes, N., 1978, Proc. Laser 78 Conf., London (March, 1978).
Harry, J.E., and Lunau, F.W., 1972, IEEE Trans. Ind. Applics., **1A-8** No. 4, July/August, p. 418.
Lunau, F.W. and Paine, E.W., 1969, Weld. Metal Fabr., Jan., p. 9.
Lunau, F.W., 1978, Proc. Laser 78 Conf., London (March, 1978).
Russel, J.D., 1969, Weld. Inst. Res. Bull., Dec., p. 345.

this is focal length/incident beam diameter. The depth of focus, Z_s, associated with this spot size is defined as the distance along the optical axis over which the central power intensity is greater than half the peak intensity at the focal point, i.e.

$$Z_s = \pm 4 \times 1.39 \left(\lambda L^2 / 2\pi a^2 \right)$$

$$= \pm 4 \times 1.39 \left(\frac{2\lambda}{\pi} F^2 \right)$$

$$= \pm 37.5 F^2 \ \mu\text{m for } 10.6 \ \mu\text{m radiation,}$$

where λ = wavelength (m); L = focal length of lens (m); $2a$ = beam diameter (m); and $F = L/2a$ = Fresnel number. Both of these values are only correct for a pure parallel Gaussian beam with infinite diameter optics. In reality the beam is not pure gaussian, it is truncated by finite sized optics, and the optics are not the required perfect parabolic shape. Therefore, the actual spot size achieved is a multiple of the theoretical by a factor of around 4 depending on the mode structure from the laser, optics and apertures used.

Whatever the correction it is true to state that the shorter the focal length of the focusing optics the smaller the F number and hence the smaller the spot size and depth of focus.

However for an F number below 5 spherical aberration starts to enlarge the focused spot size, minimising this effect. Experimentally Clarke and Steen (1978) found little variation in optimum or maximum cutting speed using 75 mm, 100 mm and 150 mm focal length lenses while cutting 4 mm mild steel. Their results are shown in fig. 2.20.

These results are supported by those published by Spectra Physics for cutting 1.57 mm 6061-T6 aluminium with a 63.5 mm and 127 mm focal length lenses. Both lenses cut with 2 kW power at 12 cm/s. Another effect of a low F number is the shallow depth of focus (varying with F^2) causing poor focal tolerance which makes cutting operations more difficult to reproduce.

At higher F numbers Lunau (1970) shows a distinct difference in penetration in perspex for a given processing speed using F numbers of 10 and 8. The greater penetration was achieved with the higher F number. This is perhaps unexpected since the higher F number would have the lower power density. At only 145 W plasma effects would not be causing this effect, but the greater depth of focus (± 3.7 mm to ± 2.4 mm) for the large penetrations recorded (~ 30 mm) could account for the observations. To penetrate 30 mm with a depth of focus of 4 mm means several internal reflections are going to occur resulting in less power at the cavity base for the smaller F

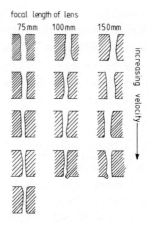

focal length of lens

75mm　　100mm　　150mm

increasing velocity →

Fig. 2.20. 4.0 mm mild steel oxy-laser cut geometries: to show the change in shape with velocity and lens focal length (Clarke and Steen 1978).

number lens. This sort of result was also recorded by Seaman (1976) in welding with very high powered lasers. An $F21$ optic gave increased welding speed relative to an $F7$ optic, provided the threshold value of the power density for the cavity formation was exceeded ($\sim 10^6$ W/cm^2).

The focal position relative to the workpiece surface is critical and is a function of the focal depth, more operating tolerance being possible with larger F number optics.

It is not easy to measure the exact place of tightest focus and so it is not possible to be positive in stating that the focus should be just below the surface, on it or above it. The most systematic record of such work is that of Seaman (1976) using his bench mark method. Fig. 2.21 shows both the decreased welding speed he recorded and the narrower focal tolerance to 90% bench mark speed with reduced F number. He also found that the optimal bench mark focal position varied with laser power as in fig. 2.22.

It was shown by Uglov et al. (1978) that for maximum penetration the focal point should lie within the workpiece by a distance that is almost equal to the achieved penetration (fig. 2.23). The cutting conditions were as listed below:

material	leather-simulating rubber
gauge	2 mm
CO_2 laser power	500 W
lens focal length	280 mm
cutting gas	air at 2 atm pressure.

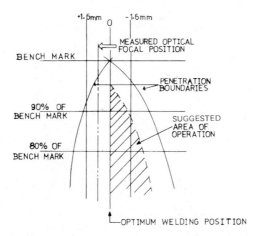

Fig. 2.21. Suggested area for process optimisation in the welding of stainless steel (Seaman 1976).

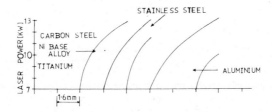

Fig. 2.22. Variation of optimum welding position with laser power for different materials (Seaman 1976).

It is quite possible, however, that their results are peculiar to leather-simulating rubber.

The conclusion from these results and others is that the optics F number should be matched to the workpiece material, thickness and laser power. But by optimising the F number only some 10% variation in cutting speed should be anticipated provided the F number stays in the range 5–20. The focal tolerance will decrease with decreasing F number.

(c) *Beam coupling with the workpiece.*

(i) *Reflectance.* The workpiece reflectivity becomes significant only as new material enters the beam. Theoretically, one would anticipate that the lower the reflectivity the faster the cutting rate. Absorption ($= 1 -$ reflectivity

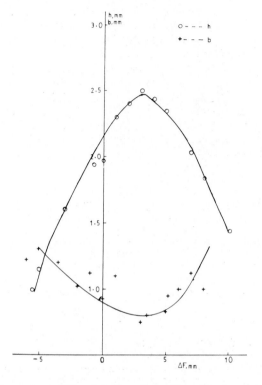

Fig. 2.23. Dependence of cut width and depth on the position of the focusing lens (Uglov et al. 1978).

for an opaque substrate) is a strong function of surface shape, surface oxidation, plane of polarisation and plasma coupling mechanisms (fig. 2.24) and is time dependent (Duley 1979).

Forbes (1976) shows a significant variation in cutting rate with surface finish between a polished surface and either an untreated or shot blasted surface (table 2.4). His results do not show much difference in cutting speed between the two unpolished surfaces. This is possibly because their reflectance does not differ much or, because at a lower value of reflectance, the beam couples more strongly resulting in a fast warm up with consequent surface upset which further enhances coupling. Thus with low reflectivity materials little power is lost because the surface alteration is very fast. Although Forbes did not quote values for the reflectivity of his samples, a considerable difference in the reflectivity between the polished samples and

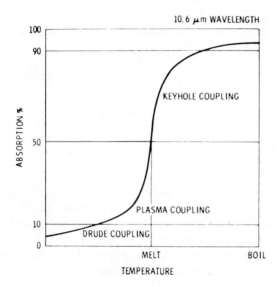

Fig. 2.24. The temperature dependence of reflectivity at 10.6 μm (courtesy of GTE Sylvania).

Table 2.4
Effects of surface treatment of cutting speed (Forbes 1976)

Material	Polished		Untreated		Shot blasted	
	Vel. (mm/s)	Power (W)	Vel. (mm/s)	Power (W)	Vel. (mm/s)	Power (W)
C263 Nickel alloy	12.7	600	12.7	600	21.1	600
N80A Nickel alloy	12.7	400	16.9	400	21.1	400
L2% Cr Steel	12.7	200	25.4	200	25.4	200

the others would be expected as shown by the values quoted by Shkarofsky (1975) (table 2.5).

(ii) *Beam polarisation*. Nearly all high powered lasers are plane polarised; that is the electromagnetic waves within the emergent beam are all oscillating in the same plane. There is a small difference in coupling between a wave oscillating at right angles to a surface and a wave oscillating in the plane of the surface. (An extreme example is reflection of a beam from a transparent material surface which is at the Brewster angle. In this case only the waveform oscillating in the plane of the surface is reflected.)

Table 2.5
Effects of surface treatment on reflectivity (Shkarofsky 1975)

Material	Reflectivity	
	As received	Polished
316 S/S	61	91
416 S/S	48	90
Al	88	98
Cu	88	98

In many laser applications polarisation is not important e.g. surface treatment, and possibly welding. In cutting, however, there is a reflection off the cutting face within the kerf. The beam would be absorbed optimally if it were vibrating along the direction of the kerf.

Variations in cutting speed with direction of cut have been observed. These effects are most noticeable on numerically controlled tables (Olsen 1981).

6.1.3. Nozzle design and gas flow effects

In gas-assisted laser cutting, the gas is introduced either off-axis or coaxially to the laser beam through a nozzle and the gas flow rate is varied by changing the pressure of the gas. Forbes (1975) has shown that nozzle design and flow characteristics will affect the cutting performance at a given laser output. Although there is no standard design for laser cutting nozzles, the common features of any coaxial nozzle are:

(1) it must be big enough to pass the beam without it touching the nozzle (the smaller the nozzle, the more difficult it is to align it with the beam);

(2) the flow from the nozzle must couple effectively with the kerf to remove the molten dross and enhance the cutting action.

In order to optimise the nozzle design it would be necessary to relate such factors as nozzle diameter, height above the workpiece, and exit pressure/momentum distribution.

(a) *Effect of nozzle diameter.* Figs. 2.25 and 2.26 show the relationships between cutting speed and nozzle diameter at constant laser power and supply pressures for both inert gas and oxygen-assisted cutting of 2 mm gauge mild steel. At a given supply pressure there is an optimum nozzle diameter. This result is explained by mapping out the stagnation pressure distributions from the various nozzles as a function of height above the workpiece and distance from the centre of the nozzle. The pressure values illustrated in fig. 2.27 were determined by using a pitot plate which was connected to a mercury manometer.

(b) *Effect of nozzle gas pressure.* Let us assume that p_0 represents the minimum pressure required to remove molten material from the kerf in the time available. The maximum cutting speed for a given nozzle (pressure distribution) will be proportional to the distance between the pressure axis and the point of intersection the horizontal line through p_0 and the nozzle pressure distribution curve. It is therefore easy to see from fig. 2.27 why the cutting speed variation with nozzle diameter passes through a maximum. The effect of increasing the gas pressure would be to expand all the pressure distribution curves as well as move them upwards, thus making it possible to increase the cutting speed by increasing the gas pressure.

However, as has been shown by most workers (e.g. Duley and Gonsalves 1972, Lunau 1970) increasing the gas flow rate increases the cutting speed up to a maximum after which further increase in gas flow rate causes a fall

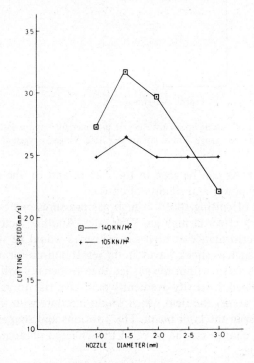

Fig. 2.25. The variation of cutting speed with nozzle diameter at constant gas pressure. (Material, mild steel; gauge, 2 mm; laser power, 1500 W; gas, argon.)

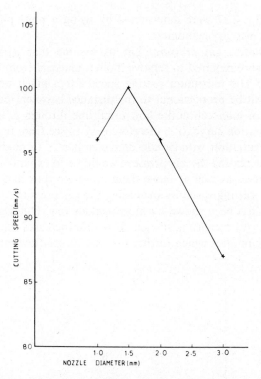

Fig. 2.26. Variation of cutting speed with nozzle diameter for oxygen-assisted laser cutting. (Material, mild steel; gauge, 2 mm; laser power, 1500 W; gas, oxygen at 140 KN/m².

in cutting speed. As can be seen in fig. 2.28 (a and b), the maximum is a function of laser power and rigidity of optics.

The lowering of cutting speed at high gas pressures can be attributed to increased cooling effect at high gas flow rates. Another factor which could affect cutting performance at high gas pressures would be the presence of discontinuities, such as shock waves, in the jet. If non-uniformity of pressure and temperature exists within the gas jet, then its density will change across the flow field. Such a density gradient would give rise to variations in the refractive index across the field which would interfere with the focus of the optical energy from the laser beam. The positions and shapes of regions of refractive index change can be revealed by Schlieren photography (Kamalu and Steen 1981).

Schlieren pictures of the gas flow through different cutting nozzles under various conditions show that at low gas flow rates (i.e. low pressures) the jet

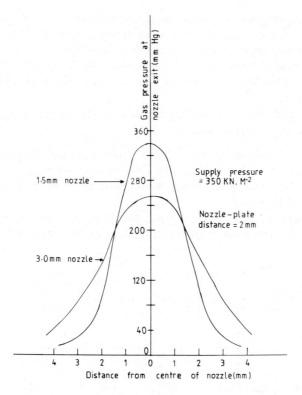

Fig. 2.27. Pressure distribution for nozzles of different diameters at a constant supply pressure.

is practically free of discontinuities (or disturbances) such as a density gradient field (DGF). At high gas pressures a DGF is formed above the workpiece and lies within the region covered by the gas jet (see fig. 2.29). The shape and size of the DGF depend on:

(1) gas pressure, p;

(2) distance, h, between the nozzle and the workpiece; and

(3) nozzle diameter, d.

We will now consider these in turn.

(1) *Gas pressure.* The DGF spreads out to cover a wider area as the gas pressure is raised. Also, the strength of the DGF (measured by its thickness and degree of darkness relative to the background) increases with pressure.

(2) *Effect of the working distance, h.* The DGF is strongest for $1.0 < h < 3.0$ mm. For $h > 3$ mm, the potential core of the jet fails to reach the workpiece,

and so the strength of the DGF decreases with increasing h. For < 1.0 mm
the shape of the DGF changes when using a stub-ended nozzle and it
consists of two regions: a high pressure central region and a low pressure
region around the periphery of the nozzle (fig. 2.30). This low pressure
region is detrimental to material removal from the kerf, thus lowering
cutting speed.

(3) *Nozzle diameter*. Increasing nozzle diameter increases the size of the
DGF, and the gas consumption.

(c) *Effect of the density gradient field (DGF)*. It is not yet possible to say
whether, and to what extent, the DGF would interfere with 10.6 μm
radiation. Any such interference would lead to either a refocusing or a
scattering of the beam.

(1) *Refocusing*. In laser processing, the beam is usually focused on the
surface of the workpiece as shown in fig. 2.31(a). Re-focusing can only shift

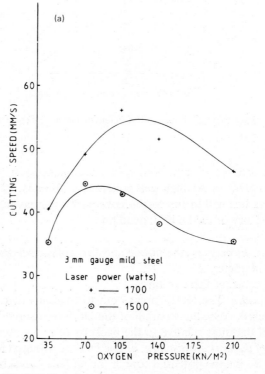

Fig. 2.28. Variation of cutting speed with oxygen pressure.

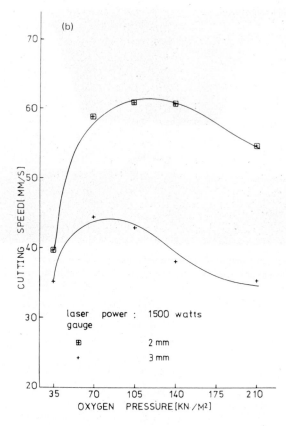

Fig. 2.28 continued.

the focal plane to lie above the workpiece and this would lead to a re-focused beam diameter, D_r, incident on the workpiece. D_r will, invariably, be greater than the original focused beam diameter. Melting efficiency will thus be decreased and we shall end up with a kerf width that is considerably greater than the original beam diameter.

(2) *Beam scattering*. Beam scattering by the DGF will also give rise to an increased incident beam diameter on the surface of the workpiece and will, therefore, result in decreased cutting performance.

Another possible effect of the presence of the DGF would be a change in the beam mode structure on which the focused beam properties depend.

If the DGF were to affect the beam as shown in fig. 2.31(c), the resulting beam diameter would be too large to be effective in cutting. For a re-focused

Fig. 2.29. Schlieren photograph of cutting nozzle and workpiece showing the presence of a density gradient field (DGF) on the workpiece surface. Nozzle pressure, 20 psig.

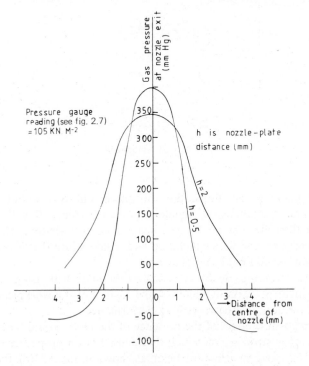

Fig. 2.30. Change of pressure distribution with nozzle–plate distance.

beam, it would be possible to initially place the focal position below the material surface and then have it moved closer to the surface by the DGF as shown in fig. 2.32. We shall, therefore, have a situation where maximum penetration will be achieved by having the focal position below the surface of the workpiece (fig. 2.33). This may be the explanation for the results of Uglov et al. (1978) and Seaman (1976), (see figs. 2.21 and 2.23).

(d) *Effect of nozzle/plate distance.* Coupling of the nozzle jet to the workpiece kerf is an aerodynamic problem. The exit flow pattern and the nozzle/plate distance are important variables. If the jet is too close to the plate severe back pressures on the lens will be created apart from mechanical jamming against splattered dross particles. If the jet is too far away there is an unnecessary loss of kinetic energy. A working distance of 0.89–2 mm is usual. For profile cutting, some automatic height adjustment device is required (e.g. feeler, jet back pressure, variable inductance). Grumman (1976) used a potential flow oxygen jet surrounded by a helium sheath to achieve 850 mm/s cutting speed in 1.27 mm thick titanium. Such a jet should allow greater nozzle-to-plate distance tolerance.

(e) *Effect of gas type.* The type of gas used in the jet affects how much heat is added to the cutting action. There is a large difference between using oxygen and using argon in cutting any metal e.g. titanium as can be seen in fig. 2.18 or mild steel such as in figs. 2.25 and 2.26. In fact Forbes (1975) has estimated that in the oxygen cutting of steel only 30% of the energy comes from the laser and 70% comes from the oxygen exothermic reaction. Fig. 2.18 indicates a reduction in severing energy from 10 J/mm^2 to 2.3 J/mm^2 in good agreement with Forbes estimate.

However with some materials oxygen is too reactive and causes a ragged edge. Thus 20–50% oxygen is recommended (Lunau 1970) for niobium and tantalum, cutting and inert gas cutting of titanium is necessary if edge

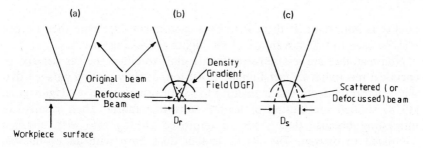

Fig. 2.31. Possible interference effects of a density gradient field.

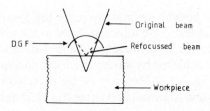

Fig. 2.32. Focal position can be moved to the material surface by a density gradient field.

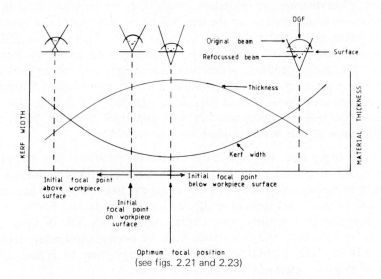

Fig. 2.33. Hypothetical curve showing possible variation of cutting performance with position of focus.

quality is important. If there is to be a subsequent edge trim then oxygen may be used to take advantage of the higher processing speeds.

Non-metallics are less sensitive than metals to either the gas density, or chemical reactivity. In fact Lunau (1970) found when cutting perspex that the penetration depth was insensitive to gas pressure (fig. 2.34). Uglov et al. (1978) studied the cutting of leather substitute fabric. Their results are interesting because they show an improved cutting rate with nitrogen compared to oxygen. The plastic leather does burn with an exothermic reaction, but probably generates considerable smoke which blocks the laser beam. If so this would be a new consideration in choosing a cutting gas.

Increasing the jet velocity increases the kerf width as discussed in section 6.1.5.

6.1.4. *Arc-augmented laser cutting*

In arc-augmented laser cutting, the power from an electric arc is coupled to that of the laser to give an economic improvement in cutting speed (Clarke and Steen 1978).

Fig. 2.35 shows the experimental set-up that is currently used in the arc-augmented cutting of mild steel. The oxygen pressure is measured as the line static pressure, as shown in fig. 2.7, and although this gauge reading and the pressure behind the nozzle are not equal, yet the two pressures are related. As shown in figs. 2.36 and 2.37 the limits of oxygen-assisted laser cutting can be considerably extended by arc-augmentation.

At low arc currents and high speeds when the arc would not be stable on its own, the arc is stabilised by the laser-generated hot spot because this is a region of increased electron concentration. Steen and Eboo (1979) have

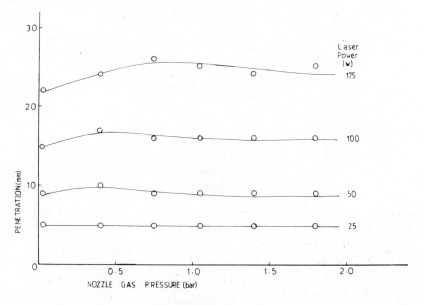

Fig. 2.34. Effect of gas flow rate on penetration into perspex (Lunau 1970); cutting speed, 130 mm/min; focal length, 200 mm.

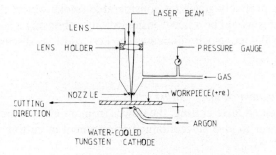

Fig. 2.35. Arrangement for arc-augmented laser cutting.

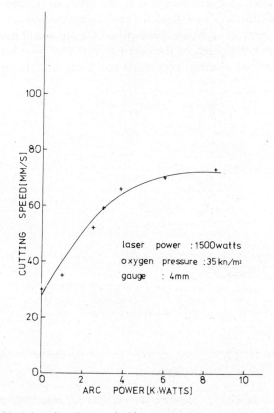

laser power : 1500 watts

oxygen pressure : 35 kn/m²

gauge : 4mm

Fig. 2.36. Variation of cutting speed with arc power in arc-augmented laser cutting.

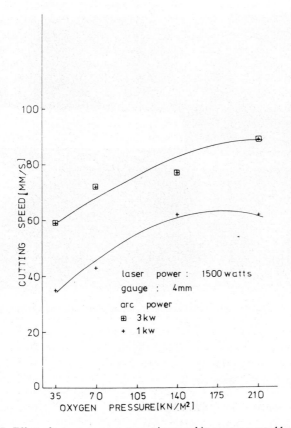

Fig. 2.37. Effect of oxygen pressure on cutting speed in arc-augmented laser cutting.

found that the arc root is contracted by the presence of a localised hot spot so that some intensification of the arc action may be expected in the cutting process, and the two energy sources can be considered to cooperate rather than simply add. The physics of the process is discussed by Steen (1980).

The arc is not of much help when it is used on the same side as the laser because it flickers from one side of the kerf to the other and so damages the cut edges. The arc is much more stable at the underside of the cut and initially keeps the dross more fluid and thus extends the velocity. At low arc currents, considerable improvement in cutting speed can be obtained without any significant deterioration in cut quality. For example, fig. 2.38 shows an arc-augmented cut on 4 mm gauge mild steel under the following

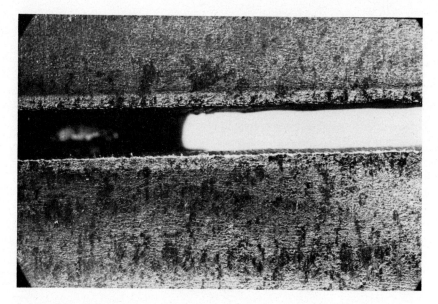

Fig. 2.38. Arc-augmentation effect on laser cutting. Left: laser only, right; laser + arc.

conditions:

laser power	1500 W;
arc power	3000 W;
oxygen pressure	5×6.895 kN m^{-2};
cutting speed	52 mm s^{-1};
magnification	$\times 15$.

It is very significant to note that neither the kerf width nor the heat-affected zone has been affected by the addition of the arc and yet there is an improvement in cutting speed of more than 70%.

6.1.5. Cut quality
The qualities sought in a cut are:
  (a) sharp corners at entry;
  (b) narrow cut width;
  (c) minimal thermal damage;
  (d) parallel sides;
  (e) non-adherent dross;
  (f) smooth cut surfaces.

These are affected by laser variables and materials variables which are now discussed in turn.

(a) *Laser parameters affecting cut quality – kerf width and HAZ.* A typical plot of kerf width versus traverse speed in 4 mm mild steel is shown in fig. 2.39 (see also fig. 2.40 for the variation of cut geometry with cutting speed). Up to the velocity V_a (fig. 2.16), there is considerable side burning due to the oxygen burning front moving faster or as fast as the laser beam traverse speed, an effect elegantly filmed by Arata et al. (1979). This results in a wide rough kerf with more burning at the bottom than the top. Between the speeds V_a and V_b the laser traverse speed is dominant and the kerf remains reasonably constant and parallel-sided with a fine striation pattern. In this region the kerf width is defined by the material thickness and the incident

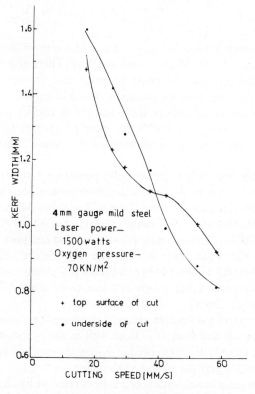

Fig. 2.39. Variation of kerf width with cutting speed.

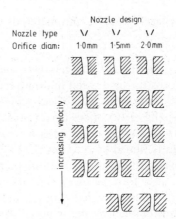

Fig. 2.40. Variation of cut geometry with cutting speed and nozzle diameter for *F*-10 optics, 1600 W, 4 mm mild steel (Clarke and Steen 1978).

laser spot size which for a focused beam is a function of the wavelength of the laser radiation and the beam mode structure. There is an optimum kerf width for a given material thickness in order that the dross should clear easily. The change over from the dominant speed being the burning reaction to the laser traverse speed can be quite sudden as can be seen in fig. 2.41. The region V_a to V_b (see fig. 2.16)[a] is the optimum operating region in which the conditions stated at the beginning of this section are most nearly fulfilled.

Beyond V_b the laser beam fails to fully penetrate without some internal reflections. This is seen in the angled striation pattern on the cut edge, an example of which can be seen in fig. 2.42(c). The dross fails to clear quickly, if at all from the bottom edge, which results in a sharp rise in the heat affected zone. The kerf loses its near parallelism and becomes wedge-shaped. Finally, at V_c, the dross fails to clear the kerf before resolidifying and cutting is lost. Beyond V_c the melting isotherm still penetrates the workpiece allowing in some cases a fairly easy fracture to be made. The penetration continues to the velocity V_d.

It is interesting to note that the thermal conduction models of a line heat source by Swifthook and Gick (1973), as well as the cylindrical heat source by Bunting and Cornfield (1975) predict a similar trend in kerf width versus speed but do not suggest the observed shoulder.

The change in parallelism with speed is illustrated in fig. 2.20 showing the diagrammatic cross section of a series of cuts in mild steel. The variation of

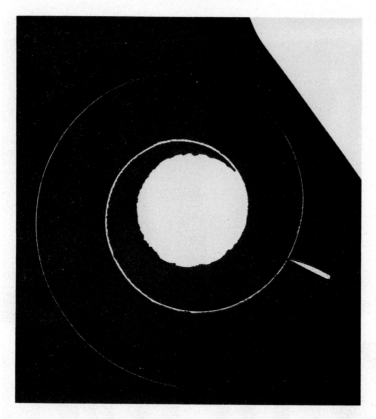

Fig. 2.41. Laser cut in mild steel illustrating the change in kerf width with speed and a sudden change in kerf width at arrowed point.

heat-affected zone in oxygen assisted laser cutting of mild steel is illustrated in fig. 2.43. This figure also shows the cooling effect of the coaxial gas jet which not only increases the optimum speed up to a certain maximum as previously noted but also reduces the HAZ.

The kerf width increases with increasing gas pressure as would be expected due to the greater drag forces from the faster flowing gas (fig. 2.44). Fig. 2.45 shows the top and bottom surfaces of laser cuts in mild steel in the four regions illustrated in fig. 2.16. The HAZ was measured by surface oxidation marks. The metallurgically affected zone was narrower. The micrograph of a cut shows how small the zone is in the optimum cutting region (fig. 2.46). Rolling and bending trials on this edge have been

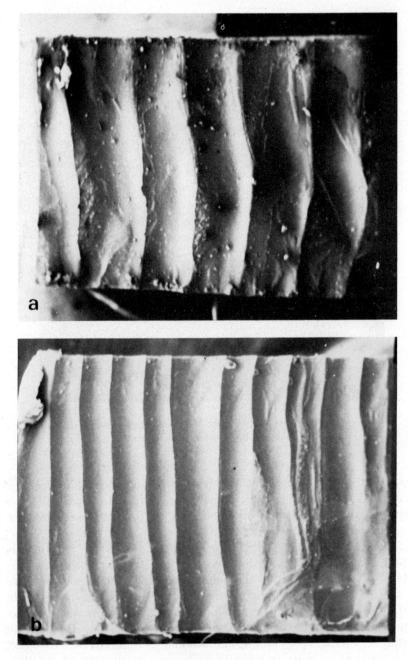

Fig. 2.42. SEM pictures of cut surface showing (a) slow, (b) optimum, and (c) fast cutting speeds.

Fig. 2.42 continued.

done for mild steel and show that no detrimental embrittlement has occurred.

The smoothness of the cut edge is a function of the striation heights and spacing (frequency). Forbes (1975) notes the presence of 300 Hz striations on optimum cuts using a 500 W laser on 2.3 mm low C steel. This may be due to ripple on his rectified 3 phase voltage since the authors' work has not shown such a steady striation frequency.

Adams (1970) and Forbes (1975) and Arata et al. (1979) have shown, using high speed photography that the striations are due to an intermittent flow of the molten product. While Shinada et al. (1980) suggest that intermittent plasma blockage could also play a part.

Adams (1970) took a film at 2000 frames/s showing the laser as it machined a thin slice from the edge. The camera was placed perpendicular to the cut edge. A 450 W CW laser was used to cut mild steel and austenitic stainless steel. At low speeds he noted that the beam heated a small area. The oxidation reaction started and molten metal oxide formed. The oxygen jet blew this away exposing new metal. In cutting 6 mm mild steel at 4 mm/s with 450 W, the oxide flowed down and out for a short period after which the interface moved forward at about 4 to 6 times faster than the traverse speed. There was then a pause until the laser caught up. Stainless

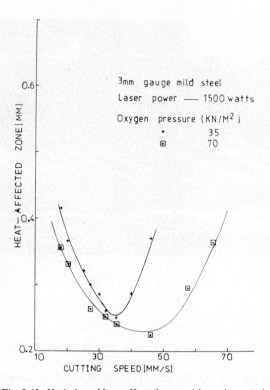

Fig. 2.43. Variation of heat affected zone with cutting speed.

steel behaved the same way but the advancing interface did not move as fast as was the case with mild steel.

At the optimum cutting region flow was continuous but dross formation as droplets on exit caused a pulsation in the melt thickness resulting in the rear edge of the melt moving sharply forward and leaving a striation mark.

The leading edge of the melt was vertical but the oxide flow rear edge was inclined 15° to the vertical (the drag angle); with a curve towards the bottom of the cut due to resolidification.

A pulsed CO_2 laser (50 Hz with peak power of 250 W) showed a cutting action occurring only during peak powers.

Similarly Arata et al. (1979) filmed the cutting action of 400 frames/s viewing into the kerf and observed the intermittent burning at speeds below V_a. This explanation was an intermittent melting due to the variation in

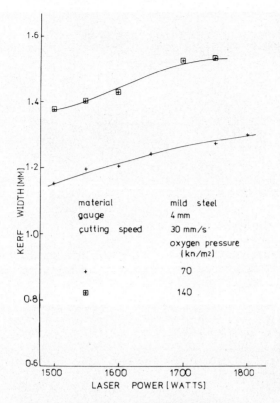

Fig. 2.44. Dependence of kerf width on laser power.

power density on the steeply sloping cutting face. They found the edge roughness varied with speed as shown in figs. 2.47a and b.

Under-bead formation can be of crucial importance, for example in cutting laminates for transformers, etc. Control of underbead formation is possible by the use of auxiliary jets above or beneath the workpiece. Lawton and Mayo (1974) showed this was useful in gas or plasma cutting.

The use of auxilliary or secondary gas jets can give rise to substantial improvements in cutting speed, depending on such factors as nature of gas, its flow rate, auxilliary nozzle design and its alignment and position relative to the primary gas jet. Three possible auxiliary jet configurations are shown in fig. 2.48. In oxygen-assisted laser cutting, fig. 2.48(b) gives the best improvement in cutting speed while the first configuration is best suited to arc-augmented cutting.

Fig. 2.45. Top (T) and Bottom (B) surfaces of cuts in mild steel illustrating the four regions shown in fig. 2.52.

Fig. 2.46. Typical laser cut profile in mild steel using 1.5 kW laser power at 25 mm/s.

Some materials are more prone to underbead formation than others, as discussed in the next section, but for mild steel there is minimal attachment when cutting at half the maximum cutting speed, V_c (Arata et al. 1979).

(b) *Material properties affecting cut quality.* The ease with which a material can be cut and the quality of the resulting cut depend not only on the laser parameters but also on material parameters. In particular, the absorptivity of the material to the laser radiation used, its thermal conductivity and coefficient of thermal expansion are very important factors which influence cutting performance.

Ulmer (1974) drew up a classification of materials on these three basic properties as shown in fig. 2.49. The successful cutting of materials in column 1 depends on adequate laser power to generate the initial keyhole required to achieve penetration. Thus column 1 materials will require lasers of greater than around 3.5 kW. Columns 2 and 3 materials have just been discussed. The high absorptivity materials are easily affected by even low

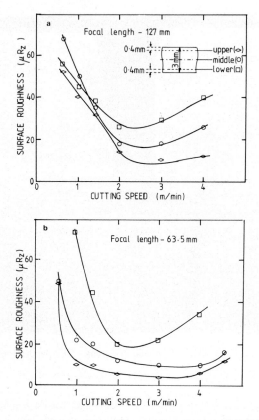

Fig. 2.47. Variation of roughness of cut surface with cutting speed. Mild steel, 1 kW laser power, 1 kg/cm^2 oxygen pressure (Arata et al. 1979).

powered lasers. The results, though, are not always useful. Some (e.g. column 6) tend to crack during cutting and will thus need special treatment such as preheating and post-cut tempering.

Other materials, column 7, will respond in a similar way to columns 2 and 3. Organic materials will usually vapourise at the beam centre and either have an invisible or undetectable effect at the kerf edge. They can also melt, char or burn. Ulmer (1974) suggests classifying these organic materials against their "incandescence resistance" and their "Vicat temperature", both being standard tests in the plastics industry. The incandescence resistance which is a measure of a materials resistance to high temperatures is found by placing a ceramic stick at 945°C against the material and

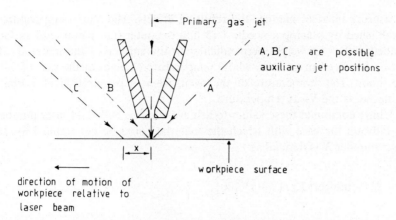

Fig. 2.48. Possible auxilliary jet configurations.

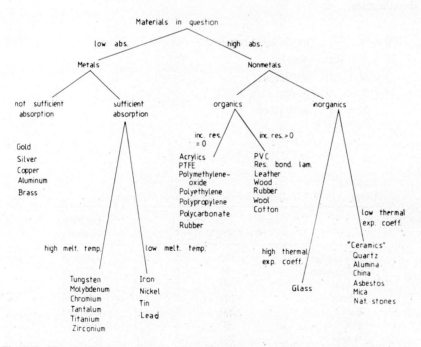

Fig. 2.49. Ulmer's (1974) classification of materials in terms of their absorptivity and incandescence resistance.

measuring in a set manner the resulting effects. The Vicat temperature is established by placing a needle 1.13 mm diameter (i.e. 1 mm^2 end surface) loaded with 5 kg weight perpendicular to the surface of the material. The set-up is placed in an oven whose temperature changes at the rate of 50°C per hour. The temperature of the oven, when a penetration of 1 mm is achieved, is the Vicat temperature.

Ulmer combined these values to achieve what he called a "laser number" to indicate the ease with which the material could be cut (table 2.6). The laser number N is defined as:

$$N = \text{integer}\left[T_V^2/(I+1)^3 100\right],$$

where T_V = Vicat temperature (°C); and I = incandescence resistance; N is expressed as an integer. However, since Vicat temperature measurement is only valid at values below 200°C, this analysis is only of limited use.

The value of N is used to predict cuttability according to the following classification:

$$
\begin{aligned}
100 &\leqslant N & &\text{extra good results,} \\
20 &< N < 100 & &\text{good,} \\
9 &< N < 20 & &\text{sufficient,} \\
4 &< N < 9 & &\text{poor,} \\
&N < 4 & &\text{very poor.}
\end{aligned}
$$

Table 2.6
Ulmer's (1974) "cuttability" forecast for different materials

	T_V(°C)	I	N	Cut ability forecast
Polyurethane (pure)	100	1	12	sufficient
Polystyrene	88	1	9	sufficient
Wood	200	2	14	sufficient
Cotton	200	2	14	sufficient
Paper	200	2	14	sufficient
PVC	80	1	8	poor
Wool	200	3	6	poor
Leather	200	3	6	poor

It is found that a material having a high Vicat temperature, T_V, and incandescence resistance, $I = 0$, will cut with a smooth edge, free of burrs, but the material is usually difficult to weld. A material that has a low Vicat temperature and $I = 0$ may have a cut with a burr. A material with a value of $I = 0$ would be expected to have a charred edge and be impossible to weld. If the value of $I > 2$ this problem is acute.

For example paper has a value of $I = 2$ and is classified as "sufficient" by its laser number. Ruffler and Gurs (1972) show laser cuts in paper which appear like those from blunt scissors. Ainsworth (1978) using a 400 W laser, found that paper could be cut without edge charring, provided the side gas jet used was properly aligned. He also found that the laser cut edge was stronger than from a conventionally slit edge, and the edge material could be repulped. In fact the laser produced what he described as a cut of superb quality, with less dust (but more smoke) than a slitting machine.

Thus, materials ranked by Ulmer as "sufficient" can, with care, be satisfactorily cut by laser.

Arata et al. (1979) describe the fairly reasonable theory that dross formation is dependent upon the difference in melting point of the oxide formed during any laser cutting and the metal,

$$\Delta T = \left(T_{m_{oxide}} - T_{m_{metal}} \right)$$

for low dross formation ΔT should be a minimum.

6.2. Hole drilling

Drilling has been one of the rapidly growing areas in the use of lasers. It is an attractive way of making small holes particularly at high production rates. The distinct advantages of the laser are:

(i) It is a non-contact process.

(ii) Precision location of holes is easily attained, even if location is difficult to reach or if the hole is to be drilled at a difficult angle – such as 10° to the surface.

(iii) The material is removed as a vapour so that there is no chip problem.

(iv) The holes are drilled quickly.

(v) Aspect ratios up to 100 : 1 are attainable.

(vi) Hard or soft materials can be drilled.

(vii) Hole diameters from 0.2 μm (Kocher 1975, using a frequency-quadrupled YAG laser) or even less (as in the latest video recording discs) up to 1.5 mm in single or repeated shots (or larger by trepanning) are available.

(viii) The drilling process can be observed with auxiliary optics and a television monitor.

The limitations of laser drilling are:

(a) limited depth (typically up to 13 mm for 0.2–0.7 mm diameter holes).

(b) Recondensed material and ejecta may be present at the lip of the hole.

(c) Wall roughness – sometimes with resolidified material on the wall.

(d) Hole taper – present to some extent in all laser holes, but controllable by mode and shape of pulse or multiple pulses. It can be minimal in thickness, < 250 μm.

(e) Limiting aspect ratio: this is a function of the depth of focus, thermal diffusivity of the workpiece and internal reflection effects. Normal values for metals of depth to width ratio are around 8–12 : 1; and for ceramics it is higher ~ 25 : 1 while in special cases 100 : 1 has been attained.

Some typical hole depths and drilling times are given in table 2.7. The most important parameters in hole drilling are: the hole diameter, depth, parallelism and edge quality. We will now consider these characteristics in turn.

Table 2.7
Typical drilling parameters from Laser Inc.

Material	Energy per pulse 50 J				
	Thickness	Hole diameter	Tolerance	Drilling rate	Application
Brass	15	1	0.1	60	controlled orifice for leak
Stainless steel		3	0.5	30	flush hole for wire EDM process (spinnerette)
Cold rolled steel	250	30	1	12	lubricating hole for conveyor system
Cobalt alloy	700	20	1.5	*	transpiration hole drilling (jet engine vanes platform holes
Steel	35(49 angle)	20	1	45	automotive air conditioner bleed hole
Stainless steel	6	4	0.5	60	hypo-needle
Stainless steel	185	10	–	25	suture needle
Copper	175	12	1	10	wire feed hole
Stainless steel	18	5	0.5	30	jet engine nozzle
Steel	18	7	0.5	30	fuel pump plate (automotive)
Stainless steel	5	4	0.1	60	spark advance hole (automotive)
Stainless steel	200	18	1	20	hydraulic nozzle (jet aircraft)
Tungsten carbide	250	25	1.5	14	tooling plate
Aluminium	–	9	0.5	30	electric circuit
Brass	20	2 to 30	0.25 to 1	60	nozzle for gauge
Inconel	100	50	2	10	feed hole (automotive)
Brass	10	0.5	–	30	optical aperture

The laser operating variables which affect these qualities are energy level/pulse, the energy spatial spread (mode structure), the energy temporal spread (pulse shape), wavelength of radiation, number of pulses/hole and the focal position.

The drilling mechanism varies from vaporization to fusion-controlled, depending on the pulse duration (fig. 2.50, Chun and Rose 1970).

The theory of hole drilling considered by means of energy balance is discussed by: Ready (1971); Dabby and Paek (1972); Anisimov et al. (1967); Wagner (1974); and Bar-Issac and Korn (1974). Stress analysis around a hole is adequately covered by Gagliano and Paek (1971).

6.2.1. Hole diameter

The diameter of the hole is dependent on the incident beam diameter. This diameter is determined by the laser cavity optics, the focusing optics and the laser wavelength. The diffraction limited spot size is given by $d_{min} = m2.4F\lambda$,

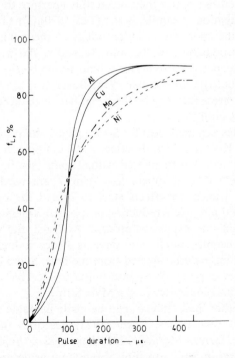

Fig. 2.50. Percentage material removed in liquid phase as a function of laser pulse duration (Chun and Rose 1970).

where: F is the focal length/incident beam diameter on the lens, and λ the wavelength; m is a multiple for the deviation in the beam mode structure from a pure parallel gaussian beam; $m = 4$ or 5 for high powered CO_2 lasers, but can be as low as 2 or 3 for a CO_2 laser with a carefully designed optical resonator. By mode control using apertures within the laser cavity, and carefully calculated cavity optics which reduce the effect of thermal drift during extended use of the laser, a value of m of about 2 can be achieved (Herziger et al. 1971). However Kato and Yamaguchi (1968) used two apparently similar ruby rods in the same laser drive system and found they produced spot sizes which differed by almost a factor of two.

As the laser rod is excited in an optically pumped solid state laser a radial temperature gradient may develop at high repetition rates. This causes thermal lensing which upsets the beam geometry. A variation in pumping power would thus cause a variation in spot size of the focused beam.

To overcome this effect, which could be a nuisance in a production situation, some manufacturers pass the beam from the laser through an aperture. The focusing optics then image this aperture to the required size (method of conjugation of pupils, Boyer et al. 1970). This has two beneficial effects. Firstly, the spot size is independent of the pumping power, thus giving more reproducible results, and, secondly, the power distribution across the heated spot is more uniform than would be the case with simple beam focusing (Battista and Shiner 1976; Boyer et al. 1970). The entrance cone created by direct focusing of the beam is also reduced by this method of conjugation of pupils.

Very small holes can be drilled by the frequency doubling or quadrupling of a YAG laser. Kocher (1975) describes a 0.2 μm hole in organic tissue.

The lens F number can be altered within limits (see fig. 2.51, Kato and Yamaguchi 1968). If F numbers less than 7 are used, then spherical abberration and truncation effects start to spread the beam. Thus an F number $\simeq 3$ would not give a reduction in spot size unless the lens was very carefully corrected – an expensive process! At $F = 3$, the working distance from the lens becomes small and there is hence a higher risk of lens contamination from material ejected from the hole. Some lens protection is possible by gas jet purging, or by inserting a replaceable transparent sheet between the lens and workpiece (e.g. a Myla Strip).

Hole shapes, other than circular, can be made in single or multiple shots by using Fresnel zone plates. Engel (1974) describes the production of square and ring forms. Moran (1971) has used a holographic lens to machine twin spots in a 0.2 μm film of tantalum using an argon laser. He records a diffraction efficiency of 50% which must be near the limit for holographic imaging.

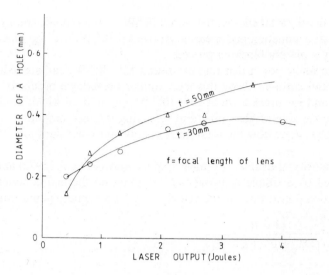

Fig. 2.51. Effects of lens focal length and laser power on hole diameter (Kato and Yamaguchi 1968).

Alternative ways of shaping holes have been suggested by Scott and Hodget (1968) who imaged a shaped aperture. The aperture was 1–3 cm in diameter and the image was 0.03 cm in diameter in order to obtain sufficient power density. The resolution was quoted at ±15 μm. Boyer et al. (1970) reports the engraving of quite complex shapes this way – machining a pattern within an area of 1 mm^2 in 250 μs. However, drilling by this method must be done with sufficient power to evaporate the workpiece, a process that is dependent on P/D^2, where P is the laser power and D is the focused beam diameter. If thermal conduction enters the process then it would depend upon P/D and the machined shape would be different from the aperture shape (Steen 1978). The hole diameter also varies with laser power. A thermal profile is generated on the workpiece by a gaussian beam and drilling will proceed at points which are heated above their boiling point. Thus it can be seen that the hole diameter will vary with laser power, and material thermal diffusivity (fig. 2.52).

6.2.2. Hole depth
The hole depth increases with input energy density up to a limiting value. For example, Hamilton and James (1976) found that the drilling efficiency was a maximum at 5×10^8 W/cm^2 when drilling stainless steel using a 250 mJ TEA laser. They measured efficiency as a specific depth where specific

depth = depth/total energy. Below 5×10^8 W/cm^2 too much energy is spent in thermal conduction and melting. Above 5×10^8 W/cm^2 too much energy is lost by a plasma blocking process.

On the whole power densities of around 10^6–10^8 W/cm^2 are required for drilling and cutting. Above this level vapour breakdown occurs at 10^8–10^9 W/cm^2 and gas breakdown at 10^9–10^{10} W/cm^2 both of which would create a plasma capable of absorbing and reflecting the incoming laser radiation.

Fig. 2.53 shows how the total pulse energy and pulse duration affect hole depth.

The velocity of drilling was measured by Andreev et al. (1972) and can be calculated by a simple heat balance assuming no thermal conduction and unidirectional heat flow as the velocity, V_{ss}, of a vapour/liquid interface,

$$V_{ss} = \frac{(1-R)I}{|\rho H_v + \rho C_p (T_v - T_0)|}.$$

Here:

I = power/cm^2;
ρ = density (g/cm^3);

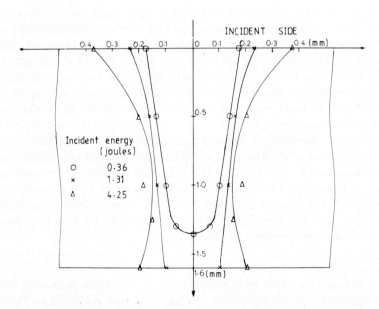

Fig. 2.52. Change in hole geometry with laser output (Kato and Yamaguchi 1968).

H_v = latent heat of vapourisation (J/g);
C_p = specific heat capacity (J/g °C);
T_v = vapourisation temperature (°C);
T_0 = initial temperature (°C); and
R = reflectivity.

In their experiments on aluminium, Andreev et al. (1972) used $I = 3 \times 10^7$, 1.5×10^8, 3×10^8 W/cm² therefore the expected velocity would be 17 m/s, assuming $R = 98\%$ up to 855 m/s if $R = 0$. (For Aluminium $H_v + C_p\rho(T_v - T_o) = 35\ 060$ J/cm²).

Penetration velocities of around 12 m/s were measured in aluminium films on glass by timing the appearance of plasma beneath the film. However, the graphs of Andreev et al. of time versus thickness suggest acceleration with increasing depth which must be wrong.

The depth of drilling can be increased by ultrasonic vibration. Mori and Kumehara (1976) obtained a 10% increase in hole depth by having a 25 μm amplitude ultrasonic vibration shaking the workpiece while drilling with a pulsed 5 J ruby laser.

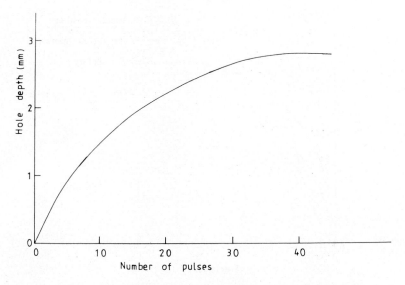

Fig. 2.53. Dependence of hole depth on number of pulses in drilling of saphire by a ruby laser (Charscham et al. 1977).

6.2.3. Aspect ratio

The aspect ratio depth/diameter of a hole is dependent on the depth of focus and internal reflections within the hole cavity. The depth of focus, z, is a function of the F number, $z = \pm 4 \times 1.39 \, (2\lambda \, F^2/\pi)$; where: λ = wavelength (m); F = Fresnel number = $L/2a$; L = focal length (m); and a = beam diameter (m); as seen in section 6.1.2.

Thus a low F number would give a smaller diameter and therefore greater aspect ratio in thin films. If, however, the thickness is greater than the depth of focus, then parallelism may be seriously reduced.

The aspect ratio also varies with the focal position (fig. 2.54). This is the single most important factor in controlling taper.

Unfortunately the same effects can be produced by varying the pulse length or peak intensity. Therefore it is not always clear how to change a particular hole geometry (Scott and Hodgett 1968). The use of multiple pulses into the same hole allows higher aspect ratios to be obtained with minimum taper. Typically a hole diameter of 0.1 mm through 0.5 mm material would have minimum taper if a low energy pulse was used about three to five times, whereas had the hole diameter been 0.5 mm through the

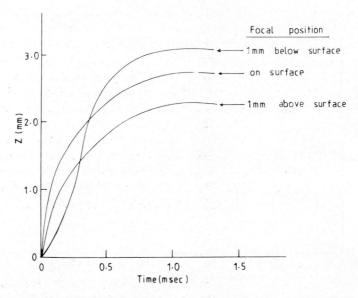

Fig. 2.54. Motion of the evaporation surface as a function of focusing conditions for Al with 5 J of laser energy (Bar-Isaac and Korn, 1974).

same material the best results would be expected with a single more powerful pulse.

The shape of the hole is due to the variations in the incident beam intensity, both transverse and along the axis of the hole, together with the scouring effects of the fast flowing vapour leaving the crater. Andrews and Atthey (1975) have calculated the gas velocity from a laser-generated keyhole but more simply, if the drilling velocity is around 10 m/s (Bar-Isaac and Korn 1974, Andreev et al. 1972) and the ratio of density of solid/gas is 1000 : 1 the exit velocity of gases would be around 10 000 m/s if it were not for sonic effects altering the density, and the possibility of three-dimensional flow.

The tapered shape is expected by a consideration of the need for a near parallel laser beam to maintain the hole walls liquid and is usually found to be from 1–10° for aspect ratios of 25 : 1 calculated on the diameter at the half depth point. Bottle shapes occur if the laser beam is focused within the material.

Edge quality. Edge quality is discussed further under the section on trimming and machining. It is however worth mentioning here some specific techniques for improving the edge and rim of a hole. They all depend on either blowing the material away or reducing the wettability of the hole lip. Mori and Kumehara (1976) used ultrasonic vibration to reduce the resolidified layer within a hole, the ultrasonic vibration increasing the fluidity of the dross. Hamilton and James (1976) blew CCl_4 into the interaction zone, thus increasing plasma formation and generating a blast wave which left a hole in a copper workpiece with no lip. They noted that this technique does not work with all materials. Other methods of improving edge quality have been reported using a gas blast effect from blowing on the underside with a strong gas jet (~ 40 psig) or coating the hole exit side with epoxy resin (Charschan 1972). Alternatively the lip surface can be adjusted to reduce lip formation. Williams (1966) suggests cutting through two layers, the top layer being removed with the offending resolidified lip material. Otstot et al. (1969) patented the idea of coating the surface with paraffin wax or silicone grease.

6.3. Machining and resistance trimming

Some of the current machining operations which utilise pulsed lasers are: (1) platemaking for the printing industry (Webster 1976); (2) resistance trimming of microelectronic circuits – both thick and thin film resistors and monolithic circuits; (3) trimming quartz oscillators; and (4) machining of conduction paths in printed circuitry.

Resistance trimming was first started in 1967 using CO_2 lasers on thick film circuits. By 1969 Q switched solid state lasers had taken over this application and by 1971 laser trimming of thick films was widely accepted. An interest in trimming the more sensitive and smaller thin film resistors also developed at this same time. In 1973 the trimming of quartz oscillators and filters for the watch and telecommunications industry was practised. The trimming of thin films on silicon and functional trimming of these circuits began in 1974. In 1976 the laser trimming of operating circuits became very important (Cozzens 1976). Today the laser has largely replaced the previous methods of sand abrasion and diamond wheel scribing. The trend is towards shorter wavelength lasers to produce final detail.

The normal method of producing thick film resistors is to silk screen the resistance pattern onto an alumina substrate in a resistor paste. The resistor paste may be a mixture of metals (AgPd) and metal oxides (Ag/Pd/PdO), glass frit, organic binders and solvents. After printing, the solvent evaporates. Firing removes the binder and melts the glass. The resulting sheet is firmly held on the alumina substrate. The best accuracy that this printing process can achieve in a given resistor is between $\pm15\%$ and 20%. A tolerance of less than $\pm1\%$ is usually required so that the resistances have to be adjusted after they are laid down.

This can be carried out by using a laser to machine a series of overlapping craters through the resistors until the desired value is achieved. The machined groove can take any of the forms illustrated in fig. 2.55. Since the laser process is a non-contact process that requires no chemical etchants it is possible to monitor the resistance, while machining, to achieve good accuracy and reliability to $(\pm0.1-0.01)\%$ (Rosen 1975). It is also possible to

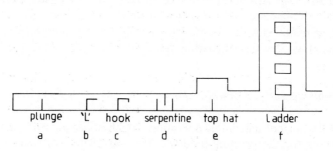

Fig. 2.55. Resistance trimming patterns (Rosen 1975).

carry out the trimming late in the production process, thus avoiding any alterations to the circuit trim by subsequent operations. Typical parameters for trimming thin films (~ 1 μm) are:

(Charschan et al. 1977)

pulse repetition rate	10 kHz,
peak power	1200 W,
pulse width	250 ns,
cut width	25 μm,
overlap of spots	65%,
post trim drift	0.1%.

Typical parameters for trimming thick films (~ 25 μm) are:

	(Charschan et al. 1977)	(Rosen 1975)
pulse repetition rate	3 KHz,	500–30 kHz,
peak power	4 kW	10^8 W/cm^2,
pulse width	250 ns,	150 ns,
cut	50 μm,	30–80 μm,
overlap	95%,	
post trim drift	1%,	
typical trimming speed	25 mm/s.	

There are five particular problems which need to be considered in this process.

(a) material recondensation in trimmed track;

(b) post trim drift in the resistance value due mainly to thermal stress microcracks appearing in the trim edge – this drifting in value ceases after 48 h if the laser parameters are correctly chosen and also if subsequent thermal shock due to soldering etc. causes no further drift (Rosen 1975);

(c) sputtered material forming a lip to the trim groove;

(d) thermal damage to the substrate; and

(e) finish at end of a cut (dependent on switching off time) (Sickman 1968).

Recondensation can be eliminated by use of an auxiliary thin jet of inert gas (Tomlinson 1978). Both (a) and (c) are minimised or eliminated by using a high peak power which avoids fused transitions in the cut and removes material as a vapour, with little thermal penetration. However, a high peak power must be as low as possible and still do the job if thermal damage to the substrate is to be avoided. Weick (1972) calculates this minimum power to be e times the threshold power needed to cause evaporation. However, a value of $4.5 \times$ threshold power is normally required to reduce the lip formation which is quite large at the threshold power.

To help solve this problem whereby low peak power is necessary to avoid substrate damage but high peak power becomes important if recondensation and lip formation are to be avoided, special resistor paints for laser trimming have been developed. In fact by careful adjustment of crystal sizes (i.e. controlled nucleation) the refractive index of a paste can be altered as predicted by the Maxwell–Garnett theory (Chopra 1969, Lou 1977). This controlled nucleation affects the reflectivity and reduces the threshold power substantially. According to Lou (1977), the threshold energy for trimming the bismuth films that are used in display and memory devices has been reduced from 22 nJ for untreated films to 7 nJ for treated films.

Post trim drift is due to microcrack formation within the thick or thin film. "Plunge" cutting as in fig. 2.55(a) has a current restriction at the tip of the plunge and a microcrack at this point would cause a significant resistance change. By cutting an "L" shape fig. 2.55(b) there would be several microcracks at the edge of the trim in the current restricted area and this would give a more reliable product since it is based on an expected average number of cracks instead of a single crack or relatively few cracks. "Hook" cutting fig. 2.55(c) takes the resistance rate to 10% of required accuracy on the plunge part, to 1% on the cross part and the final 1% on the cross cut. This has been shown to give a statistically significant improvement in resistor stability (Bube et al. 1978). In fact Bube et al. (1978) found less than 0.01% drift during thermal cycling between 50–150°C in a small hybrid microcircuit resistor.

Ladder cutting [fig. 2.55(f)] can avoid trim edge being totally located in a current-restricted zone. Post trim drift is predictable for "L" type cuts and hence can be allowed for.

Functional trimming. Laser trimming to accuracies of 0.25% in thin films deposited directly on to the integrated circuit silicon chip has removed the need for external adjustment potentiometers (Wagner 1978). The only restriction on this process is the heat sensitivity of certain elements on the chip. If the resistors are located on the thermal centre line between these devices measurement errors during trimming will be avoided. Optoelectronic effects can interfere with these measurements and this is avoided by syncopating measurements with laser pulses.

A resistor is usually present in order to control some functional output of an electrical circuit, e.g. voltage, ac or dc, frequency, or overall circuit resistance. These functional outputs can be monitored during laser trimming and this results in a considerable improvement in the cost effectiveness of the entire production process (Abenaim 1973).

The case for functional trimming is summarised as follows:

(a) there is no need to rely on calculated resistance values;

(b) tolerance on the parameters of other components can in consequence be relaxed – this results in considerable cost saving;

(c) functional trimming is usually the last operation so that changes in resistance value due to the subsequent bonding of other components is avoided; and

(d) final testing is carried out during trimming.

Engraving. The same micro-machining for resistance trimming can be applied to any material producing a clearly visible permanent line. It thus forms a useful marking system for identification, production information or theft protection. The patterns can be burnt onto the surface with minimal damage by engraving, dot matrix marking or mask imaging. Rastor machining with a laser allows the direct dry preparation of offset printing plates. The pattern to be printed is scanned with a laser, such as an He–Ne laser. The intensity of the reflected light is used to modulate the power of the machining laser (e.g. YAG, CO_2 etc) which scans the print roll in exact sequence with the He–Ne laser.

Line densities of up to 600 lines/cm with engraving speeds of up to 5 m/s with a minimum line thickness (which can be read and engraved) of around 85 μm were obtained using a 20 W CO_2 laser with 40 μm spot size engraving a polymer coating laid on aluminium. The modulation of the cutting laser power can be achieved by using a plane polarised laser beam and modulating the plane of polarisation.

Scribing. Scribing is a process whereby a groove is formed in a surface and a bending stress subsequently allows fracture of brittle material along the line of the groove.

Conventionally the groove was formed with a diamond tool. The laser offers an attractive alternative not only because it is a chemically clean non-contact process but also because the groove shape has a reasonably sharp-angled base suitable for a stress raiser and it can be made as deep as is required for reliable separation.

The scribing of silicon wafers cannot be done with a CO_2 laser since silicon is partially transparent to 10.6 μm radiation. Q-switched Nd/YAG lasers are usual for this application. The scribed groove is made with a power density of around 10^9 W/cm^2 and 0.1 μs pulses to ensure good vapourisation and low heat affected zone. 70% penetration and 87% overlap of spots gives optimum fracturing characteristics with minimum force (Charschan 1972). But in any case, a penetration of at least one fourth of the substrate thickness is required for good separation (Charschan et al. 1977). The scribed groove may be continuous or a series of blind holes acting as perforations. The use of a laser allows closer spacing of components than with diamond point scribing. It is also quicker, typical speeds being around 7.5 cm/s (Eleccion 1972).

Typical scribing parameters and rates are shown in table 2.8.

Hot machining. Heating the workpiece to the softening point just prior to machining mechanically can reduce tool wear by around 50%.

Homach Ltd, part of BOC Ltd new ventures, has successfully operated this process using plasma torches since 1973. More recently Bass et al. (1979) have reported the use of a laser for the required localised heating.

There is a problem however with heating transformation-hardening materials this way. Either the material has to be removed before it has cooled sufficiently, to harden or the depth of hardness is less than the mechanical machining depth.

Table 2.8
Scribing parameters and rates for different materials (Lumley 1969)

Material	Material thickness	Laser	Power	Pulse length	Groove depth	Width	Rate
Al_2O_3	0.635 mm	CO_2	100 W				25.4 mm/s[a]
Si Transition							0.254 mm/s[a]
Al_2O_3	0.71 mm	Nd/YAG?	300 W	100 μs 1 kHz	25%	0.127 mm	2.54 mm/s[b]
Si			500 W	300 ns 35 kHz	80% overlap	0.025 mm	2.54–3.4 mm/s[c]
Si		Nd/YAG	10^9 W/cm^2	0.1 μs	70%–87% overlap		[c]
Al_2O_3	0.64 mm	CO_2	20 W	3 ms	0.125 mm	0.1 mm	50 mm/s[c]

[a] Gagliano et al. (1969).
[b] Charschan et al. (1977).
[c] Shkarofsky (1975).

Surgical uses of lasers. Cutting biological material with a laser is a specialist subject area. It is mentioned here because the use of lasers for surgical purposes is becoming a major cutting application.

Consider the following advantages of laser in the field of surgery.

(i) The laser is an absolutely sterile cutting blade that creates a biologically clean wound which is protected by coagulated tissue.

(ii) A laser cut is self cauterising and gives substantial reductions in bleeding and oozing.

(iii) There is no pressure on the wound area. Therefore there is less risk of spreading poisonous material, less trauma and cuts can be made very close to malignant tissue.

(iv) The cuts are so clean they are able in some cases to heal without a visible scar.

(v) Certain radiation (e.g. 1.06 μm from Nd/glass or YAG lasers or 0.4880 or 0.5145 μm from an argon laser) can be passéd down glass fibres and thus used for internal surgery by means of an endoscope (Brunetaud et al., 1975).

(vi) Certain radiation (e.g. an argon laser) will pass through the vitreous humour and serous tissues (membranes) of the eye with little absorption and so allow retinal surgery in the eye without having to remove the eye from its socket.

(vii) The laser can cut bone or flesh with ease.

Many people think of the laser as having unlimited penetration and therefore as being a highly dangerous surgical tool. This is not the case. A short focal length lens and low powered laser < 200 W will only be able to cut biological material for a short depth approximately equal to the depth of focus. Some surgeons use lasers with a short stroking action to cut a depth of around 1 mm per stroke depending on the case. In order to get this flexibility of action the laser beam is passed into a "pen-like" lens holder, which the surgeon manipulates. The beam is transmitted from the laser by a bundle of glass fibres if the wavelength allows or via a series of articulated mirrors.

For eye surgery the laser is precisely located through a microscope objective down which the surgeon can simultaneously view the operation.

The biological threshold is 10^{-3} W/cm^2 at which power density thermo-chemical reactions would start; at 1 W/cm^2 coagulation is possible and at 10^2 W/cm^2 charring and evaporation can occur (Eichler et al 1977).

For coagulation processes the argon ion laser is preferred to the Nd/YAG since the blue–green light from the argon laser is highly absorbed by body tissues and haemoglobin whereas the YAG laser radiation has considerable penetration depth which could lead to complications.

In neurosurgery a 20 W CO_2 laser with a He/Ne alignment laser has been used by Ascher (1979) who observed that at these low powers there was no explosive evaporation of brain tissue, as was feared, and that the heat affected zone around a laser hole was minimal.

Penetration rates of about 1 mm/s at 30 W produced well sealed surfaces. The laser effect was better than the tearing or heating action of a lancet or electric needle. Ascher (1979) also observed that nerves cut using a laser in a rabbit did not continue to grow due to the better sealing action. The laser would thus appear to have advantages in amputations to stop nerve resuscitation.

"Lasers in medicine" is really the subject of a separate book and so the interested reader is referred to the conference proceedings on Lasers in Medicine in Laser '75 and Laser '79 Opto-Electronic Conferences.

7. Conclusion

What has been shown is that the laser is one of the most versatile cutting tools, capable of producing high quality, high speed cuts in most materials with little thermal damage or chemical contamination. However, the success of the laser cutting process depends upon many variables, principally the absorbed laser power, the focused spot size and the cutting gas flow characteristics.

The process has only been seriously studied since 1971 and during this time it has been used in a wide variety of applications from metal working to medicine and from woodworking to microelectronics. Undoubtedly many new applications will be added in the years to come, as well as many significant advances in the technology of the process. The present review has been an attempt to describe the current state of the art.

References

Abenaim D., 1973, Microelectronics 5 No. 1, 33.
Adams M.J., 1970, Metal Construction and Brit. Weld. J. (Jan.) 2 No. 1, 1.
Addison G.M., 1971, Machinery and Prod. Eng. (Feb.) 17 256.
Ainsworth H.S., 1978, Paper Technology and Industry 19 No. 7, 220.
Anderson D.G., 1978, Mech. Eng. (USA) 100 No. 6, 44.
Andreev S.I., et al., 1972, Sov. Phys.-Tech.Phys. (Oct.) 17 No. 4, 705.
Andrews J.G. and Atthey D.R., 1975, J. Inst. Math. Applics. 15 59.
Andrews J.G. and Atthey D.R., 1976, J. Phys. D: Appl. Phys. 9 2181.
Anisimov, S.I., et al., 1967, Sov.Phys.-Tech. Phys. 11 No. 7, 945.
Arata Y. and Miyamoto I., 1972, Trans. JWRI 1 No. 1.
Arata Y., Mario M., Miyamoto I. and Takachi S., 1979, Trans. JWRI 8 No. 2, 15.

Ascher P.W., 1979, Laser applications in neurosurgery, in: Laser 79 Opto-Electronics Conf., Munich, FRG, 2–6 July, 1979, ed., W. Waidelich (IPC Science and Tech. Press).

Atkey M., 1978, Machinery and Production Eng. **133** No. 3441, 23.

Bar-Isaac C. and Korn U., 1974, Appl. Phys. (Jan.) **3** 45.

Bass M., Copley S., Beck D. and Wallace R.J., 1979, Laser-assisted hot spot machining, in: AIP Conf. Proc. **50** No. 1, 205.

Battista A.D. and Shiner W.H., 1976, SME Laser Conf., Dec. 1.

Berry M., 1978, Sheet Metal Industries, **55** No. 1, 24, 26.

Boehme D., and Herbrich N., 1978, Metalwork production (GB) **22** No. 10, 137.

Boyer G., Huriet J.R. and Lamouroux B., 1970, Opt. Laser Tech. **2** No. 4, 196.

Brunetaud J.M., et al., 1975, Endoscopic argon-ion laser coagulation in the digestive tract: development of a photo-coagulator and experimental study, in: Laser 75 Opto-Electronics Conf., Munich, FRG, June 1975 (IPC Science and Tech. press, 1976).

Bube K.R., et al., 1978, Sol. St. Tech., (Nov.) **21** No. 11, 55.

Bunting K.A. and Cornfield G., 1975, Trans. ASME, J. Heat Trans. (Feb.) 116.

Carslaw H.S. and Jaeger J.C., 1959, Conduction of heat in solids, (Oxford University Press, New York).

Charschan S.S., ed., 1972, Lasers in industry (Van-Nostrand Reinhold, New York).

Charschan S.S., et al., 1977, Guide for material processing by lasers (the Laser Institute of America) ch. 8.

Chopra K.L., 1969, Thin film phenomena (McGraw Hill, New York).

Chun M.K. and Rose K., 1970, J. Appl. Phys. (Feb.) **41** No. 2, 614.

Clarke J. and Steen W.M., 1978, Arc augmented laser cutting, Proc. Laser 78 Conf., London, March.

Courtney C. and Steen W.M., 1978, Measurement of the diameter of a laser beam, Appl. Phys. **17** 303.

Cozzens W.B., 1976, Electron. Eng. (Feb.) **48** No. 576, 58.

Dabby F.W. and Paek U.C., 1972, IEEE J. Quantum Electron. **QE-8** No. 2, 106.

Diels J.-C., 1976, Laser Focus 1976, pp. 12–33.

Doxey B., 1976, Laser engraving of flexographic printing cylinders, PIRA/EFTA Flexographic Technology Conf., London, Hilton.

Duley W.W., 1976, CO_2 lasers: effects and applications (Academic, New York).

ℓ Duley W.W., 1979, Optics and laser technology, **11** 331.

Duley W.W. and Gonsalves J.N., 1972, Can. J. Phys. **50** 215.

Duley W.W. and Young A., 1973, J. Appl. Phys. **44** No. 9, 4236.

Dutta M., 1974, PhD Thesis, Birmingham University.

Eichler J., et al., 1977, Kvantovaya Elektron (Moscow) (Dec.) **4** 2610; Sov. J. Quantum Electron. (Dec.) **7** No. 12, 1492.

Eleccion M., 1972, IEEE Spectrum **9** 62.

Engel S.L., 1974, Appl. Opt. **13** No. 2, 269.

Engel S.L., 1978, Opt. Eng. (USA) (May–June) **17** No. 3, 235.

Evans L.J., 1976, Plastics Tech., October.

Fletcher M.J., 1977, Eng. Mat. Design, **21** No. 6, 52.

Forbes N., 1975, The role of the gas nozzle in metal cutting with CO_2 Lasers, 1976, in: Laser 75 Opto-Electronics Conf., Munich, FRG, June 1975 (IPC Science and Tech.) p. 93.

Forbes N., 1976, Fabricator **6** No. 5.

Ford (Cologne) 2/3, 1974: Reprint.

Gagliano F.P. and Paek U.C., 1971, IEEE J. Quantum Electron. **QE-7** No. 6, paper 3.3, p. 277.

Gagliano F.P., Lumley R.M. and Watkins L.S., 1969, Lasers in industry, Proc. IEEE, **57** 114.

Grumman, 1976, Iron Age **217** No. 24, 43.
Haller W. and Winogradoff N.W., 1971, J. Am. Ceram. Soc. – Discussions and Notes **54** 314.
Hamilton D.C. and James D.J., 1976, J. Phys. D: Appl. Phys. **9** No. 4, L41.
Herbrich H., 1975, Advanced Weld Tech. (Japan Weld. Soc., Tokyo) p. 131.
Herziger G.I., Loebtscher J.P. and Steffen J., 1971, IEEE J. Quantum Electron. **QE-7** No. 6, paper 3.7, p. 279.
Herziger G., Stemmep R. and Webber H., 1974, IEEE J. Quantum Electron. (Conf. paper) **QE-10** No. 2, pt. 2.
Hitachi K.K., 1974, Japanese Patent No. 75052349.
Hoffman M., 1974, Weld. Metal. Fabr. **42** No. 1, 6.
Houldcroft P.T., 1968, Optics Tech., (Nov.) **1** 37.
Huber J., 1977, in: Proc. 5th N. Am. Metalworking Res. Conf., Amherst, Mass., 23–25 May (Dearbon, Mi., SME., 1977) p. 57.
Hultgren R., et al., 1963, Selected values of thermodynamic properties of metals and alloys (John Wiley, New York).
Kahaner L., 1977, Am. Metal Market, Dec. 12.
Kamalu, J.N. and Steen W.M., 1981, TMS paper A81-38 publ. AIME.
Karr L.A., 1975, Proc. Laser 75 Optoelectronics Conf., Munich, FRG, June 1975 (IPC Science and Tech., 1976) p. 96.
Kato, T. and Tamaguchi T., 1968, Laser machining, NEC research and development, No. 12.
Klemens P.G., 1976, J. Appl. Phys. (May) **47** No. 5, 2165.
Kocher E., 1975, Proc. Laser 75 Opto-Electronics Conf., Munich, FRG, June 1975 (IPC Science and Tech., 1976).
Kovalenko V.S., Arata Y., Maruo H. and Miyamoto J., 1978, Trans. JWRI, **7** No. 2, 101.
Lawton J. and Mayo P.J., 1974, J. Eng. Mat. Tech. (Trans. ASME) (July) **96** No. 3, 168.
Locke E.V. and Hella R.A., 1974, IEEE J. Quantum Electron. (Feb.) **QE-10** No. 2.
Longfellow L., 1973, Sol. St. Tech. **16** No. 8, 45.
Lou D.Y., 1977, J. Appl. Phys. (May) **48** No. 5.
Lumley R.M., 1969, Am. Ceram. Bull. **48** 850.
Lunau F.W., 1970, Conf. paper: in: Conf. on Advances in welding processes, Weld. Inst., Cambridge, GB.
Masters, J.I., 1956, Problem of intense surface heating of a slab accompanied by a change of phase, J. Appl. Phys. **27** 477.
Mazumder J. and Steen W.M., 1980, Heat transfer model for CW laser processing, J. Appl. Phys. **51**(2) 941.
Megaw H.P.C. and Spalding I.J., 1976, Phys. Tech. **7** No. 5, 187.
Moran J.M., 1971, Appl. Oct. (Feb.) **10** No. 2, 412.
Mori M. and Kumehara H., 1976, Bull. Japan Soc. Precision Eng. **10** No. 4, 177.
Moss A.R. and Sheward J.A., 1970, IEE Conf. on Electrical methods of machining forming and coating, March 1970, 41.
Olsen, F., 1981, Proc. Laser 81 Optoelectronics in Engineering Conf. (Springer, Berlin) pp. 227–231.
Otstot R.S., et al., 1969, Method for machining with laser beams, US Patent No. 3 440 388.
Peters C.C. and Banas C.M., 1977, Forestry Prod. J. (Nov.) **27** No. 11, 41.
Prifti W.E., 1971, Laser industrial application, Notes 1–71.
Ready J.F., 1965, J. Appl. Phys. **36** 1522.
Ready J.F., 1971, Effects of high power laser radiation (Academic, New York).
Ready J.F., 1978, Industrial applications of lasers (Academic, New York).

Rosen H.G., 1975, Conf. Proc., Laser 75 Opto-Electronics Conf., Munich, FRG, June 1975 (IPC Science and Tech., 1976).

Rosenthal D., 1946, Trans. Inst. Min. Mech. Eng. **68** 849.

Ruffler C., and Gurs K., 1972, Optics and Laser Tech., (Dec.) **4** pt. A, 265.

Saunders R.J., 1977, in: paper presented at: Conf. on Industrial applications of high power laser tech., San Diego, Calif., 24–25 August 1976 (Palos Verdes Estates, Calif., Soc. of Photo-Optical Engrs. 1977) p. 32.

Schawlow A.L., 1977, Laser interactions with materials, in: Laser 77 Opto-Electronics Conf. Proc., Munich, FRG, 20–24 June, ed. W. Waidelich (IPC Science and Tech., 1977).

Scott B.F. and Hodgett D.L., 1968, Proc. Instn. Mech. Engrs. 1968–69, **183**, pt. 3D, 75.

Seaman F.D., 1976, AFML-TR-76-158 Tech. Report.

Serchuk, A., 1978, Mod. Packag., **51**, No. 1, 29.

Sharp C.M., 1978, Laser welding and drilling, in: SPIE **164** Fourth European Electro-Optics Conf. (Sira, Oct., Ultrech, Netherlands) p. 271.

Shinada K., et al., 1980, in: Proc. Int. Conf. on Weld. Res. in the 1980s, Osaka, Japan, Oct. 27–29.

Shkarofsky I.P., 1975, RCA Rev., **36** No. 2, 336.

Sickman, J.G., 1968, Cutting of thin film metals with a CO_2-gas laser beam, in: Microelectronics and reliability, Vol. 7 (Pergamon) pp. 305–311.

Slivinsky S.H. and Ogle N.E., 1977, J. Appl. Phys. (Sept.) **48** No. 9, 360.

Smithells C.J., 1976, ed., Metals reference book (Butterworths) 5th ed.

Steen W.M., 1977, Thermal history of a spot heated by a laser, Heat and Mass Trans. Lett. **4** 167.

Steen W.M., 1978, UK Patent Application No. 7 944 188.

Steen W.M., 1980, J. Appl. Phys. **51** 5636.

Steen W.M. and Eboo M., 1979, Metal Construction **11** No. 7, 332.

Steverding B., 1970, Pressure and impulse generation by Q-switched lasers, J. Phys. D: Appl. Phys. **3** 358.

Steverding B., 1971, Thermomechanical change by pulsed lasers, J. Phys. D: Appl. Phys. **4** 787.

Swift-Hook B.D. and Gick A.E.F., 1973, (Nov.), Weld, Res. Suppl., 492S.

Tandler W.S.W., 1971, Laser Focus **7** No. 3, 24.

Tomlinson J., 1978, Radio and Electron. Eng. (GB) (Jan.–Feb.) **48** No. 1–2, 43.

Uglov A.A., Kokora A.N. and Berlin N.V., 1978, Cutting of non-metallic materials by a CO_2 laser beam, Sov. J. Quant. Electron. **8**(7), 884–887.

Ulmer W., 1974, Electro. Optical Systems design, (July) **6** 33.

Wagner R.E., 1974, J. Appl. Phys. (Oct.) **45** No. 10, 4631.

Wagner R., 1978, New Electron., (June) **11** No. 13, 26, 29.

Webster J.M., 1976, Prod. Engr. **55** No. 7/8, 373.

Weick W.W., 1972, IEEE, J. Quantum Electron. (Feb.) **QE-8** No. 126.

Wessling J., 1977, Am. Metal Market/Metalworking News (July 11) p. 12.

Williams D., 1966, Eng. Proc., (May), Penn. State Univ., p. 44

Withers P.B. and Wilshaw T.R., 1973, J. Phys. D: Appl. Phys. **6**.

Yessick M. and Schmatz D.J., 1975, Met. Prog., (May) **107** No. 5, 61.

Young A., 1978, Metal Construction, (Jan.) 34.

CHAPTER 3

LASER WELDING

J. MAZUMDER

University of Illinois at Urbana-Champaign
Urbana, Illinois 61801, USA

Laser Materials Processing, edited by M. Bass
© *North-Holland Publishing Company, 1983*

Contents

1. Introduction

Lasers have been hailed as a potentially useful welding tool for a variety of applications since their initial development. Until the seventies, however, laser welding was restricted to relatively thin materials and low speeds because of the limited continuous power available. A variety of laser systems had already been developed by 1965 for the production of micro-welds in the fabrication of electronic circuit boards, inside vacuum tubes, and in other specialized applications where conventional technology was unable to provide reliable joining. The availability of high power CW CO_2 lasers and the limitations of current welding technology have stimulated considerable interest in deep penetration welding in the past ten years using these devices. The rapid technological developments have brought many of these laboratory curiosities into daily use.

2. Principles of laser welding

The potential advantages (Spalding 1971, Arata and Miyamoto 1978) of laser welding are the following.

(1) Light is inertialess; hence high processing speeds with very rapid stopping and starting becomes possible.

(2) Focused laser light can have high energy density.

(3) Laser welding can be used in room atmosphere.

(4) Difficult materials (e.g. titanium, quartz, etc.) can be handled.

(5) The workpiece need not be rigidly held.

(6) No electrode or filler materials are required.

(7) Narrow welds can be made.

(8) Very accurate welds are possible.

(9) Welds with little or no contamination can be produced.

(10) The heat affected zone adjacent to the cut or weld is very narrow.

(11) Intricate shapes can be cut or welded at high speed, using automatically controlled light deflection techniques.

(12) Time sharing of the laser beam can be achieved.

The localized heating obtained with laser sources was soon realized to be an important advantage. Anderson and Jackson (1965b) reported an interesting comparison between heating effects produced with a conventional arc source and those occurring with a pulsed laser. Their result is shown in fig. 3.1. They show that not only is the heat affected zone small but also the laser source is more efficient since it requires only 37.8 J/cm^2 to produce melting to the required depth, while the arc source must deliver 246 J/cm^2 to the workpiece. Localized heating makes the laser ideal for welding electronic components such as printed circuit boards where high average temperatures even in small volumes surrounding the weld region cannot be tolerated (Duley 1976).

The scope for technically and commercially feasible laser welding applications has increased greatly since the development of multikilowatt CW CO_2 lasers around 1970 (Brown 1970, Brown and Davis 1970, Patel 1964). The laser's capability for generating a power density greater than 10^6 W/cm^2 is a primary factor in establishing its potential for welding in general and for deep penetration welding in particular.

The mechanism of deep penetration welding by a laser beam is very similar to that encountered with an electron beam, i.e. energy transfer via "keyhole" formation (Duley 1976, Meleka, 1971, Swifthook and Gick 1973, Klemens 1976). This keyhole may be produced when a beam of sufficiently high power density causes vaporization of the substrate and the pressure

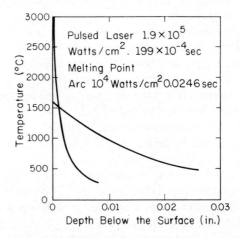

Fig. 3.1. Maximum temperature reached as a function of depth in iron for heating with a laser pulse and the pulse from a conventional arc source (Brown and Davis 1970).

produced by the vapor in the crater causes displacement of the molten metal upwards along the walls of the hole. This hole acts as a blackbody and aids the absorption of the laser beam as well as distributing the heat deep in the material. On the other hand, in most conventional welding processes the energy is deposited at the surface of the workpiece and is brought into the interior by conduction.

The conditions of energy and material flow during beam welding were investigated theoretically by Klemens (1976). According to Klemens the "keyhole" or cavity is formed only if the beam has sufficient power density. The keyhole is filled with gas or vapor created by continuous vaporization of wall material by the beam. This cavity is surrounded by liquid which in turn is surrounded by solid as shown in fig. 3.2.

The flow of the liquid and surface tension tend to obliterate the cavity while the vapor which is continuously generated tends to maintain the cavity. There is a continuous flow of material out of the cavity at the point where the beam enters. For a moving beam, this "keyhole" achieves a

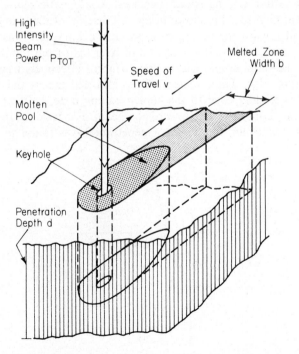

Fig. 3.2. Penetration weld geometry (Swifthook and Gick 1973).

steady state, i.e. the cavity and the beam associated molten zone move forward at the speed set by the advance of the beam, while the material lost by vaporization shows up as a depression in the solidified melt, as porosity or as an inward deformation of the workpiece, or possibly as a combination of these effects. The requirement that sufficient vapor be produced to maintain a steady state leads to a minimum advance speed for a steady state. While the cavity moves through the solid and liquid material at a speed determined by the motion of the beam, material must be moved continuously from the region ahead of the cavity to the region behind it. This is confirmed by experimental works of Sickman et al. (Sickman and Morijn 1968a, b) and Arata et al. (1977).

An interesting experiment on the mechanism of deep penetration welding with a CW CO_2 laser has been reported by Sickman and Morijn (1968a, b). This experiment was performed on transparent fused quartz so that the development of the weld in time could be directly followed photographically. The laser was seen first to drill a hole in the quartz, which was then translated through the material. However, they found that the profile of the hole varies as the welding speed increases. A schematic representation of this profile at high speed is shown in fig. 3.3. The tip of the hole was seen to bend around toward the direction in which the workpiece was translated. This process is caused by the reflection of laser light from the leading edge of the hole. Material evaporated at this surface is effectively trapped by the cooler trailing edge. Thus, material is transported across the laser beam from the hot leading edge to the cooler trailing edge, without significant ejection of material back out toward the beam. For welding speeds in the range 10–45 mm/min, the depth of penetration was found to be linearly

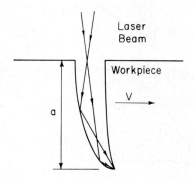

Fig. 3.3. Shape of "keyhole" during laser welding of quartz (Sickman and Morijn (1968b).

related to welding speed. As expected, the penetration was least for the highest welding speed.

Fluid flow during penetration welding was also studied by Arata et al. (1977) using laser beam irradiation of a glass with low viscosity at high temperature. High speed photography of the phenomenon, taken at a speed of 8000 frames per second, clearly showed a melt flow and the motion of the cavity. Molten fluid, formed at the front wall of the cavity, was accelerated at an angle along the wall by the violent driving force of the laser driven vaporization. In the process, a large vortex was formed behind the cavity near the surface (fig. 3.4). This is considered to be the cause of the generation of so called "wine-glass" beads.

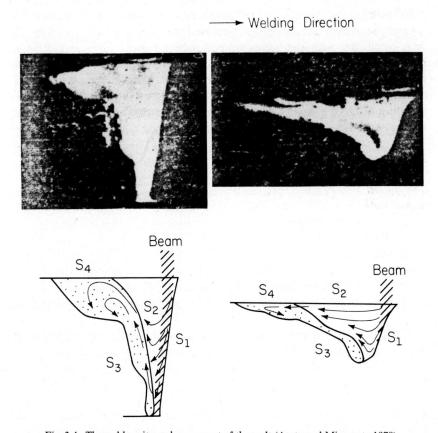

Fig. 3.4. The weld cavity and movement of the melt (Arata and Miyamoto 1978).

The transport of material is mainly by flow in the liquid. However, part of the transport is in the vapor phase, and the vapor transport provides the excess pressure which drives the liquid flow.

The extent to which the beam can penetrate the cavity and the absorption of energy depend on the degree of ionization and hence on temperature (Klemens 1976). Once the temperature is high enough for ionization to be complete, the absorption decreases with increasing temperature, but faster than in the case of electron beam welding. The absorption is least on the outside, and increases towards the center. However, if the central temperature is high enough, the absorption drops off again towards the center (Klemens 1976). For low beam intensities, the beam penetration is limited, but at high beam intensities penetration grows with beam power, as in the case of electron beam welding. An oversimplified model by Klemens (1976) indicates that an increase in welding depth can only be achieved by a quadratic increase in beam power provided the total flux exceeds 1 kW. He also noted that with the increased power requirement there would be an increase in the power absorbed per unit length, so that it would be possible to increase welding speeds and thus recover at least some of the disadvantage of the beam power requirement. Although the quantative estimates resulting from the model are uncertain, the increased penetration with increased power should be qualitatively correct.

3. Laser welding variables

The major independent process variables for laser welding include the following:
  (1) incident laser beam power;
  (2) incident laser beam diameter;
  (3) absorptivity;
  (4) traverse speed of the laser beam across the substrate surface; and
  (5) other parameters such as: shielding gas, depth of focus with respect to the substrate, weld design and gap size for butt welds also play important roles.

The dependent variables are considered to be (a) depth of penetration, (b) microstructure and metallurgical properties of the laser welded joints.

3.1. Laser beam power

The depth of penetration during laser welding is directly related to the power density of the laser beam and is a function of incident beam power and beam diameter. Generally, for a constant beam diameter, penetration increases with the increasing beam power. Lock et al. (1974) and Baardsen et al. (1973) report that penetration increases almost linearly with incident laser power as can be seen in figs. 3.5 and 3.6. Generally it is observed that for a laser welding of a particular thickness, a minimum threshold power is required.

The total power produced by the laser is measured by various techniques. For example, a precalibrated infrared detector is used for the continuous monitoring of the output power in the AVCO 15 kW CO_2 laser and in Photon Sources lasers. A flowing water copper cone calorimeter is used in 2 kW CO_2 laser manufactured by Control Lasers (UK) Ltd. Flowing water copper cone calorimeters are also being used by AVCO metal working lasers to measure the total output power outside the laser cavity. Blackbody calorimeters are also commercially available to measure output powers up to several kilowatts. Since the weld penetration depends on the power delivered to the workpiece surface, care should be taken to measure the power incident on the workpiece. Generally, the power delivered to the workpiece is less than the output power generated at the laser cavity due to the reflection losses from the steering mirrors.

3.2. Laser beam diameter

This parameter is one of the most important variables because it determines the power density. However, it is very difficult to measure for high power

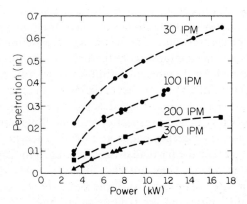

Fig. 3.5. Relation between penetration depth versus laser power for welding with the AVCO CO_2 laser; material: 304 stainless steel (Locke and Hella 1974).

J. Mazumder

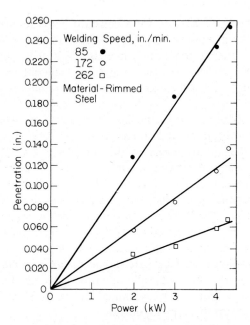

Fig. 3.6. Relation between penetration depth versus laser power for welding with the United
Technologies CO_2 laser (Baardsen et al. 1973).

laser beams. This is partly due to the nature of the beam diameter and
partly due to the definition of what is to be measured. A gaussian beam
diameter may be defined as the diameter where the power has dropped to
$1/e^2$ or $1/e$ of the central value. The beam diameter defined on the basis of
$1/e^2$ of the central value contains more than 80% of the total power whereas
the power contained for $1/e$ beam definition is slightly over 60% (Harry
1974) (fig. 3.7). Therefore, the $1/e^2$ beam diameter is recommended.

Many techniques have been employed to measure the beam diameter, but
most are unsatisfactory in some respect or another. Single isotherm contouring techniques such as charring paper and drilling acrylic or metal plates,
suffer from the fact that the particular isotherm they plot is both power and
exposure time dependent. They are also highly unlikely to coincide with
either the $1/e$ or $1/e^2$ diameters. Multiple isotherm contouring techniques
overcome these difficulties but are tedious to interpret.

One of the better methods for the measurement of beam diameter
employs a beam chopper and a photon drag detector (Mazumder 1978,

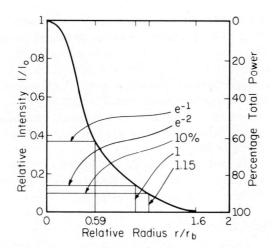

Fig. 3.7. Variation of relative intensity and percentage total power with radius for a Gaussian beam (Harry 1974).

Gibson et al., 1970, Danismevskii et al. 1970, Gibson and Walker 1971). Generally, a photon drag detector consists of a bar of p-type germanium which generates a longitudinal emf under open circuit conditions when exposed to 10.6 μm radiation from a CO_2 laser. The emf generation is due to the transfer of photon momentum to mobile holes and this phenomenon is known as photon drag effect (Gibson et al. 1970, Danismevskii et al. 1970, Gibson and Walker 1971). A beam chopper with a small slit width is used to scan the beam so that a part of the total power is incident on the photon drag detector as a function of the position of the slit within the beam. This technique gives an idea of the spatial distribution of power within the beam. To convert the signal from this combination to a measure of the beam's spatial distribution, it is necessary to assume the beam is axisymmetric and this might not always be strictly correct. However, this assumption is only critical if one makes an attempt to interpret the spatial mode distribution of the beam. For simple beam diameter measurement, this is not a problem. Although to date this device has only been used for TEM_{00} or lower order mode beams it may also be used for beams with other types of spatial distributions. Previous work by the author (Mazumder 1978) revealed that the response time for the photon drag detector is fast enough to determine an accurate beam diameter.

The equivalent gaussian beam diameter for any laser beam may be determined in terms of its heating effect as reported by Courtney and Steen

(1978a, b). In this method the time taken for a spot heated by the laser to reach 90% of its equilibrium temperature is measured using an infrared thermometer. This time is then used in a mathematical model of the heating arising from a gaussian power distributed beam to calculate the gaussian beam diameter equivalent of the beam used. This indirect method of measuring the beam diameter has the benefit of measuring it by the heating effect, not geometrically, and thus removes much speculation on mode structure. The overall accuracy for this method is considered to be $\pm 30\%$ (Courtney and Steen 1978b).

The diffraction limited spot size at the focal point of a laser beam can be calculated on the basis of the diffraction theory of light, assuming there are no aberrations formed by the lens. This gives

$$D_b = 2.44 \frac{\lambda f}{D} (2M + 1)^{1/2}, \tag{3.1}$$

where: D_b = beam diameter; λ = wavelength of the laser beam; f = focal length of the lens; D = diameter of the unfocused laser beam; and M = number of oscillating modes.

It is evident from eq. (3.1) that the smaller the f/D ratio the smaller the D_b. Smaller values of M will also give the smaller beam diameter. Therefore, when a high power density is required, a simple low output which is in the TEM_{00} mode may be more advantageous than a higher power output oscillating in higher order modes. For efficient welding a tightly focused beam with small f/D value (F number) and a low order mode beam (i.e. small M) is desirable.

It should be noted that, in practice, the beam diameter is larger than calculated using eq. (3.1). This is due to the aberrations formed by the focusing optics. However, the relationship between λ, f, D, M and D_b described by eq. (3.1) is useful to estimate D_b, but experimental determination of the beam diameter is essential for high quality work.

3.3. Absorptivity

Absorptivity is one of the important parameters for laser welding. The efficiency of laser welding depends on the absorption of light energy by the workpiece. Any heat transfer calculation for laser processing is based on the energy absorbed by the workpiece.

The infrared absorption of metals largely depends on conductive absorption by free electrons. Therefore, absorptivity is a function of the electrical resistivity of the substrate material. Arata et al. (1977, 1978) measured the absorptivity of polished surfaces of various materials and concluded that absorptivity is proportional to the square root of the electrical resistivity. This agrees closely with the following formula:

$$A = 112.2\rho_r^{1/2}, \tag{3.2}$$

where: A = absorptivity; and ρ_r = electrical resistivity.

A temperature dependent relationship between the electrical resistivity and emissivity of the metal was derived by Bramson (1968). Bramson's formula can be used for theoretical calculation of the absorptivity from the electrical resistivity since absorptivity equals emissivity. However, such a calculation will be valid only for metals heated in vacuum without a surface oxide layer. The presence of an oxide layer will increase the absorptivity. The relationship between the emissivity and the electrical resistivity of a substrate, for perpendicular incidence of radiation of long wavelength, derived by Bramson is,

$$\varepsilon_\lambda(T) = 0.365\left(\frac{\rho_r(T)}{\lambda}\right)^{1/2} - 0.667\frac{\rho_r(T)}{\lambda} + 0.006\left(\frac{P_r(T)}{\lambda}\right)^{3/2}, \tag{3.3}$$

where $\rho_r(T)$ is the electrical resistivity at absolute temperature T expressed in Ω cm. $\varepsilon_\lambda(T)$ is the emissivity of the substrate at $T°C$ temperature for radiation having a wavelength of λ.

The estimated absorptivity for Ti–6Al–4V at 300°C using Bramson's formula is around 15%. Experimental data published by Arata et al. (1977,

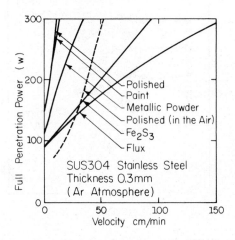

Fig. 3.8. Velocity versus full penetration power (Arata and Miyamoto 1978).

1978), indicate that the absorptivities of Al, Ag and Cu are 2 and 3% and those of stainless steel (304), Fe and Zr are below 15% even in the molten state. This means that reflection losses are particularly great. Therefore, when sheet metal is welded various measures must be taken when required to avoid reflection. Applying absorbent powder or forming an anodized film, for example, are considered to be very effective (Arata et al. 1978). Fig. 3.8 shows the relationship between the surface treatment and the power required for welding (Arata and Miyamoto 1972b).

Absorptivity can also be increased by the use of reactive gas. Jørgensen (1980) reported that the addition of 10% of oxygen to argon shielding gas gave an increase of up to 100% in welding depth. Jørgensen (1980) also found that gas flow had no significant effect on weld depth but an increase in depth was associated with a decrease in reflectivity which was achieved

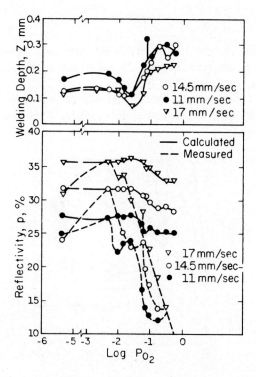

Fig. 3.9. Weld depth z, and reflection coefficient p, as function of partial pressure of oxygen, for different velocities (Jørgensen 1980).

by the addition of a small amount of oxygen. Data obtained by Jørgensen (1980) is illustrated in fig. 3.9.

Although metals are poor absorbers of infrared energy at room temperature, above a certain threshold (approximately $10^6 - 10^7$ W/cm^2) energy transfer via the "keyhole" leads to much higher effective absorptivity. Once a keyhole has formed absorptivity increases rapidly as shown in fig. 3.10. Multiple internal reflections inside a "keyhole" as illustrated in fig. 3.11 are responsible for the deep penetration welding by the laser in spite of the large convergence angles for laser beams compared to electron beams (Arata and Miyamoto 1978). However, the threshold energy required for keyhole formation in laser welding is higher than that required for an electron beam (1.5×10^5 W/cm^2) because of the poor absorptivity. Nevertheless, energy transfer by this "keyhole" mechanism permits lasers to perform efficient welding of even highly reflective materials such as aluminum.

3.4. Traverse speed

The correlation of penetration depth with laser welding speed is discussed by Duley (1976) (p. 241) and Locke et al. (1974) in comparison with that of electron beam welding. It is evident from fig. 3.12 that the penetration in the laser weld is consistently less than that possible with an electron beam, but the difference between the two penetration depths diminishes as the welding speed increases. But Duley (1976) (p. 241) found that this was

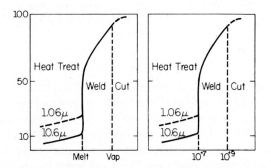

Fig. 3.10. Simplified model of the change in percentage absorption by a metal, at left as a function of its surface temperature and at right as a function of power density required to achieve melting and vaporization. Power density is shown in W/in^2 for tungsten; for aluminum and copper, melting and vaporization points are 10^6 and 10^7 W/in^2, respectively, while for carbon and stainless steel's melting and vaporization points are 10^5 and 10^7 W/in^2, respectively (Engel 1976).

(a)

Laser Beam

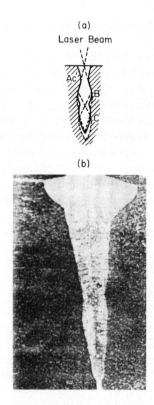

(b)

Fig. 3.11. Schematic diagram illustrating multiple internal reflections of the incident beam. Based on 3/8 in plate of stainless steel made at 23 ipm with a 4 kW CO_2 laser (Arata and Miyamoto 1978).

somewhat surprising, since, as pointed out by Baarsden et al. (1973), the time to form a void or "keyhole" may become comparable to the illumination time for a particular area on the surface of the workpiece as the welding speed increases. When this occurs, the average power dissipated in the sheet is expected to drop because the keyhole is no longer a completely effective trap for the incident laser radiation. For an electron beam, the absorptivity of the material is independent of the shape and extent of the keyhole and hence the total power dissipated in the workpiece will be less strongly dependent on the welding speed. However, Crafer (1976b) reports that the keyhold penetration threshold for laser or electron beams of radius 0.1 mm incident on a steel surface of thermal diffusivity $\simeq 10$ mm^2/s is achieved in

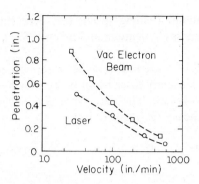

Fig. 3.12. Weld penetration versus weld speed for type 304 stainless steel welded with a 10 kW CO_2 laser (Locke and Hella 1974).

1 ms. This could be regarded as instantaneous. Again, for very low welding speeds, the penetration depth of laser welds becomes significantly less than that attainable with the electron beam. According to Locke et al. (1972a,b) this can be attributed for the formation of a plasma cloud, which attenuates the incident beam.

Fig. 3.13 shows that for a particular power, welding can be performed over a range of thickness (Mazumder and Steen 1980b). Beyond that range full penetration welding is not possible. Higher speeds may lead to improper penetration whereas lower speeds lead to excessive melting, loss of material and weld perforation. The range of welding speed (i.e. range between

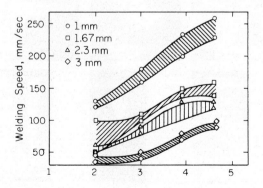

Fig. 3.13. Welding speed against power for Culham/Ferranti 5 kW CO_2 Laser; material: 6Al–4V–Ti (Mazumder and Steen 1980b).

J. Mazumder

maximum and minimum welding speed) for successful welding decreases with increasing thickness (Mazumder and Steen 1980b), as previously noted by Adams (1970).

A laser beam may be traversed over the workpiece either by moving the workpiece under a stationary laser beam or by moving the focusing optics over a stationary workpiece. The latter system is particularly useful when welding an awkard, heavy or fragile object.

3.5. Other parameters

3.5.1. Shielding gas

The plasma produced during laser welding absorbs and scatters the laser beam. It is necessary, therefore, to remove plasma. The higher the power is, the more clearly the phenomenon can be observed. Fig. 3.14 is an example obtained with a 20 kW laser, showing the way in which the depth of penetration decreases with the amount of beam absorption caused by plasma. Plasmas of this type can be removed by supplying a shielding gas

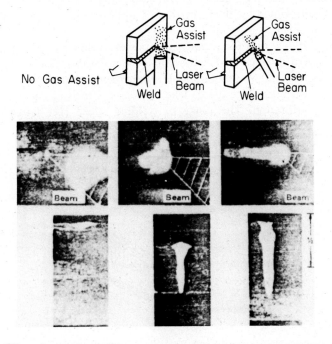

Fig. 3.14. Effect of gas assist on the cross section of the weld (Locke et al. 1972a). Marker at the right represents 1/2″.

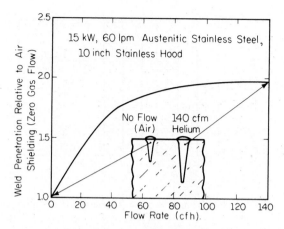

Fig. 3.15. Flow rate effects (Seaman 1977).

such as He. Shielding gas is also required to protect the weld surfaces from oxidation. Both the composition and flow rate of the shielding gas influence the depth of penetration (Seaman 1977, Rein et al. 1972).

Seaman (1977) studied the role of shielding gas in high power CO_2(CW) laser welding. Cross sections of welds made with various shielding gases and gas mixtures show a 60% difference in penetration. However, gases that permit the greatest penetration do not necessarily "blanket"* the weld effectively at characteristically high laser welding speeds. Therefore, compromises are necessary to permit sound, deep penetration laser welds (Seaman 1977).

The effect of the composition of the shielding gas on depth of penetration was studied by Seaman (1977) and Rein et al (1972). Generally, helium is used as shielding gas for high power laser welding. As shown in fig. 3.15 helium seems to improve transmission whereas argon cause severe beam blockage (Seaman 1977). This is probably due to the lower ionization potential of argon (15 eV) compared to that of helium (25 eV). This is further substantiated by the data obtained from the addition of 1% hydrogen to helium as shielding gas (fig. 3.16). Since hydrogen has even higher ionization potential than helium, the hydrogen–helium mixture improves the transmission of the beam (Seaman 1977). The beam transmission effects of air and carbon dioxide lie between the extremes represented by argon

*Blanketing – the ability to displace air from above the weld rapidly.

J. Mazumder

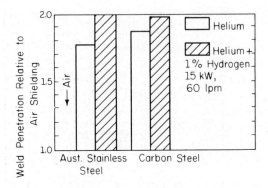

Fig. 3.16. Adding hydrogen improves beam transmission in helium (Seaman 1977).

and helium (fig. 3.17). Rein et al. (1972) have shown that addition of small quantities of hydrogen, sulfur hexafluoride and carbon dioxide to the helium enhances penetration at welding speeds below 40 mm/s.

The ionization potential of the shielding gas is not the only consideration in choosing one for laser welding, especially at higher speeds. This is because gases with higher ionization potentials have lower atomic numbers and thus lower masses. These lighter gases are less effective in displacing air

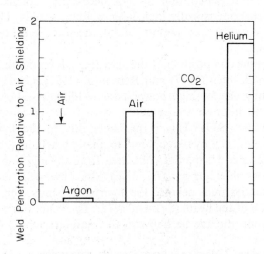

Fig. 3.17. Achievement of welds in different gases (Seaman 1977).

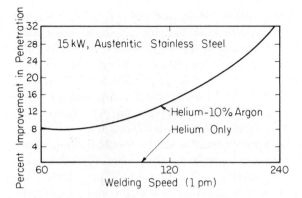

Fig. 3.18. Improvement resulting from the blanketing effect of an addition of 10% argon to a helium shielding gas (Seaman 1977).

from the laser–material interaction area in the short time available in high speed welding. Heavier gases are better able to displace air in a short time. Therefore, a mixture of heavier and lighter gases will result in optimum penetration. Fig. 3.18 [after Seaman (1977)] shows that, as speed increases, the improvement resulting from the addition of a small amount of argon to helium (10% Ar and 90% He) becomes more noticeable.

3.5.2. Position of focus

The optimum position for the focal point of the laser beam, with respect to the substrate was investigated by Wilgoss et al. (1979). Fig. 3.19 shows the transverse profiles generated by moving the focus point in 2.5 mm steps perpendicular to the plane of the workpiece (Wilgoss et al. 1979). The reported plate thickness for this study was 6 mm. When the focal point was

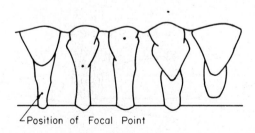

Fig. 3.19. Transverse profiles as a function of focus position (material stainless steel 310, beam power 5 kW, traverse speed 16 mm s^{-1}) (Wilgoss et al. 1979).

positioned deep inside the workpiece, a V-shaped weld resulted necessitating more precise alignment than that for a parallel sided weld of the same cross-section. When the beam was focused well above the level of the workpiece surface, a large "nail head" with a consequent loss of penetration was observed.

Wilgoss et al. (1979) concluded that the optimum focus is approximately 1 mm below the level of the workpiece surface. This produced welds with little or no "nail head" and nearly parallel sides.

Engel (1976) reported that for several metals the optimum position of the focus is 1.25 mm to 2.5 mm below the surface. Research work by Alexander (1981) using the University of Illinois 10 kW laser and $\frac{3}{4}''$ thick AISI 1020 steel revealed that the optimum position of focus is 2 mm below the surface. When the beam was focused 4 mm below the surface the depth of penetration was almost the same as that for the focus 2 mm below the surface. The general consensus is that the optimum position for the focal point is below the surface but the exact distance is dependent on the thickness of the workpiece and the laser power used.

There are four principal ways in which a laser beam can be focused on to a substrate, these are shown in fig. 3.20.

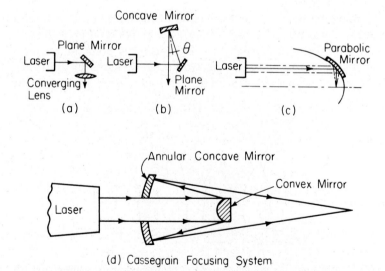

Fig. 3.20. Beam focusing method (Arata and Miyamoto 1978).

Generally, polished copper or gold coated copper is used for mirrors for high power lasers. Several infrared transmission materials such as Ge, KCl, GaAs, ZnSe and CdTe are used for converging lens. A laser beam focused by a lens has excellent light convergence and symmetry properties. However, at laser power in excess of 2 kW transmissive optics cannot be used without major risk of optical failure. Reflective beam optics which can be efficiently cooled are recommended for high power laser beam handling.

3.5.3. *Joint preparation and fit up*
Most weld-joint geometries used in conventional fusion-welding processes (e.g. autogeneous automatic GTAW or electron beam welding) are suitable for laser beam welding (LBW). However, it must be remembered that the laser beam is focused to a spot of few hundred microns in diameter and thus fit-up tolerances and alignment requirements are also of that order of magnitude.

Different types of joint design reported for laser beam welding were reviewed by Shewell (1977). Some of the important joint geometries are illustrated in fig. 3.21 (Shewell 1977).

3.5.4. *Weld design*
Butt joint – full penetration type. Material to be laser welded does not need bevelled edges. Sheared edges are acceptable if square and straight (Engel 1978, Schwartz 1979). The fit-up tolerance should be within 15% of the workpiece thickness. Misalignment and out-of-flatness should be less than 25% of material thickness as shown in fig. 3.22A, to keep the laser beam from wandering. The transverse alignment should be kept within half of the focused beam diameter (Mazumder 1978, Mazumder and Steen 1980b). Clamping is recommended and buckling is not troublesome except when welding material thinner than 0.10 in (0.25 mm) (Engel 1978).

Lap joints. Air gaps severely limit penetration and welding speeds (Engel 1978). Compressive clamping as shown in fig. 3.22B should be used to maintain a separation of less than 25% of the material thickness (Schwartz 1979). For welding of dissimilar thicknesses, the thinner material should be welded to the thicker (Schwartz 1979).

Pulsed laser spot lap joints demand tighter tolerances. The maximum gap for ferrous and nickel alloys is 15% of the material thickness and it is less for higher conductive materials (Shewell 1977). Baranov et al. (1976), studied the influence of gap-size on the formation and strength of spot lap joints in laser welding. They reported for laser spot lap-joints of 0.4 mm

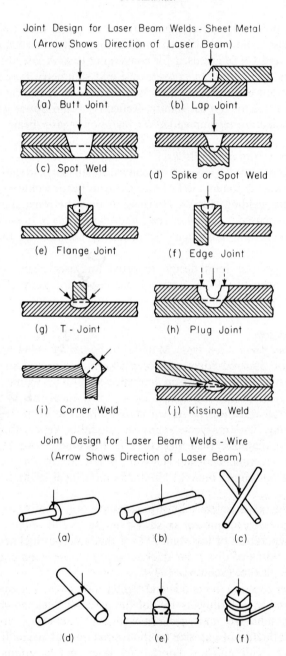

Fig. 3.21. Joint designs for laser beam welds (Shewell 1977).

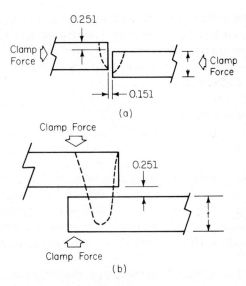

Fig. 3.22. Clamping and tolerances for (a) butt joint and (b) lap joint. Dimensions are in units of material thickness t (Schwartz 1979).

thick steel and 0.3 mm thick nickel, a gap of up to 0.1 mm has hardly any influence on joint strength as there is almost no change in the diameter of the fusion zone. On the other hand, the shape of the weld pool changed with gap size. It has also been observed by Baranov et al. (1976), that the presence of air in the gap may lead to the formation of porosity and non-fusion. Micropores sometimes form in the fusion zone and the number of pores and their size increase with increasing gap size.

Flange joints. Flange joints, as butt joints, require straight, square edges, good fit up, clamping and precise transverse alignment. Flange joints are suitable for welding high shrinkage metals such as aluminum (Engel 1978).

Kissing weld. This type of joint is called a kissing weld because the puddle forms where two pieces just "kiss". This offers a small angle between the two parts that traps most of the energy of the laser beam. Very little pressure, if any, is required during welding, but the facing surfaces must fit well (Shewell 1977). A gap between the sheets will allow radiation to escape. This joint successfully joins thin foils in cases where spot welding would burn through.

The joint configurations for wires as shown in fig. 3.21 were developed by the electronics industry. For wire-to-wire joints, the two wires must share the incident laser energy. In a cross joint, for example, the laser beam should be directed at the intersection of the wires so that both wires are exposed directly to the beam.

4. Welding of various materials

4.1. Microwelds

A large variety of metals and alloys have been welded with pulsed or CW lasers. Until the first reported demonstration of deep penetration LBW by Brown and Banas (1971), most laser welds were performed with pulsed lasers. Several reviews of pulsed laser welding are available (Duley 1976, Ready 1971, Nichols 1969, Schwartz 1971). The more important of these are noted here.

Pfluger and Maas (1965) have discussed the problems associated with making wire–wire and wire–ribbon welds. Application of laser-formed wire–wire welds to the generation of a plated-wire magnetic memory device has been described by Cohen et al. (1969). Laser welding of fine wires without the need to remove insulation was reported by Anderson and Jackson (1965a) and Lebeder and Granista (1972). Lebeder and Granista (1972) reported that the strengths of lap welds of enamel-coated wires of 0.1 to 0.2 mm in diameter were found to be 70 to 90% that of the wire.

A quantitative study of the strength of laser spot welds in sheets made with a gap between the sheets has been reported by Velichko et al. (1972a). They report that gaps of 0.1 mm before welding do not seem to affect the subsequent strength of the bond. In this case, however, the laser power and pulse duration must be tailored to produce optimum melt penetration.

Some specialized laser welding has been discussed by Miller and Nunnikhoven (1965) and Miller (1966). One involved the welding of a bearing assembly made of type 321 stainless steel to a massive type 440C stainless bearing housing. GTAW, resistance spot welding, and EBW had previously been used without satisfactory results whereas a circumferential weld was made with a ruby laser by overlapping individual spot welds which satisfied all design requirements.

Schmidt et al. (1965) studied the feasibility of the welding 302 stainless steel and 18% Ni maraging steel. The welding variables investigated by Schmidt et al. (1965), include sheet thickness, incident laser beam diameter, and the beam energy. A typical result for 0.011 in thick 18% Ni maraging steel sheet is shown in fig. 3.23. Optimum penetration is reported to be obtained for a laser energy of 4 J with slight defocusing of the spot on the workpiece.

A laser beam may be directed to a remote location to effect a weld. Moorhead used a pulsed ruby laser to weld thermocouples to the inside wall of a tube as shown in fig. 3.24A (note that either a CW or a pulsed laser may be used for this purpose). Alternatively, the laser beam may be directed at the thermocouple through a small hole drilled in the tube as shown in fig. 3.24B. The hole may be resealed by laser welding a small plug into the hole after the thermocouple has been attached. Some data reported by Moorhead (1971) on the welding of thermocouples to various metals are given in table 3.1.

4.2. Penetration welding

The scope of technically and commercially feasible laser welding applications has increased greatly since the demonstration of penetration welding using a multikilowatt CW CO_2 laser by Brown and Banas (1971). Numerous experiments have shown that the laser permits manufacture of precision weld joints of high quality rivalled only by those made by electron

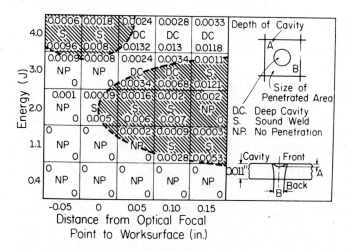

Fig. 3.23. Weldability test results with 0.011 in thick, 18% Ni maraging steel sheet; the cross-hatched areas indicate the conditions for sound welding (Schmidt et al. 1965).

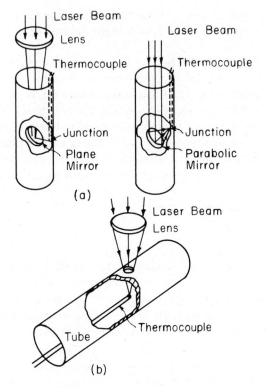

Fig. 3.24. Laser welding of thermocouples to the inside of tubes (Moorhead 1971).

Table 3.1
Thermocouples welded with a ruby laser to some metals[a]

Base metal	Thickness ($\times 10^{-3}$ in)	Thermocouple	Diameter ($\times 10^{-3}$ in)	Pulse length (ms)	Energy (J)
Molybdenum	62	Pt versus Pt–10% Rh	10	4.0	3.2
	62	W versus W–26% Re	10	4.35	4.2
Tantalum	0.5	Pt versus Pt–10% Rh	3	2.75	0.03
	62	W versus W–26% Re	20	5.2	6.5
304 Stainless steel	20	Pt versus Pt–10% Rh	3	3.0	0.04
Zircaloy-4	–	chromel–P versus alumel	13	2.75	0.78
Niobium	1	Pt versus Pt–10% Rh	3	3.50	0.06
Tungsten (arc cast)	125	Pt versus Pt–10% Rh	10	4.50	4.6

[a]After Duley (1976), Moorhead (1971).

beam welding (EBW). Data describing the plate thicknesses of different laser welded materials are given in table 3.2.

4.3. Aluminum and its alloys

Although the use of high power, CO_2 lasers for welding titanium, stainless steel, alloy steels is well researched, their application to aluminum welding did not attract the attention it deserved. To date, very limited data are available in the open literature (Crafer 1976a, Metzbower and Moon 1979, Snow and Breinan 1978). One of the reasons is that aluminum alloys have proven to be very difficult to weld because of the high initial surface reflectivity for 10.6 μm radiation from CO_2 lasers.

The laser welding study of aluminum alloys 5456 and 5086 by Snow and Breinan (1978) revealed that these two aluminum alloys appear to differ in their welding response to a significant degree, both in ability to penetrate under a given set of conditions and also in bead appearance. Penetration in 5456 was substantially greater, the difference being primarily attributed to the 1.5% greater magnesium content of this alloy. Porosity was present in unacceptable amounts in all weld specimens as shown in fig. 3.25 (5086) and fig. 3.26 (5456). Excessive "drop-through" of the bead was encountered in all full penetration welds (Snow and Breinan 1978). This problem is related to liquid metal viscosity and surface tension. The interaction of shielding gas and/or plasma with the beam and the workpiece is very much a part of this phenomenon. Snow and Breinan (1978) concluded that aluminum alloys are very sensitive to the intensity of the input energy and the welding variables.

Breinan et al. achieved some success in laser welding of aluminum alloys 2219 (fig. 3.27) and 5456 (fig. 3.28). Both of these materials have been welded in thicknesses of up to 9.5 mm with acceptable microstructure and bead profile but porosity has been a problem. Tensile tests on these welds resulted in failure by diagonal shear through the welds at an average strength of 342 MPa (49.7 ksi) whereas the ultimate tensile strength value for the parent metal was 345 MPa (50 ksi) (Breinan et al. 1975). Bend test results of laser butt-welds were acceptable (Schwartz 1979). Bend test results for laser fillet weld joints of aluminum alloy 5456 (4.7 mm thick) exhibited sharper-bend radii without fracture of the weld (Snow and Breinan 1978) when compared with those of similar gas metal welds (GMAW).

Metzbower and Moon (1979) studied the mechanical properties of laser welded 12 mm thick aluminum alloy (5456). Tensile tests were performed on a standard ASTM round specimen 6.4 mm in diameter. The weld was

Table 3.2
Materials welded by high-power lasers (Duley 1976, Mazumder 1978, Locke and Hella 1974, Schwartz 1979, Metzbower and Moon 1979, Breinan et al. 1975, Nagler 1976)

Material	Thickness		Laser power
	(in)	(mm)	(kW)
Ship steel, grades A, B, C	1.125 butt	28.6	12.8
	1.0 butt	25.4	12.0
	0.75 butt	19.0	12.0
	0.625 butt	15.9	12.0
	0.50 butt	12.7	12.0
	0.375 butt	9.5	10.8
	0.375 to 0.5 tee	9.5 to 12.7	11.9
Low-alloy carbon steel	0.375 to 0.5 tee	9.5 to 12.7	7.5
AISI 4130	0.60 butt	15.2	14
Low-alloy high-strength steel 300M	0.75 butt	19.0	14
Arctic pipeline steel X-80	0.52 butt	13.2	12
D6AC steel	0.25 butt	6.4	15
	0.50 butt	12.7	15
HY-80	0.49	12.5	10.6
HY-130 steel	0.25 butt	6.4	5.5
HY-180 steel (HP9-4-20)	0.062 butt	1.6	5.5
	0.062	1.6	10.5
	0.64	16.3	5.5
Nickel-base alloy – Inco 718	0.57 butt	14.5	14
Stainless steel (AISI 304)	0.24	6	17
	0.49	12.5	17
	0.67	17	17
	0.65	16.5	17
	0.15	3.8	11.5
	0.22	5.6	11.5
	0.35	8.9	11.5
	0.48	12.3	9.5
	0.35	8.9	8
	0.8	20.3	20
	0.5	12.7	20
Stainless steel (AISI 321)	0.57 butt,	14.5	14
Aluminum alloy:			
2219	0.50 butt	12.7	13
2219	0.25 butt	6.4	5
2219	0.25 butt	6.4	16
5456	0.125 butt	3.2	5.5
5456	0.187 butt	4.7	5.5
5456	0.49 butt	12.5	8
Titanium alloy 6A14V	0.60 butt	15.2	13.5
	0.125, 0.25	3.2, 6.4	5.5
	0.12	3	4.7
	0.49	12.5	11

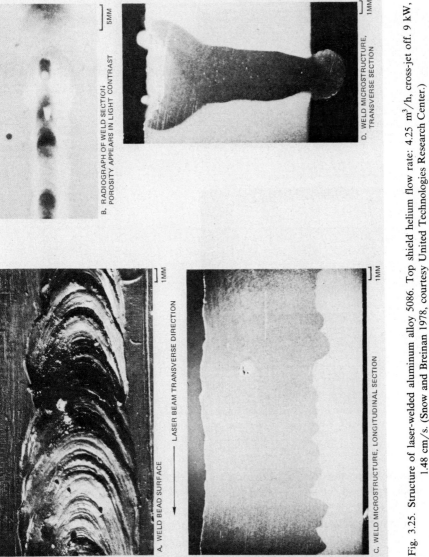

A. WELD BEAD SURFACE

LASER BEAM TRANSVERSE DIRECTION

C. WELD MICROSTRUCTURE, LONGITUDINAL SECTION

1MM

1MM

B. RADIOGRAPH OF WELD SECTION; POROSITY APPEARS IN LIGHT CONTRAST

5MM

D. WELD MICROSTRUCTURE, TRANSVERSE SECTION

1MM

Fig. 3.25. Structure of laser-welded aluminum alloy 5086. Top shield helium flow rate: 4.25 m³/h, cross-jet off. 9 kW, 1.48 cm/s. (Snow and Breinan 1978, courtesy United Technologies Research Center.)

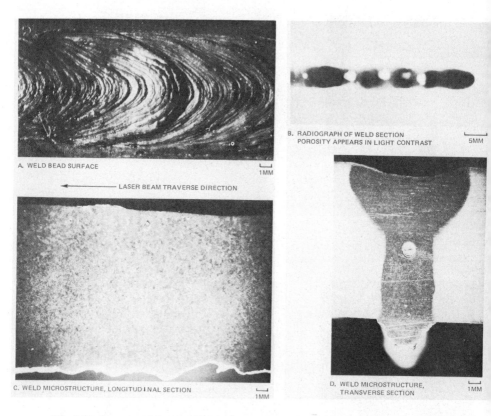

A. WELD BEAD SURFACE   1MM

LASER BEAM TRAVERSE DIRECTION

B. RADIOGRAPH OF WELD SECTION
POROSITY APPEARS IN LIGHT CONTRAST   5MM

C. WELD MICROSTRUCTURE, LONGITUDINAL SECTION   1MM

D. WELD MICROSTRUCTURE,
TRANSVERSE SECTION   1MM

Fig. 3.26. Structure of laser-welded aluminum alloy 5456. Top shield helium flow rate: 2.83 m³/h, cross-jet off. 8 kW, 1.19 cm/s. (Snow and Breinan 1978, courtesy United Technologies Research Center.)

transverse to the loading direction in the tests. All of the aluminum laser welds failed in the weld zone. The ductility of these specimens was low and the amount of porosity visible on the fracture surfaces was high. The mechanical properties data obtained by Metzbower and Moon (1979) are given in table 3.3. They used a non-standard 12 mm thick dynamic tear (DT) test specimen of laser welded 12 mm thick aluminum alloy (5456). Tensile tests were performed on a standard ASTM round specimen 6.4 mm in diameter. The weld was transverse to the loading direction in the tests. All of the aluminum laser welds failed in the weld zone. The ductility of these specimens was low and the amount of porosity visible on the fracture surfaces was high. The mechanical properties data obtained by Metzbower and Moon (1979) are also given in table 3.3. They used a non-standard 12

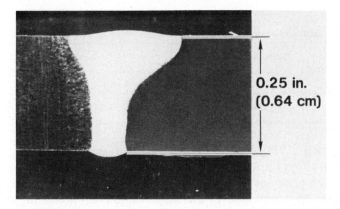

Fig. 3.27. Laser weld in 2219 aluminum alloy. Power: 5 kW, speed: 60 in/min (2.54 cm/s).

mm thick dynamic tear (DT) test specimen with the notch at the weld for their fracture toughness tests (the standard size is 16 mm). Porosity was also observed in DR fracture surfaces.

The variation of the microstructure of the laser welded AL-5456 is shown in fig. 3.29 (Metzbower and Moon 1979). The parent material consists of a fine distribution of the precipitate Mg_2Al_3 in a matrix of Al. Large insoluble particles of Mg_2Si and $(Fe, Mn)Al_6$ are also present. The fusion zone consists solely of α-Al and the precipitate Mg_2Al_3 (fig. 3.29B). The HAZ microstructure (fig. 3.29C) reveals elongated grains and an increase in density of insoluble constituents (Mg_2Si) and a second phase (Mg_2Al_3), both in the matrix and at the grain boundaries, as compared to the weld zone (Metzbower and Moon 1979).

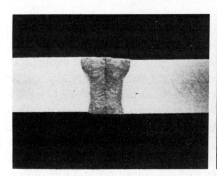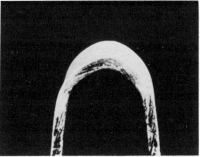

Fig. 3.28. Laser weld and bend test specimen. Material: 5456 aluminum alloy, thickness: 0.110 in (0.279 cm).

Table 3.3
Mechanical properties of a laser welded aluminum alloy (Metzbower and Moon 1979)

0.2% yield strength (MPa)	Ultimate tensile strength (MPa)	% elongation	% reduction in area	Average dynamic tear (DT) energy (N m)	Range of DT energy (N m)
193	289	9.2	16.9	249	176–311

Mazumder (1978) studied the feasibility of the laser welding of three aluminum alloys (2036, 5182, 6009). A 1350 W CW CO_2 laser was used for the bead-on-plate and lap welding of 1 mm thick aluminum plates. It was found that the composition of the alloy determined whether the irradiation parameters were critical. Mazumder (unpublished) reported that it is possible to produce crack-free laser welds with optimum irradiation parameters for aluminum alloy 2036. Two types of precipitates (disc shaped and ribbon shaped) as shown in fig. 3.30A, were observed in the weld pool. Fig. 3.30B represents the copper X-ray image corresponding to fig. 3.30A showing higher copper concentration at the disc shaped precipitate. The ribbon-like precipitate consists of Cu–Al–Mg–Si. The precipitate composition was found to be a function of irradiation parameters. Aluminum alloy 5182 was found to be the easiest to weld compared to Al 2036 and Al 6009. Laser irradiation parameters were not very critical for successful welds but a pronounced loss of magnesium by evaporation was observed. Aluminum alloy 6009 demonstrated very poor weldability. Laser irradiation parameters required for successful weld seemed to be extremely critical. An excessive precipitation of Al–Mg–Si was observed and may be responsible for the cracking of these welds. However, additional cooling of the irradiated material by a chill plate facilitated successful welding.

4.4. Heat resistant alloys

Schwartz (1979) reviewed the laser beam welding studies on heat-resistant alloys. Many nickel and iron base alloys have been successfully welded using CO_2 lasers. INCO 718 alloy was welded using a CW CO_2 laser at power levels ranging from 8 to 14 kW for a specimen thickness of 0.25 in (6.4 mm). Mechanical properties for the 0.25 in thick welded joints (Nagler 1976) are given in table 3.4.

Adams (1973) conducted extensive research on LBW of high-temperature alloys using a fast-axial flow CO_2 laser with an output up to 2 kW. Adams (1973) studied a ferrous base alloy Jethete M152 and a number of nickel base alloys which included PK33, N75, and C263 of thicknesses 1.0 and 2.0

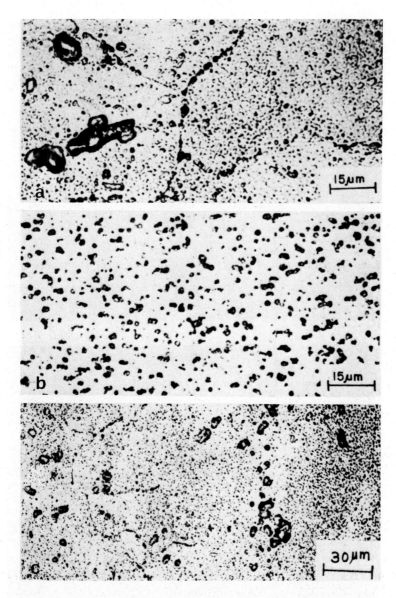

Fig. 3.29. The variation of the microstructure of the laser welded Al 5456. (a) Parent matrix; (b) fusion zone; (c) heat affected zone (HAZ) (after Metzbower and Moon 1979).

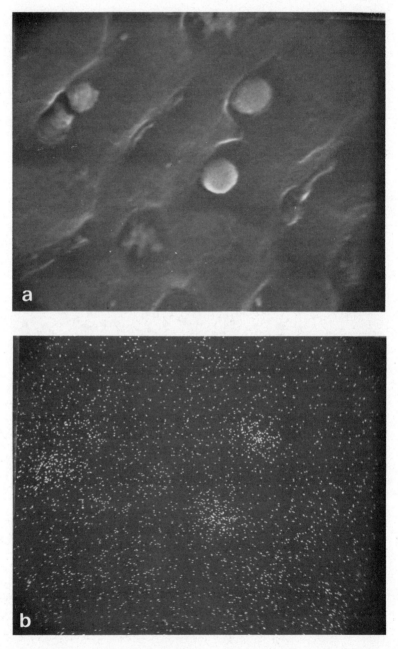

Fig. 3.30. (a) Scanning electron micrograph for Al 2036 showing disc-type Cu–Al precipitate (Mazumder 1978). (b) Copper X-ray image corresponding to (a) (Mazumder 1978).

Table 3.4
Inco 718 nickel-base alloy tensile results (Schwartz 1979, Nagler 1976). Specimen thickness 0.25 in (6.4 mm)

		Elongation (%)		Ultimate strength			
				Tensile		Yield	
Description	Identification	1 in (25.4 mm)	2 in (50.8 mm)	(ksi)	(MPa)	(ksi)	(MPa)
Transverse	TTB-1N		16.4	201.1	1387	175.7	1211
base metal	TTB-2N		16.4	201.0	1386	177.6	1225
Longitudinal	LTB-1N		17.0	204.1	1407	177.9	1227
base metal	LTB-2N		20.0	204.3	1409	178.6	1231
Longitudinal	ALT-1N		5.2	200.2	1380	174.6	1204
welded	ALT-2N		10.5	200.0	1379	173.2	1194
	ALT-3N		6.0	198.5	1369	170.2	1174
Transverse	ATT-1N	5.7	3.9	198.8	1371	179.3	1236
welded	ATT-2N	3.1	2.2	194.9	1358	180.9	1247
	ATT-3N	5.0	3.2	197.3	1360	180.4	1244

mm. In the tensile specimens of Jethete 152, failure occurred in the base metal away from the weld zone whereas the nimonic 75 failed preferentially at the edge of the fusion zone at about 90% of the strength of the parent metal. Elongation of the welded material was 12 to 13% compared with 33 to 45% for the parent material. Tensile-test data for the laser welded ferrous and nickel base alloys are summarized by Schwartz (1979) and given in table 3.5.

Alwang et al. (1969), reported successful laser welding of alloys AMS 5525 and 5544 with thicknesses of up to 0.8 mm. These materials are used for applications requiring high strengths at temperatures up to 700 and 815°C and oxidation resistance up to 800 and 950°C.

4.5. Steels

A great deal of laser welding effort has been expended on ferrous alloys. Most of the initial parametric laser welding studies were carried out on stainless steel (Locke and Hella 1974, Locke et al. 1972a, Brown and Banas 1971). This was investigated by various researchers (Wilgoss et al. 1979, Crafer 1978) due to its importance in the power plant and chemical industries. Laser welding of rimmed steel sheet was evaluated for automotive applications by Baardsen et al. (1973). High speed laser welding of tin

J. Mazumder

Table 3.5
Tensile properties of laser welds in nickel-base and iron-base high-temperature alloys (Schwartz 1979, Breinan et al. 1975)

Alloy and weld or parent material	Nominal thickness		Average UTS		Elongation in 1.98 in (50 mm), (%)	Fracture location
	(in)	(mm)	(ksi)	(MPa)		
PK33-W	0.039	1.0	142	982	20.0	weld
PK33-P	0.039	1.0	147	1014	23.0	–
PK33-W	0.079	2.0	155	1067	26.0	parent
PK33-P	0.079	2.0	151	1041	27.0	–
C263-W	0.039	1.0	145	1000	31.0	parent
C263-P	0.039	1.0	141	973	30.0	–
C263-W	0.079	2.0	141	973	23.0	weld
C263-P	0.079	2.0	143	986	35.0	–
N75-W	0.039	1.0	104	714	13.0	weld
N75-P	0.039	1.0	117	807	34.0	–
N75-W	0.079	2.0	106	729	18.0	weld
N75-P	0.079	2.0	117	807	32.0	–
JM152-W	0.039	1.0	129	887	15.0	parent
JM152-P	0.039	1.0	133	917	16.0	–
JM152-W	0.079	2.0	126	869	11.0	parent
JM152-P	0.079	2.0	126	869	16.0	–

plate and tin free steel for container industries has been reported by Mazumder and Steen (Mazumder 1978, Mazumder and Steen 1981).

Results in HY-alloy steels were evaluated and reported by Banas (1972b) and Metzbower and Moon (1979). Laser welding of modified 4340 alloy has been reported by Seaman and Hella (1976). Several other ferrous alloys studied for the applicability of laser welding include X-80 Arctic pipeline steel (Breinan and Banas 1971), tanker construction steels (Banas and Peters 1974), D6AC (Banas 1979), and high strength, low alloy (HSLA) grade steels (Yessik and Schmatz 1974).

High-power laser welding experiments with 300 series stainless steel have been conducted by Banas (1972b). He reported that welds were formed in stainless steel with aspect ratios (depth-to-width ratios) as large as 12 : 1. The observed variation in weld penetration in stainless steel with welding speed and laser power level is shown in fig. 3.31 (Banas 1972b). Data were obtained in a series of bead-on-plate penetration tests conducted on samples

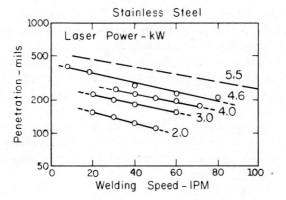

Fig. 3.31. Effect of laser welding speed on penetration (Banas 1972b).

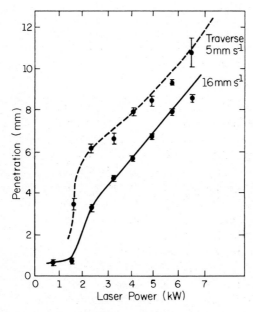

Fig. 3.32. Penetration depth of laser welds as a function of beam power for 316 stainless steel (Wilgoss et al. 1979).

exposed to the ambient atmosphere using laser powers up to 5.5 kW. It is evident from fig. 3.31 that the depth-of-penetration at constant laser power is a relatively weak function of welding speed.

The variation in weld penetration in stainless steel (316) with laser powers up to 7 kW is given in fig. 3.32 (Wilgoss et al. 1979). Similar data for 304 stainless steel (Locke and Hella 1974) for laser powers from 4 to 16 kW is shown in fig. 3.5.

The effect of power (up to 77 kW) on butt weld penetration for 304 stainless steel is given in fig. 3.33 (Banas 1977). The qualitative observation by Banas (1977) is that the maximum penetration is porportional to the laser power to the 0.7 power.

The effect of weld speed on weld shape for type 304 stainless steel is shown in fig. 3.34 (Crafer 1978).

Radiographic inspection of selected laser welds in stainless steel has shown that high-density, non-porous welds can be achieved (Banas 1972b). Tensile tests of stainless steel welds performed by Banas (1972b) have shown that the joint strength can, with appropriate selection of weld parameters, equal that of the parent material. Similar observations were

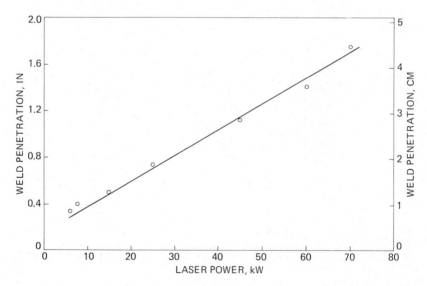

Fig. 3.33. Effect of power on butt weld penetration; material 304SS; weld speed: 63 in/min (Banas 1977).

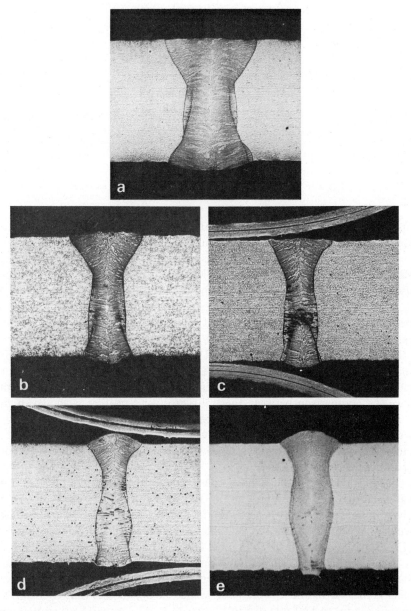

Fig. 3.34. The effect of weld speed on weld shape for stainless steel (304). (After Crafer 1978, courtesy of British Welding Institute, Oxford, England).

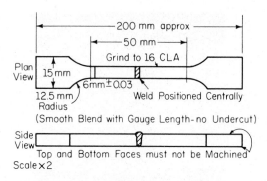

Fig. 3.35. Tensile test pieces specification (Wilgoss et al. 1979).

made by Wilgoss et al. (1979) for stainless steel 316 but data obtained for stainless steel 310 were less encouraging.

Wilgoss et al. (1979) performed tensile tests on laser butt welded stainless steel 316, 310, and ferritic steel DUCOL W30. Flat tensile specimens were used with the plane of the weld running transversely across the gauge length at its center (fig. 3.35). The room temperature uniaxial tensile tests were performed at a Crosshead velocity of 1.6 mm/s corresponding to a strain rate of 2.8×10^{-2} s^{-1}. Results from Wilgoss et al. (1979) for the tensile tests are given in table 3.6. Wilgoss et al. (1979), evaluated laser welding of 6 mm thick stainless steel 316 and DUCOL W30 against TIG, plasma arc, and electron beam welding methods. Criteria for the comparison is given in table 3.7. The comparison between the transverse weld profiles is shown in fig. 3.36. It is evident from the comparison that for high productivity and better weld quality electron beam and laser welding techniques are to be preferred. However, one should keep it in mind that electron beams require a vacuum chamber which lasers do not.

Laser welding parameters for HY alloys are given in table 3.8. The variation of weld penetration in HY-130 with laser power up to 3.6 kW is given in fig. 3.37 (Banas 1972b) and the mechanical properties of some laser welded HY alloys are given in table 3.9. The results of impact tests on HY-130 laser welds are given in table 3.10.

Schwartz reviewed the laser welding capability of tanker construction steels (Nagler 1976, Banas and Peters 1974). Merchant ship construction steel grades A, B, and C were laser welded in thicknesses ranging from 9.5 mm (0.375 in) to 12.7 mm (0.5 in) for grade A; 12.7 mm (0.5 in) to 19.0 mm (0.75 in) for grade B; and 25.4 mm (1.0 in) to 28.6 mm (1.125 in) for grade

3.6
le test results (Wilgoss et al. 1979)

rial	Welding conditions	Welding speed (mm s^{-1}), power = 5 kW	Stress at ultimate tensile strength (MPa)			Strain (%)	
			Failure in parent (P) or weld (W)	Mean			Mean
tainless steel	optimum	16	P 599	598	50.0		50.0
		16	P 598		50.0		
	non-	25	W 634	594	40.4		45.7
	optimum	25	W 554		51.0		
		16	W 715	636	57.6		56.8
		16	W 557		56.0		
	parent	–	P 592	591	54.0		54.0
	plate	–	P 590		54.0		
tainless steel	non-	25	W 410	428	16.8		19.7
	optimum	25	W 446		22.6		
	(but fully	16	W 506	508	31.0		28.7
	penetrated)	16	W 510		26.4		
l W30 steel		16	P 766	767	16.0		15.5
	optimum	16	P 769		15.0		
	non-	25	W 666	710	13.2		10.6
	optimum	25	W 754		8.0		
		16	P 774	757	22.8		21.9
		16	P 740		21.0		
	parent	–	P 769	775	18.0		17.5
	plate	–	P 780		17.0		
501 Standard Data							
ness)	parent plate (143–212)	–	–	at least 520	–		at least 40
ness)	parent plate (149–217)	–	–	at least 540	–		at least 40
l W30 ness)	parent plate (not quoted)	–	–	at least 590–700	–		at least 16

J. Mazumder

Table 3.7
Comparison of processes (plate thickness 6 mm) (Wilgoss et al. 1979)

Process	Laser	TIG	Plasma	Electron Beam
Power absorbed by workpiece	4 kW	2 kW	4 kW	5 kW
Total power used	5 kW	3 kW	6 kW	6 kW
Typical traverse speed	16 mm s^{-1}	2 mm s^{-1}	6.7 mm s^{-1}	40 mm s^{-1}
Alignment accuracy required	±0.5 mm	±1.0 mm	±1.0 mm	±0.3 mm
Energy input per unit length (absorbed by workpiece)	250 J mm^{-1}	1000 J mm^{-1}	600 J mm^{-1}	125 J mm^{-1}
Capital cost and availability	$250K for 5 kW output, available on a year's delivery in UK and USA. $750K for 15 kW output, available on a year's delivery in USA	> $16K per welding set, immediate delivery	~ $20K per welding set, immediate delivery	~ $100K for 6 kW machine, on short delivery

Possibility of all positional welding	YES but requires optics to manipulate the beam. Optimum when moving workpiece	serious penetration characteristic changes with attitude	serious penetration characteristic changes with attitude	YES but requires mechanics to move gun. Optimum when moving workpiece
Distortion { axial shrinkage	Minimal (small HAZ)	significant ~ 1 mm on 5 mm plate	significant ~ 1 mm on 5 mm plate	minimal (small HAZ)
{ angular	minimal parallel sidedness	significant V-shaped weld	significant V-shaped weld	minimal parallel sidedness
Surface profile defects	Very fine flow lines	underside protrusion held in by surface tension	underside protrusion held in by surface tension	produces ruffled swarf on backface
Special requirements for process operation	safety interlock to guard against misplaced beam reflections	normal light screening	normal light screening	vacuum chamber, local vacuum, X-ray screening
Process END effects { start	slight surface protrusion	smooth	slight surface protrusion	slight surface protrusion
{ finish	smooth		slight surface protrusion	slight surface protrusion

J. Mazumder

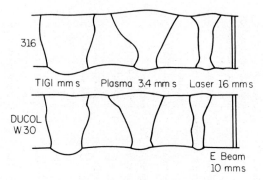

Fig. 3.36. Comparison of transverse profiles produced by welding processes (Wilgoss et al. 1979).

Table 3.8

Material	Thickness (in)	(mm)	Weld type	Laser power (kW)	Weld speed (in/min)	(mm/s)	Refs. and remarks
HY 80	0.5	12	butt	10.6	30	12.7	(Metzbower and Moon 1979)
HY 80	1.5	38	bead on plate	90	100–120	42–50	not radiographically acceptable (Banas 1977)
HY130	0.25	6.4	butt	5.5	50	21.2	(Nagler 1976)
HY130	0.5	12	butt	11	30	12.7	(Metzbower and Moon 1979)
HY180	0.062	1.6	butt	5.5	160	67.7	(Nagler 1976)
HY180	0.062	1.6	lap	5.5	140	59.2	(Nagler 1976)

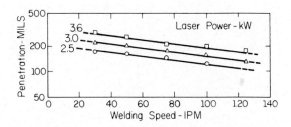

Fig. 3.37. Effect of laser welding speed on penetration (Banas 1972b).

Table 3.9 (Metzbower and Moon 1979)

	0.2% yield stress		UTS				Dynamic T bar energies			
					Elongation (%)	Reduction in area (%)	Av. DT energy		Range of DT energy	
Materials	(MPa)	(ksi)	(MPa)	(ksi)			(N m)	(ft lb)	(Nm)	(ft lb)
HY 80	627	91	752	109	23.5	75.8	76	56	612–176	38–130
HY130	952	138	987	143	18.5	68.4	707	530	836–66	459–627

Table 3.10
Results of impact tests on HY-130 laser welds (Nagler 1976)

Welding speed		Temperature		Weld impact strength		Base-metal impact strength	
(in/min)	(mm/s)	(°F)	(°C)	(ft lb)	(J)	(ft lb)	(J)
45	19.0	30	−1.1	39.0	52.9	26.4	35.8
45	19.0	75	24	39.0	52.9	27.0	36.6
35	14.8	75	24	28.3	38.4	24.0	32.5
20	8.5	75	24	27.0	36.6	25.0	33.9

Table 3.11
Demonstration butt welds (Nagler 1976)

Thickness		Laser power (kW)	Weld speed		Comment
(in)	(mm)		(in/min)	(mm/s)	
0.375	9.5	10.8	50	21.2	single pass
0.375	9.5	10.8	45	19.0	single pass
0.5	12.7	12.0	27	11.4	single pass
0.5	12.7	12.0	30	12.7	single pass
0.625	15.9	12.0	24	10.2	single pass
0.75	19.1	12.0	45	19.0	dual pass
1.0	25.4	12.0	30	12.7	dual pass
1.0	25.4	12.0	30	12.7	dual pass
1.125	28.6	12.8	27	11.4	dual pass
0.375–0.5	9.5–12.7	11.0	90	38.1	tee joint
0.375–0.5	9.5–12.7	7.5	65	27.5	tee joint
1.0	25.4	12.0	27	11.4	dual pass[a]
1.0	25.4	12.0	25	10.6	dual pass[a]

[a] 0.001 in (0.03 mm) aluminum foil preplaced at weld interface.

C. Some of the laser welding parameters are given in table 3.11. These steels are all in the 0.23% maximum carbon, 0.60 to 1.03% manganese class, and increase in deoxidation from grade A through to C (Schwartz 1979). The welding was accomplished using a convectively cooled CW CO_2 laser at power level from 5.5 kW to 12.8 kW. At these power levels, two passes were necessary to weld 19 mm to 25.4 mm (0.75 in to 1 in) thick materials.

Mechanical test data for the laser welded ship construction steels are summarized (Breinan and Banas 1975) in table 3.12. Although all welds made within the thickness capability of the equipment passed X-ray radiograph examination, root porosity was observed in the welds where penetra-

Table 3.12
Weld properties of tanker construction steels (Schwartz 1979, Breinan et al. 1975)

Number and grade	Thickness (in)	(mm)	X-ray	Tensile	Side bend	Charpy impact energy at 0°F (−15°C) Weld (ft lb)	(J)	HAZ (ft lb)	(J)
1-A	0.375	9.5	OK	failure in base plate	OK	14.3	19.4	12.7	17.2
2-A	0.375	9.5	OK	failure in base plate	OK	8.0	10.8	10.3	14.0
3-A	0.5	12.7	OK	failure in base plate	OK	9.7	13.1	7.3	9.9
4-A	0.5	12.7	OK	failure in base plate	OK	11.7	15.8	11.7	15.8
5-B	0.625	15.9	OK	failure in base plate	OK	36.0	48.8	14.0	19.0
6-B	0.75	19.0	OK	failure in base plate	OK	10.7	14.5	5.0	6.8
7-C	1.0	25.4	aligned root porosity	failure in base plate	OK	12.7	17.2	11.7	15.8
8-C	1.0	25.4	aligned root porosity	failure in base plate	OK	26.7	36.2	15.3	20.8
9-C	1.125	28.6	large scale porosity	no mechanical tests					
12-C	1.0	25.4	OK	failure in base plate	OK	8.8	11.9	5.3	7.1
13-C	1.0	25.4	OK	failure in base plate	OK	8.0	10.8	11.5	15.6

tion was marginal. Root porosity was also observed in dual pass welds. However, porosity may be reduced by appropriate selection of welding parameters (Schwartz 1979). The tensile properties of the welds in grades A, B, and C were found to be very good. Fractures mainly occurred at the boneplate. Adequate ductility was observed by Banas et al. (1974), who concluded that higher power laser systems offer significant potential for shipyard applications. Compared to conventional arc welding processes, laser welding produces welds with minimum distortion.

Successful, autogenous, square-butt welds of X-80 Arctic pipeline steel using a high power CW CO_2 were reported by Nagler (1976) and Breinan and Banas (1971). Both single pass and dual pass techniques are used to weld 13.2 mm (0.52 in) material. Dual-pass welds exhibited smaller grain structure than single-pass welds; the upper shelf for dual-pass welds was also greater than 358 J (264 ft lb) and the transition temperature was below $-51°C$ ($-60°F$). The impact test data for dual pass laser welds of X-80 Arctic pipeline steel are summarized in table 3.13. The mechanical properties of the laser welds appear to be better than those of the base metal. Therefore, laser welding promises to be suitable for high-quality large diameter pipe welding application.

Successful laser welding of D6AC steel in thicknesses of 6.5 mm and 12.7 mm (0.25 in and 0.5 in) using a CW 15 kW laser has been reported (Banas 1979). However, preheating up to 232°C (450°F) was found to be necessary to eliminate porosity. The resultant welds exhibited consistent and reproduceable tensile test data with fracture appearing in the base metal demonstrating the quality of the weld.

The laser welds in HSLA steels exhibited properties equivalent to those of the base metal (Yessik and Schmatz 1974). This is very difficult to obtain using conventional arc welding techniques.

Laser welded rimmed steel sheet was found to be acceptable for automobile sheet joining applications (Baardsen et al. 1973). However, the welds possessed some gas porosity.

One of the principal commercial areas for the welding of thin material is that of can manufacture. Mazumder and Steen (Mazumder 1978, Mazumder and Steen 1981) evaluated the possibility of the high speed welding of steels used in can making (tin plate and tin free steels) with a 2 kW CW CO_2 laser. A welding speed in excess of 19 m/min was achieved for bead on plate welding of 0.2 mm tin plate using 1950 W of laser output energy. The laser welding process was compared with other can making processes (table 3.14) and found to be the only method capable of welding tin free steel (with a 0.01 μm layer of chromium and a 0.04 μm layer of chromic oxide as a

Table 3.13
Impact test results for dual-pass laser welds (Schwartz 1979, Nagler 1976)

Specimen number	Test temperature		Notch configuration and direction[a]	Charpy energy		Comment
	(°F)	(°C)		(ft lb)	(J)	
1	−100	−73.3	side i	12.5	16.9	visible pores in fracture
2	−24	−31.1	side o	245.0	332.2	one pore in fracture
3	70	21.1	side i	185.0	250.8	grouped porosity at one corner
4	−18	−27.8	side o	160.0	216.9	small pores only
5	−62	−52.2	side i	> 264.0	358.0	fine root porosity
6	−60	−51.1	side o	> 264.0	358.0	fine root porosity
7	−105	−76.1	side o	16.0	21.7	considerable porosity in plate
8	−21	−29.4	side i	44.0	59.7	considerable porosity in plate
9	18	27.8	side i	163.0	221.0	considerable porosity in plate
10	103	75.0	side o	38.5	52.2	considerable porosity in plate
11	−100	−73.3	side i	96.0	130.2	fine root porosity
12	−24	−31.1	side o	39.0	52.9	fine root porosity
13	−100	−73.3	side o	34.5	46.8	planar defect and pores
14	70	21.1	side i	239.5	324.7	considerable porosity
15	−80	−62.2	side o	22.0	29.8	considerable porosity
16	−100	−73.3	side i	23.5	31.9	large flaw
17	−40	−40.0	side o	178.0	241.3	
18	−20	−28.9	side i	226.0	306.4	
19	−20	−28.9	side o	> 246.0	358.0	
20	75	23.9	side i	247.0	334.9	small pore
21	−100	−73.3	side o	198.0	268.4	
22	−80	−62.2	side i	7.5	10.2	
23	−38	−38.9	side o	> 264.0	358.0	
24	−80	−62.2	side i	6.0	8.1	
25	−80	−62.2	side o	68.0	92.2	

[a] i, in welding direction; o, opposite welding direction.

Table 5.14 (Mazumder and Steen 1981)

Method	Speed (m/min)	Advantages	Disadvantages
1. Lock seam – solder (L.S.S.)	60–75	(a) fast (b) well tried (c) relatively cheap equipment	(a) uses expensive Sn and Pb (b) health hazard (c) poor appearance (d) cannot use tin free steel (e) cans cannot be recycled
2. Draw wall – ironing	comparable to ≈ 70 m/min	(a) pleasing appearance (b) thin walls; saves metal for beverage cans (c) recyclable cans (d) no filler material required	(a) tool wear (b) cannot make thick wall cans required for food cans. Because food cans have to withstand the pressure due to vacuum sealing (c) cannot use tin free steel
3. Soudronic	25–30 m/min for lap welding 40–50 m/min for H.F. semilap welding	(a) neat (b) fast, up to 50 m/min (with frequency tripling) (c) cans can be recycled	(a) electrode problem (b) cannot use tin free steel (c) electrical oscillation problem in the can (d) porosity (ref. 81)
4. Draw re-draw (DRD)	Comparable to 70 m/min	(a) pleasing appearance (b) thicker walls than DWI (c) can use tin free steel	(a) tool wear
5. Laser weld	7–8 m/min for lap welding of 0.2 mm with 2 kW power 10–20 m/min for bead-on-plate welds 0.2 mm sheet with 2 kW	(a) no tool wear (b) pleasing appearance (c) no filler materials (d) cans can be recycled (e) can use tin free steel	
6. Laser/arc augmented welds	≈ 60 m/min bead-on-plate welds	same as for laser	(a) high speed gas shielding problem

corrosion inhibitor) without auxiliary preparation. Although a 2 kW CW CO_2 laser by itself cannot reach the required welding speeds, arc augmentation of the 2 kW laser appears to be capable to do so (Mazumder and Steen 1981, Eboo 1979).

Tensile test data for laser welded tin plate and tin free steel are summarized in table 3.15 (Mazumder and Steen 1981). Mechanical properties of laser welds seem to be at least as good as those of the base material. All fractures during tensile tests were observed in the base material (Mazumder and Steen 1981). Laser welds were also found to be radiographically sound (Mazumder and Steen 1981). Simple bending fatigue tests revealed an endurance ratio of 0.45 to 0.5 (Mazumder and Steen 1981). Data for corrosion rates of the laser welds using the Tafel extrapolation method has shown that the weld zone is at least as corrosion resistant as the base metal, if not better (Mazumder and Steen 1981). In conclusion, sound welds of good appearance and mechanical properties can be made using a laser in tin plate and tin free steel. The welding speeds in the initial experiments (Mazumder and Steen 1981) are low compared to present can making speeds; however, there are ways of increasing this speed by increasing the laser power or augmenting it, for example by using an electric arc. The laser welds have a very narrow HAZ, and could be made through painted areas; they are autogeneous and so present no recycling problem as the solder does in lock seam soldered cans.

4.6. Titanium and its alloys

Until recently, the EBW technique has been the most popular method for the welding of Ti–6Al–4V, this alloy being widely used in the aerospace industries for its remarkable strength:weight ratio. However, the deep penetration of EBW can be obtained only up to a short distance under non-vacuum conditions, and for optimum efficiency electron beam welding is carried out in an evacuated chamber (Meleka 1971). In contrast, CO_2 laser beams can be transmitted for appreciable distances through the atmosphere without serious attenuation or optical degradation. The laser thus offers an easily manoeuvered, chemically clean, high intensity, atmospheric welding process, producing deep penetration welds (aspect ratio greater than $1:1$) with narrow heat affected zone (HAZ) and subsequent low distortion.

The application of the laser technique to a metal such as a titanium alloy, which is difficult to weld, is not only of direct interest to the aerospace and chemical industries but also more generally to a study into the welding of a

Table 3.15 (Mazumder and Steen 1981)

Materials	Welded material			Original material		
	U.T.S. (MN/m^2)	0.2% proof stress (MN/m^2)	Elongation (%)	U.T.S. (MN/m^2)	0.2% Proof stress (MN/m^2)	Elongation (%)
Tin plate (tin coated mild steel)	≈ 310–355	≈ 267–316	17–27	313–319	≈ 267	≈ 28
Hi-top drum quality mild steel (tin free steel)	386–426	340–380	21–27	≈ 353–380	284 ≈ 328	≈ 30–32

Table 3.16
Typical mechanical properties of titanium alloys (Mazumder and Steen 1979)

Materials	Welded material			Original material		
	U.T.S. (MN/m^2)	0.2% proof stress (MN/m^2)	Elongation (%)	U.T.S. (MN/m^2)	0.2% proof stress (MN/m^2)	Elongation (%)
AMS 4911-TA-10 (6Al–4V–Ti)	860–923	800–860	11–14	895–1004	834–895	10–15
Commercially pure titanium	≈ 530–573	≈ 460–503	≈ 27	> 494	> 416	27–28

chemically sensitive metal with a complex temperature dependent structure. The importance and the need for better joining methods for titanium and its alloys has resulted in several investigations of laser welding techniques over various power ranges: Banas (1972a, b, 1975a, b) (up to 5.5 kW), Seaman and Hella (1976) (up to 16 kW), Adams (1973) (up to 2 kW), and Mazumder and Steen (Mazumder 1978, Mazumder and Steen 1979, 1980b) (up to 5 kW). The reader is also referred to Mazumder and Steen (1982).

Mazumder and Steen (Mazumder 1978, Mazumder and Steen 1979, 1980b) reported the relationship between laser welding parameters and the metallurgical and mechanical properties of laser welded Ti–6Al–4V and commercially pure titanium. As shown in fig. 3.13, welding speeds in excess of 15 m/min were obtained by Mazumder and Steen (1980b) for 1 mm thick Ti–6Al–4V for 4.7 kW of laser power. The effect of laser welding speed on penetration as observed by Banas (1972a) for Ti–6Al–4V is shown in fig. 3.38.

X-ray radiographs of successful laser butt-welds of Ti–6Al–4V and commercially pure (CP) titanium showed no cracks, porosity or inclusions (Mazumder and Steen 1979). Low porosity in laser welded titanium alloy was also observed by Seaman (1978) and radiographically sound welds were also produced by Banas (1975a). Undercutting was not prominent.

Tensile tests performed by Mazumder and Steen (Mazumder 1978, Mazumder and Steen 1979) on laser butt welds of titanium alloys revealed that laser welds are at least as strong as the base metal. Typical tensile test data (Mazumder and Steen 1979) are given in table 3.16.

Under simple bending fatigue, the endurance ratio for welded specimens (with a transverse central weld) was found to be 0.40 to 0.47 whereas that for unwelded specimens was 0.50 (Mazumder and Steen 1979). Adams (1973) reported that, under proper welding conditions, laser welds can be made in Ti–6Al–4V which exhibits the same fatigue characteristics as the base metal.

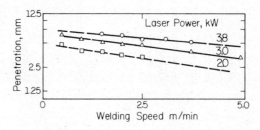

Fig. 3.38. Penetration against welding speed (Banas 1972a).

The variation of microstructure for a typical Ti–6Al–4V weld as seen in an optical photomicrograph in comparison with its macrostructure is represented in a composite (fig. 3.39) after Mazumder and Steen (1979). The parent metal (fig. 3.39A) shows dark β in a bright α matrix which represents a typical annealed structure of α–β titanium. The heat affected zone consists of a mixture of martensitic α and primary α which corresponds to a structure quenched from the range 980 to 720°C (Hochied et al. 1970). The HAZ is shown in fig. 3.39B. The right-hand side of the micrograph, fig. 3.39B, is near the fusion zone. In that area, traces of α' (martensitic α) can be observed, whereas the left-hand side of the micrograph (i.e. further away from the fusion zone) shows a relative increase of primary α. This observation is confirmed by the scanning electron microscopy study.

The fusion zone consists mainly of α' (martensitic α). Both figs. 3.39D and 3.39E represent the fusion zone of the same specimen. The abundance of α' (martensitic α) with the β-grain boundary is evident in these micrographs. This is confirmed by the structure revealed under both scanning and transmission electron microscopes. Such microstructure corresponds to a structure quenched from the β phase above 985°C (Hochied et al. 1970) and is confirmed by comparison with the hardness values of others (Hochied et al. 1970).

Fig. 3.40 shows the transmission electron micrograph (fig. 3.40A) of the weld zone. It confirms the observations using optical microscopy that the fusion zone consists mainly of α' (martensitic α) of acicular morphology. This (acicular martensitic α) is characterized by large "primary" plates and smaller "secondary" plates. The primary plates extend for large distances across the parent β grain and effectively partition the untransformed β. The partitioned β subsequently transforms to a series of short, acicular, secondary plates of martensite (figs. 3.40B and 3.40C). A similar microstructure was observed by Zaidi (1978) for water quenched 6Al–4V–Ti from above 1000°C which also correlates with Hochied et al. (1970). The selected area diffraction pattern (fig. 3.40D) of a martensite plate shown in fig. 3.40B indicates that the zone axis is $(10\bar{1}1)$ hcp. But, in another sample, the zone axis of the martensite plate was found to be (0001).

This martensitic structure of the weld zone is responsible for the good tensile properties of the laser welded titanium alloy. By comparing the microstructure with the prediction of a three-dimensional transfer model (Mazumder and Steen 1980a) the cooling rate of the weld zone was estimated to be 10^{4}°C/s (Mazumder and Steen 1979). This high cooling rate is responsible for the martensitic structure.

The total oxygen analysis of a weld sample is given in table 3.17. This indicates that there is no significant oxygen contamination taking place

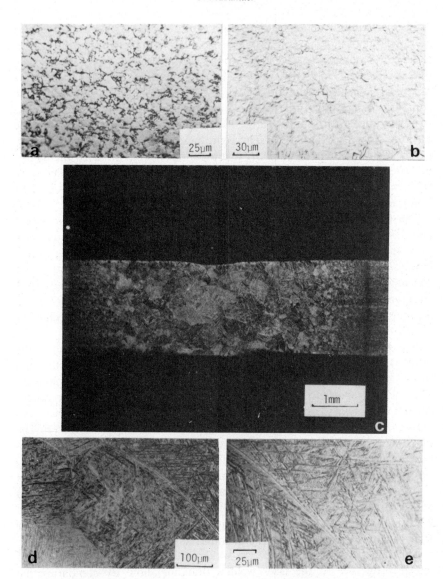

Fig. 3.39. Composite showing structural variation of 2 mm thick 6Al–4V–Ti alloy welded with CW CO_2 laser. Welding speed 7.5 mm/sec, laser power 1500 W (Mazumder and Steen 1979). (a) Parent matrix; (b) HAZ; (c) macrostructure of the weld; (d) and (e) fusion zones.

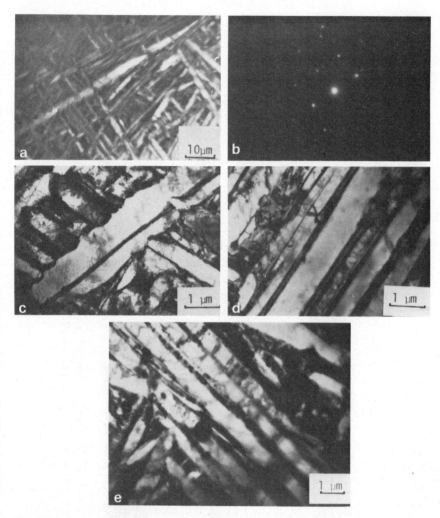

Fig. 3.40. Composite showing the transmission electron micrographs of the 6Al–4V–Ti weld
(Mazumder and Steen 1979).

during laser welding when shielding gas is used (Mazumder and Steen 1979).

A comparative study of electron beam, laser, and plasma arc welds in Ti–6Al–4V alloy has been conducted by Banas (1975a). Radiographically sound welds were produced by all three techniques (Banas 1975a). The electron beam welds were quite narrow and exhibited a somewhat non-uni-

Table 3.17
Total oxygen analysis of weld samples[a]

Material	Oxygen in ppm		
	Parent metal	Weld zone with surface oxide	Weld zone after removing surface oxide
6Al–4V–Ti	3200	3400	3250
Commercially pure titanium	1600	1900	1700

[a]After Mazumder and Steen (1979).

form radiographic appearance due to lower surface weld spatter (fig. 3.41) whereas the arc welds were considerably broader but also quite uniform in density. Laser welds were narrower than arc welds and were comparable to but more uniform than EBW weld beads. Following stress relieving for two hours at 538°C (100°F), the welds produced by all three techniques had tensile strengths equivalent to or exceeding those of the base metal.

Fig. 3.41 Radiographic weld comparison (Banas 1975a). Ti–6Al–4V, 0.64 cm (nominal). Left: electron beam, 1.61 kW, 1.27 cm/s. Center: laser, 5.50 kW, 2.12 cm/s. Right: plasma arc, 2.10 kW, 0.19 cm/s.

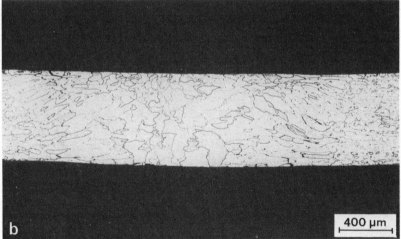

Fig. 3.42. Fusion zone microstructure of a laser weld. Welding speed: 12.5 mm/s, (a) top surface; (b) transverse section (David and Lin 1980). Material: thorium doped iridium alloy (DOP-14). (Courtesy of Dr. S. A. David, Oak Ridge National Laboratory.)

4.7. Iridium alloys

High power CW CO_2 lasers are found to be an attractive tool for welding difficult alloys such as thorium doped iridium alloys (DOP 14 and DOP 16). These alloys crack severely during GTA welding or welding with a highly defocused electron beam (David and Liu 1980). Successful laser welds free from hot cracking have been reported in DOP-14 by David and Liu (1980). This is due to the characteristics of the highly concentrated heat source available with the laser and from the refinement in the fusion zone structure. The fusion zone structure is a strong function of the welding speed (David and Liu 1980). Typical microstructures of the DOP-14 laser welds made at 12.5 mm/s are shown in fig. 3.42.

Laser welds with a refined fusion zone structure have also been reported in DOP-26 alloy (David and Liu 1980). The fusion zone structure of the laser welds compares well with the best structures obtained by arc welding using arc oscillation.

Laser welding parameters for thorium doped iridium alloys are given in table 3.18.

Table 3.18
Laser welding parameters[a]

Specimen	Power on work (kW)	Welding speed (mm/s)	Bead width (mm) top	root	Type of weld
Alloy DOP-26					
1A	4.8	8.3	1.97	1.5	melt run
2A	6.0	12.5	2.2	1.6	melt run
3A	7.0	16.7	2.0	1.5	melt run
4A	5.2	8.3	2.4	1.7	butt weld
5A	6.0	12.5	2.2	1.8	butt weld
6A	7.1	16.7	2.1	1.7	butt weld
Alloy DOP-14					
21A	5.0	8.3	2.4	1.9	melt run
22A	5.8	12.5			melt run
23A	6.0	12.5			melt run
24A	8.6	25.0			melt run
25A	5.8	12.5	2.2	1.6	butt weld
26A	8.7	25.0	1.8	1.1	butt weld

[a]After David and Liu (1980).

4.8. Dissimilar metals

Laser welds have been produced between a variety of dissimilar metals when other methods have proved unsatisfactory. Possibly due to the more rapid rate of cooling and smaller HAZ there is an alteration of formation of phases allowing more pairs of dissimilar metals to be laser welded. Miller and Nunnikoven (1965) have described successful welds between a B-113 steel feed through and type 321 steel casting in a thermistor enclosure.

Garaschuk and Molchan (1969) have reported a metallographic analysis of laser welds between nickel and copper, nickel and titanium, and copper and titanium. Garaschuk and Molchan (1969) have reported that the fusion region in a nickel–titanium weld showed a finely dispersed structure and an intermetallic phase at the boundary of the melt. But there was no comment on crack formation, whereas, Seretsky and Ryba (1976) confirm the presence of intermetallics but have reported that 75%˙ of their welds were cracked. Garaschuk and Molchan (1969) have also reported the welding of tantalum–molybdenum. Baranov et al. (1968a,b) have reported the laser welding of brass, mild steel, and stainless steel to copper.

Baranov et al. (1978) reported laser welding of KOVAR to copper. They found that the strength of the welded joints was 92% of annealed copper and the ductility was sufficiently high. The weld metal has a "whirling" structure and it was chemically heterogeneous (Baranov et al. 1978).

Laser beam welding was found to be more feasible for joining beryllium to titanium for beryllium-reinforced titanium matrix composites used in fan and compressor blades of gas turbines (Schwartz 1975). The composite blade fabricated by laser welding procedure, as shown in fig. 3.43 has acceptable geometric tolerances but small defects are present as a direct result of impurities which contaminates the composite sheet during fabrication.

Gagliano (1969) reports welding between tungsten and aluminum using a ruby laser. There are several other useful publications (Gagliano 1969, Schwartz 1975, Velichko et al. 1972b,c, Garashchuk et al. 1972, Rykalin and Uglov 1969, Baranov et al. 1972, Zhukov 1970, Duddenhagen 1963, Hice 1963, Platte and Smith 1963, Jackson 1965, Frick 1966, Orrok 1967, Kloepper 1970, Benson 1970) reporting laser welding of dissimilar metals listed in table 3.19.

4.9. Arc-augmented laser welding

No review of laser welding would be complete without the discussion of relatively new development known as "arc-augmented laser welding". In

Fig. 3.43. Composite blade after welding and prior to surface "blending". Also an enlarged
laser beam weld (Schwartz 1975).

this process an electric arc is rooted to the point where the laser interacts
with the surface. It has been shown by Steen et al. (Eboo et al. 1978, Steen
and Eboo 1979, Eboo 1979) that an electric arc will preferentially root into
the laser generated hot spot (fig. 3.44) or plasma (figs. 3.45 and 3.46). Eboo
(1979) showed by mathematical modelling that this dual heat source also
caused a contraction of the arc into the laser generated "keyhole". This
coupled power source leads to a significant increase in welding speed in
electrically conducting materials. The resulting welds have many of the
characteristics of a weld made by a laser alone, but have a larger heat
affected zone (HAZ). It is apparent that this process is a cheap method of
amplifying the effect of an expensive laser.

Steen et al. (UK Patent 12681/75, 6008/78, Eboo et al. 1978, Steen and
Eboo 1979, Eboo 1979, Alexander and Steen 1980) used a 2 kW CW CO_2
laser in conjunction with a tungsten inert gas (TIG) welding system as their
dual power sources for the arc-augmented laser welding studies. The arc can
be rooted to the laser generated hot spot on the specimen at the side
opposite to the laser as shown in fig. 3.4A. For thin materials this system

Table 3.19
Laser welding studies of dissimilar metals[a]

Metal combination[b]	Reported by
0.003 tungsten wire	Buddenhagen (1963)
0.020 nickel wire	
20 SiC fibres	Buddenhagen (1963)
0.005 nickel wire	
0.0025 nichrome wire	Hice (1963)
0.040 silver-plated brass	
0.002–0.005 gold wire	Platte and Smith (1963)
Silicon and Al-coated Si	
0.015 copper wire	Jackson (1965)
0.015 tantalum wire	
0.030 stainless steel wire	Anderson and Jackson (1965)
0.025 tantalum wire	
0.005 tungsten wire	Anderson and Jacksón (1965) ·
0.020 nickel wire	
0.020 nickel wire	Frick (1966)
0.025 tantalum wire	
0.015 copper wire	Frick (1966)
0.020 nickel wire	
0.009 phosphor bronze wire	Orrok (1967)
0.005 palladium wire	
0.020 nichrome wire	Orrok (1967)
0.140 monel rod	
0.005–0.025 rene 41	Orrok (1967)
0.125 Columbium D36	
0.002–0.003 copper	Barnov and Metashop (1968)
0.019 brass	
0.002–0.003 copper	Barnov and Metashop (1968)
0.019 mild steel	
0.002–0.019 copper	Barnov and Metashop (1968)
0.031 stainless steel	
0.025 rene 41	Kloepper (1970)
0.025 hastalloy X	
0.031 titanium wire	Benson (1970)
0.016 gold	
0.1×0.3 mm kovaralloy to	Velichko et al. (1972)
6 mm gold	
Tantalum and	Garashchuk et al. (1969)
Modybdenum	
B-1113 steel and	Miller and Nunnikhoven (1965)
321 stainless steel	

[a]After Seretsky and Ryba (1976).
[b]Decimals indicate thickness in inches unless otherwise mentioned.

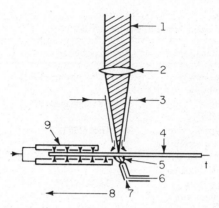

Fig. 3.44. The laser/arc welding system, 1: laser beam; 2:75 mm KCl lens; 3: He shield; 4: positive workpiece; 5: arc plasma; 6: tungsten cathode; 7: argon; 8: welding direction; 9: argon shielding jets (Steen and Eboo 1979).

produces a dramatic increase in welding speed. Eboo et al. (1979) when operating on thinner material (0.2 mm thick mild steel) observed a fourfold increase in welding speed for only 200 W of arc power (fig. 3.47). It is reported by Eboo et al. (1979) and Eboo (1979) that when the laser and arc are on opposite sides the arc fails to root in a stable manner if the undersurface temperature generated by the laser is less than 400°C above that of the surrounding material. This figure is derived from a back calculation on a mathematical model (Mazumder 1978) of an experimental run in which the decoupling speed was accurately measured.

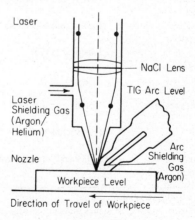

Fig. 3.45. Schematic diagram of experimental arrangement for laser and arc welding (Alexander and Steen 1980).

Fig. 3.46. The arc rooting down the laser induced plasma (Steen and Eboo 1979).

On combining the laser and arc they appear to complement so that the combined effect is more than simply additive. It was shown by Eboo et al. (1979) and Eboo (1979) using case studies of welding runs, and from records of the arc current and voltage as in fig. 3.48 that the arc roots in a stable way to the hot spot generated by the laser.

The arc can also be rooted to the laser generated "keyhole" on the specimen when the arc is on the same side as the laser as shown in figs. 3.45 and 3.46. This coupled power source resulted in a doubling of the laser alone welding speed (for 0.8 mm cp titanium) with a small current (~ 50 A as shown in fig. 3.47). Only a slight broadening of the fusion zone was noted when compared to a laser alone weld.

The arc roots down the laser generated plasma so readily that it can be attracted to this location from quite large distances (Eboo et al. 1979). It is thought that the high electron density in the laser generated plasma (Glasstone 1961) makes it more conducting and hence the favored route for the arc (Steenbeck 1932). The arc energy is thus largely located in exactly the same area as the laser energy. However, if this current were to raise the plasma temperature to the point where the electron density exceeds 10^{16} mm^3, the plasma would reflect the laser beam (Glasstone 1961). These plasma absorption or reflection effects could cause a difficulty if very large

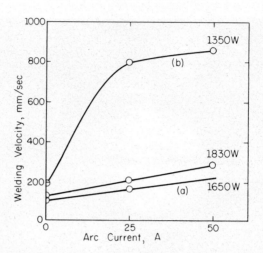

Fig. 3.47. Welding velocity against arc current; (a) 0.8 mm commercial purity titanium with both laser and arc above sheet; (b) 0.22 mm mild steel with laser above and arc below (Steen and Eboo 1979).

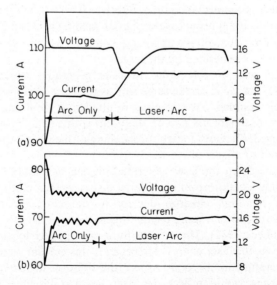

Fig. 3.48. Current–voltage characteristics of laser/arc traces; (a) decrease in arc column resistance because of laser, 3 mm thick mild steel at 22.5 mm/s; (b) stabilization of an unstable arc by laser, 2 mm thick mild steel at 45.0 mm/s (Steen and Eboo 1979).

currents were used, but they are not thought to be crucial at the small currents (< 300 A) used by Steen et al. (UK Patent 12681/75, 6008/78, Eboo et al. 1979, Steen to be published, Eboo 1979, Alexander and Steen 1980). However, they could account for the differences in welding speed enhancement seen in fig. 3.47 (Eboo et al. 1979).

Alexander (Alexander and Steen 1980) studied arc-augmented laser welding using the arc and the laser at the same side of a 12 mm thick steel (BSS En3) specimen. Alexander reported that arc-augmented laser welding doubled the welding speed for same penetration whilst maintaining an acceptable bead profile (fig. 3.49). Bead profiles become unacceptable when currents greater than 150 A or speeds in excess of 100 mm/s are used. However, it appears that augmentation of 1300 W to 2000 W laser power with up to 100 A of arc current is possible. This leads to a consequent increase in welding speed for a given penetration of 80 to 100%, and in penetration for a given speed of around 20% (Alexander and Steen 1980).

Alexander (Alexander and Steen 1980) noted that the choice of shielding gas for arc-augmented laser welding is important. Transmission of the laser beam through the plasma is facilitated by the use of a gas with higher ionization potential such as helium, whereas argon is beneficial for arc stability. However, at higher speeds argon is better for displacing the air. It was observed by Alexander that at the characteristic high speed of arc-augmented laser welding, argon shielding gas produce optimum welds.

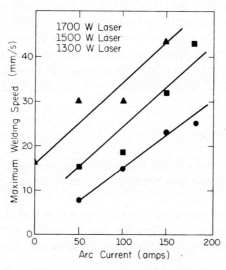

Fig. 3.49. Maximum welding speed versus arc current (Alexander and Steen 1980).

5. Heat transfer theory for laser welding

In understanding the penetration welding processes, the experimenter is faced with a multiparameter problem which is difficult to solve without extensive factorial experimentation. The principal variables are the substrate thermophysical properties, the total energy input, power distribution and diameter, and the traverse speed. Alternatively, an assumed physical picture of the process can be modeled mathematically and the model's results compared to experimental results to test the model's validity and thus, by inference, the physical model.

A model capable of predicting experimental results means that previously unmeasurable parameters can be estimated. In metal welding, Borland and Jordan (1972) summarized the most important metallurgical data to be as follows:

(1) The thermal cycle at each point in the weld zone and heat affected zone (HAZ), since this determines the type and extent of any phase change.

(2) The peak temperature distribution; since this can define the extent of the HAZ.

(3) The cooling rate through certain cooling ranges; since it indicates the likelihood of the formation for metastable products.

A large number of investigations have been made with the object of obtaining data for heat transfer during welding processes. It has been found that all these thermal factors are strongly influenced by the welding process, the welding variables, the thermal properties of the material, and the geometry of the weldment (Borland and Jordan 1972). Presumably, as a consequence attention has been concentrated on theoretical techniques for calculating the thermal history associated with welding rather than direct experimental measurement. There is a range of numerical and analytical methods which have been developed and are discussed in the following sections.

5.1. Analytical solutions

Different methods for solving the heat conduction equations for various conditions are elegantly and methodically described by Carslaw and Jaeger (1959).

Myers et al. (1967) have reviewed a large number of investigation on the theoretical techniques for calculating the thermal history associated with welding. The analytical methods discussed here are summarized in table 3.20.

Table 3.20
Analytical models for moving heat source

Type of mathematical model	Particular process aimed	Eq. no.	Ref.
Moving point heat source for three-dimensional heat flow	general welding processes	(3.4)	Borland and Jordan (1972), Rosenthal (1941, 1946), Christensen et al. (1956)
Moving line heat source for two-dimensional heat flow	general welding processes	(3.5)	Borland and Jordan (1972), Rosenthal (1941, 1946), Christensen et al. (1956)
Moving line source for two-dimensional heat flow	laser and electron beam	(3.8–19)	Swifthook and Gick (1973)
Uniform and finite heat source for one-dimensional heat flow	CO_2 laser welding	(3.20–21)	Alwang et al. (1969)
Uniform band or rectangular heat source for steady state two-dimensional heat flow	EBW and CW laser welding	–	Arata et al. (1972a)
Uniform circular heat source of a semi-infinite slab (i.e., for two-dimensional heat flow calculating transient surface isotherms)	pulsed laser	(3.28–29)	Guenot and Racinet (1970)
Stationary gaussian or uniform circular heat source for transient temperature distribution in thin film	electron beam heating	(3.22–27)	Tung-Po Lin (1967)
Gaussian heat source moving at a constant velocity for 3D heat flow	laser and electron beam	–	Cline and Anthony (1977)

Among numerous analytical methods reported, the most generally useful appears to be that involving the equations developed by Rosenthal (1941, 1946) and Rykalin (1951a). Although their equations do not give the most accurate result, according to Borland and Jordan (1972), they do show clearly the type and magnitude of the thermal changes occurring in welding and the way they are affected by material and process variables.

Two specific welding situations are considered by Rosenthal (1941, 1946):

(i) a point source of heat moving over the surface on an infinitely wide thick plate where the heat flow can be regarded as three dimensional; and

(ii) a line source of heat moving though an infinitely wide thin plate where the heat flow may be regarded as two dimensional.

These two idealized situations correspond to the final passes in a multi-run weld in thick plate and a single-run full penetration in thin plate respectively. In the first instance (Rosenthal 1941, 1946) derived equations for the temperature distribution around the moving sources.

For the three-dimensional case:

$$(T - T_0) = \frac{Q}{2\pi K} e^{-VX/2\alpha} \frac{e^{-VR/2\alpha}}{R}. \tag{3.4}$$

For the two-dimensional case:

$$(T - T_0) = \frac{Q}{2\pi K g} e^{-VX/2\alpha} K_0(VR/2\alpha), \tag{3.5}$$

where:

T = temperature at a point (X, R) (K);
T_0 = the original plate temperature (K);
Q = heat input per unit time (W);
g = plate thickness (m);
K = thermal conductivity (W m^{-1} K^{-1});
K_0 = Bessel function of second kind and zero order;
α = thermal diffusivity (m^2 s^{-1});
ρ = density (kg m^{-3});
C = specific heat (J kg^{-1} K^{-1});
V = welding speed (m s^{-1});
t = time (s);
R = distance to heat source (m);
X = distance along the weld center line (m); and
Y = distance from the weld center line (m).

From these, Rosenthal then obtained expressions for the center line cooling rate since this could be regarded as representative of the weld zone.

For the three-dimensional case

$$dT/dt = 2\pi\kappa(V/Q)(T - T_0)^2. \tag{3.6}$$

For the two-dimensional case

$$dT/dt = 2\pi\kappa\rho C(Vg/Q)^2(T_0 - T)^3. \tag{3.7}$$

Subsequently, Adams et al. (1958, 1962) produced peak temperature equations for the same two situations and also provided means for generalizing all the equations to a much wider range of practical conditions.

Christensen et al. (1965) used the Rosenthal equation for point sources moving across the surface of a semi-infinite body to construct a temperature chart for the particular set of materials properties, heat input, and welding speed under consideration. They generalized the Rosenthal equation [eq. (2.4)] by using the dimensionless parameters and produced a series of dimensionless temperature contours against different dimensionless welding parameters. This generalized chart in turn predicts temperature distribution for various welding conditions but it has limitations due to the assumptions in the point source equation.

None of the models mentioned above address the problem of finite heat source, which is often comparable to the size of weld puddle. These analytical models also do not allow for "keyholing" and therefore, predict a unsatisfactory temperature profile for laser beam and electron beam welding. However, they are quick and simple to use for a rough estimation of temperature at locations further away from the molten pool.

To allow for the keyhole effect Swifthook and Gick (1973) developed a two-dimensional analytical model on a line source moving through the substrate. Swifthook and Gick (1973) solved the moving line source model of Carslaw and Jaeger (1959) for steady state heat diffusion ignoring the transients at the beginning and end of the process. They deduced the following equations which would give the isotherms:

$$\theta = \frac{q}{2\pi} \left[\exp(Ur\cos\phi) \right] K_0(Ur), \tag{3.8}$$

where: $q = P/a$; a = penetration depth; K_0 = Bessel function of the 2nd kind and zero order; $\theta = KT$; P = power; $x = r\cos\phi$; v = velocity in positive x direction, K = thermal conductivity; $U = v/2\alpha$; α = thermal diffusivity. Now, writing the power per unit depth the above equation becomes

$$M = \left[2\pi/K_0(Ur) \right] \exp(-Ur\cos\phi), \tag{3.9}$$

where $M = P/a\theta$.

Swifthook and Gick (1973) have shown that at the point where the melt has its greatest width the following condition is satisfied.

$$\frac{\delta M/\delta\phi}{\delta M/\delta r} = \frac{\delta y/\delta\phi}{\delta y/\delta r}, \tag{3.10}$$

where y = half width of the molten pool $= r\sin\phi$. Then using eq. (3.9) in eq. (3.10) they deduced the following relationship

$$\cos\phi = -K_0(Ur)/K_0'(Ur) \tag{3.11}$$

and showed that the normalized maximum melt width

$$Y_{max} = vW/\alpha = 2UW = 4Ur\left[1 - K_0^2(Ur)/K_0^2(Ur)\right]^{1/2}, \qquad (3.12)$$

where W = total melt width = $2y$. Substituting $\cos\phi$ in eq. (3.9) one gets

$$M = \frac{2\pi}{K_0(Ur)}\exp\frac{UrK_0(Ur)}{K_0'(Ur)}. \qquad (3.13)$$

The relationship between absorbed power P and the melt width W can be found by eliminating Ur in eqs. (3.12) and (3.13). Swifthook and Gick (1973) discuss two limiting cases when this can be done analytically.

In the high speed limit (Ur large), the Bassel functions can be expanded to give

$$M \simeq 8.26(Ur)^{1/2} \qquad (3.14)$$

and

$$Y_{max} \simeq 4(Ur)^{1/2} \qquad (3.15)$$

so that

$$Y_{max} \simeq 0.484M \quad (Ur \text{ large}). \qquad (3.16)$$

When the speed is low (Ur small)

$$M \simeq \frac{2\pi}{\ln(2e^{-\gamma/Ur})} \qquad (3.17)$$

and

$$Y_{max} \simeq 4Ur \qquad (3.18)$$

so that

$$Y_{max} = \exp(1.50 - 2/M) \quad (Ur \text{ small}), \qquad (3.19)$$

where $\gamma = 0.577$ Euler's constant.

Fig. 3.50 shows a plot of Y_{max} versus M constructed from these two limiting solutions with an interpolation in the intervening region where neither solution is strictly valid (Swifthook and Gick 1973).

Arata and Miyamoto (1972a) report that a model using a line heat source provides an infinite temperature at the location of the source and leads to a noticeable error in temperature estimation at points distant from the beam axis compared with the beam diameter. In a line source model the size of the beam spot cannot be taken into account in calculating the temperature rise in spite of the fact that the weld depth is considerably affected by the

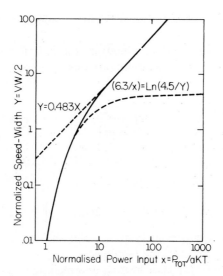

Fig. 3.50. Theoretical curve for penetration welds (Swifthook and Gick 1973).

beam diameter. Therefore, Arata et al. (1972a) introduced a band or rectangular plane heat source of uniform intensity which moves in the direction perpendicular to the source plane, the width of which corresponds to the hole diameter. They described the temperature and beam penetration depth by a set of non-dimensional variables in order to establish a universal relationship independent of parameters such as physical constants of material, welding speed, beam diameter and beam power.

Arata et al. (1972a) simplified the problem assuming that heat distributes two-dimensionally at steady state and neglected the variation of physical constants of material, radiation heat loss and convection in the molten pool.

Alwang et al. (1969) reported a one-dimensional solution for heat transfer for CW CO_2 laser welding. In such a system they had to assume that the incident power density in the beam was uniform and thus independent of position and the temperature distribution and used Carslaw and Jaeger's (1959) one-dimensional equation of linear heat flow for heat generation within the sample which is as follows:

$$\frac{\partial^2 T}{\partial z^2} - \frac{1}{\alpha}\frac{\partial T}{\partial t} = \frac{\varepsilon_a F(t)e^{\varepsilon_a z}}{K} \tag{3.20}$$

where: T = temperature; z = depth into material; α = thermal diffusivity;

t = time; $F(t)$ = incident beam power density; and ε_a = optical absorption coefficient.

According to Alwang et al. (1969) a solution for semi-infinite slab approximation of the above assuming a constant flux, i.e., $F(t) = F_0$ is as follows:

$$T(x,t) = \frac{2F_0\varepsilon_a(\alpha t)^{1/2}}{K} \text{ierfc}\left(\frac{z}{2(\alpha t)^{1/2}}\right) - (F_0/\varepsilon_a K)e^{\varepsilon_0 z}$$

$$+ (F_0/2\varepsilon_a K)\exp\left(\varepsilon_a^2\alpha t - \varepsilon_a z\right)\text{erfc}\left[\varepsilon_a(\alpha t)^{1/2} - z/2(\alpha t)^{1/2}\right]$$

$$+ (F_0/2\varepsilon_0 K)\exp\left(\varepsilon_a^2\alpha t + \varepsilon_a z\right)$$

$$\times \text{erfc}\left[\varepsilon_a\left(\varepsilon_a(\alpha t)^{1/2} + z/2(\alpha t)^{1/2}\right)\right]. \tag{3.21}$$

The equation is approximately true when the beam diameter is large compared to the thickness of the workpiece.

The thermal analysis for laser heating and melting reported by Cline and Anthony (1977) seems to be the most realistic analytical model reported so far. Cline and Anthony (1977) used a gaussian heat source moving at a constant velocity. To determine the temperature distribution the following equation was solved:

$$\frac{\partial T}{\partial t} - \alpha\nabla^2 T = \frac{Q}{C_p}, \tag{3.22}$$

where: T = temperature (°C); t = time (s); α = thermal diffusivity (cm/s); Q = power absorbed per unit volume (W/cm); and C_p = specific heat per unit volume (J/cm °C).

Cline and Anthony (1977) used a coordinate system fixed in the material in which the laser beam is parallel to the z axis and impinges on the surface $z = 0$ at time $t = 0$. The laser moves in the x-direction at a constant V (fig. 3.51). A moving gaussian beam normalized to give a total power P for a spot radius $(1/e)R$ is of the form

$$Q = P\frac{\exp\{-\left[(x - vt)^2 + y^2\right](2R^2)^{-1}\}}{2\pi R^2}\frac{h(z)}{\lambda} \tag{3.23}$$

where: P = absorbed power (W); λ = absorption depth; $h(z) = 1$ for $0 < z < \lambda$; and $h(z) = 0$ for $z > \lambda$; and v = velocity of scanning laser beam (cm/s).

To solve the above equations for the temperature, Cline and Anthony (1977) superimposed the known Green function solution for the thermal distribution of unit point source (Morse and Feshback 1953), located at

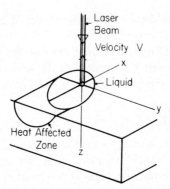

Fig. 3.51. Schematic illustrating laser-beam heating by a scanning laser beam moving in the x-direction at a constant velocity (Cline and Anthony 1977).

coordinates x', y', t' on the surface to represent the gaussian distribution. Heat flow from the laser at x', y', t' on the surface influences the temperature at x, y, z, in the material at a later time t by (Cline and Anthony 1977)

$$T = \int_{-\infty}^{t} \int_{-\infty}^{\infty} \int_{-\infty}^{\infty} \int_{-\infty}^{\infty} (Q/C_p)(x'y'z't')G(x'y'z't'|_{xyzt})\, dx'\, dy'\, dz'\, dt',$$
(3.24)

where: G = Greens function; (x', y', z', t') = coordinates of the source; x = direction of motion of laser (cm); y = direction normal to laser beam and direction of motion (cm); and z = direction of laser beam.

For the special case of a point source ($R = 0$) the simplified solution is (Cline and Anthony 1977)

$$T = \frac{P}{C_p \alpha 2\pi r} \exp \frac{-v(r+x)}{2\alpha},$$
(3.25)

where $r = (x^2 + y^2 + x^2)^{1/2}$.

For deep penetration welding the temperature distribution is given by eq. (3.24), where the absorption depth in eq. (3.23) becomes the keyhold depth, Z (Cline and Anthony 1977). The power is related to the keyhole depth Z as follows:

$$\frac{P}{P_v} = \frac{2(Z_v/Z_0)}{1 - \exp[-2(Z_v/Z_0)]},$$
(3.26)

where: Z_0 = penetration depth; P_v = power level where temperature reaches boiling point T_v; and Z_v = keyhole depth:

The keyhole depth Z_v increases from zero as the power increases above power necessary to vaporize the material at the surface P_v (fig. 3.52).

At high power levels the melt depth is a fixed distance below the keyhole depth:

$$Z_m = Z_v + Z_0 \ln(T_v / T_m), \tag{3.27}$$

where: Z_m = melt depth; and T_m = melt temperature.

Cline and Anthony (1977) reported that the agreement between the calculation and the experiment is good at penetration depths below 1 cm. At higher penetration depths the experimental points show less penetration than calculated.

Guenot and Racinet (1970) proposed a new mathematical formulation for pulsed laser welding by means of which the transient isothermal surfaces can be calculated for any sheet material as a function of heating conditions at the surface. They assumed a circular heat source of uniform intensity on a semi-infinite solid (practically a thick plate) and used the theory of instantaneous heat sources (Dennery and Guenot 1962, Guenot et al. 1966) to obtain a convenient formulation for temperature distribution. It reads as follows:

$$r > r_b, \quad \bar{\theta} = \theta_i / \theta_M = t^* \int_0^{\phi_0} \left[\psi\left(\bar{R}_1 / t^* \right) - \psi\left(\bar{R}_2 / t^* \right) \right] d\phi,$$

$$r < r_b, \quad \bar{\theta} = \theta_i / \theta_M = t^* \int_0^{\pi} \left[\psi\left(\bar{z} / t^* \right) - \psi\left(\bar{R}_2 / t^* \right) \right] d\phi,$$

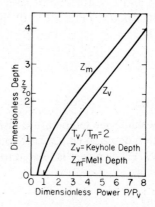

Fig. 3.52. The calculated penetration of liquid into 304 stainless steel at high power levels where the liquid–vapor interface is depressed a distance determined by the boiling point (Cline and Anthony 1977).

where $\bar{\theta}, t^*, R_i$ ($i = 1$ or 2) are dimensionless parameters;

θ_i = instantaneous temperature rise of the material;

z = depth;

r = radial distance;

$\bar{\theta} = \theta_i / \theta_M$ setting $\theta_M = q_0 r_b / k$;

q_0 = uniform intensity of circular heat source;

$t^* = (t / t_0)^{1/2}$ setting $t_0 = r_b^2 / 2\alpha$

= dimensionless time;

$\bar{R}_i = R_i / r_b$ setting $R_i^2 = l_i^2 + z^2$ ($i = 1$ or 2);

l_1, l_2 = functions of the polar angle ϕ defined as the roots of the quadratic equation $l^2 - 2l\cos\phi + r^2 - r_b^2 = 0$;

ϕ_0 = limiting value for the real roots; and

ψ = integrated complementary error function [known as "ierfc" in Carslaw and Jaeger (1959)].

Their results are presented diagrammatically using dimensionless parameters such as dimensionless temperature and time, thus providing the best working conditions for any material of any thickness.

All the studies discussed above ignored the latent heat associated with phase change, however, Andrews and Atthey (1975) calculated penetration speeds for various power ranges for the steady condition and variable thermal conductivity. Andrew and Atthey (1976) also studied the hydrodynamic limit to penetration of a material by a high power beam. For this purpose they considered a stationary beam and semi-infinite molten pool. They discussed the shape and depth of the molten pool and concluded that surface tension reduces the depth of penetration. But for simplicity they assumed that all of the incident power is used to evaporate material and totally neglect heat conduction. Therefore, their work is not particularly relevant to temperature prediction for laser welding.

5.2. Numerical solutions

Numerical calculations remove many of the limitations that apply to the analytical method. For example:

(1) the heat source does not have to be concentrated in a point, line or plane;

(2) the geometry of the workpiece may be taken into account;

(3) physical properties of the substrate may be considered temperature dependent without much difficulty;

(4) truncation error from series expansion of different functions associated with analytical solutions is absent; and

(5) the difficulty of application of the analytical solution of the heat flow equation to real boundary conditions is eliminated.

Although numerical calculation removes many limitations that apply to the analytical method, only a few numerical models for heat flow in welding have been developed so far partly due to the large computer time and storage required for the solution.

However, Westby (1968), Arata and Inoue (1973), and Paley and Hibbert (1975) have reported numerical models for the temperature distribution in conventional welding processes. Mazumder et al. (Mazumder and Steen 1980a, Chande et al. 1981) have developed a three-dimensional heat transfer model for laser welding applications. These numerical models and their results are discussed in brief. For details of the construction of numerical models, readers should study the original papers.

Arata and Inoue (1973) report that a point or line heat source approximation frequently gives unsatisfactory results for the temperature surrounding the heat source. Therefore, they have presented an intermediate case between a point heat source approximation and line heat source approximation by introducing a dimensionless quantity "β" and they have calculated the temperature distribution numerically. In a nutshell they have numerically calculated the temperature distribution in the vicinity of the heat source for a moving line heat source of non-uniform input energy. Arata and Inoue (1973) have presented their results in the term of dimensionless isothermal contours as shown in fig. 3.53. Fig. 3.53 (a), (b), (c) and (d) shows the distribution of the maximum temperature T_m^* on the Y^*-Z^* plane and its dependence on β.

Rykalin (1963) provided a numerical solution for the one and two-dimensional heat flow cases and Westby (1968) tackled the problem using a large modern computer and developed a program for both constant and variable thermal properties.

Paley and Hibbert (1975) based their computer program on Westby's (1968) work and they have added several capabilities to the program. These include (a) the capability of analyzing non-rectangular cross sections such as single and double V and U grooves and (b) the use of a non-uniform mesh in the finite difference scheme which allows a better approximation of nonrectangular cross-sections as well as permitting the placement of a finer grid size in regions of higher thermal gradients. Paley and Hibbert's (1975) program also permits the option of resuming the computation from the point left off in a previous run as well as the capability of intermediate print out in order to check the convergence of the results.

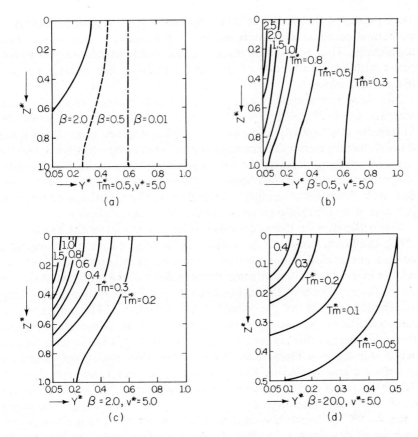

Fig. 3.53. The contour map of T^* on the Y^*-Z^* plane; $Y^* = Y/D$, $Z^* = Z/D$, D is the plate thickness; $v^* = vD/\alpha\alpha$, v is the velocity, α the thermal diffusivity (Arata and Inoue 1973).

A three-dimensional gaussian heat source has been considered by Paley and Hibbert (1975) using the following assumption to simplify their mathematical model:

(1) the size of the workpiece is infinitely long;

(2) quasi-steady state conditions are required;

(3) heat exchange with the surroundings are neglected; and

(4) the path of the electrodes is on an axis of symmetry and is parallel to the length of the plate.

They presented the results of their computations in graphical displays of isothermal contours at a given section perpendicular to one of the three coordinates. They report that there is a reasonable correlation when compared with macro-structure of the weld and have also plotted different temperature cycles.

The three-dimensional quasi-steady state heat transfer model for laser material processing with a moving gaussian heat source developed by Mazumder et al. (Mazumder and Steen 1980a, Chande et al. 1981) using finite difference numerical techniques is the most comprehensive model for the laser welding reported so far. This model allows for "keyholing", temperature dependent thermophysical properties, convective heat losses due to shielding gas and radiative heat losses. The recent updated version (Chande et al. 1981) also incorporates the latent heat of fusion.

In order to develop the model (Mazumder and Steen 1980a, Chande et al. 1981), the process is physically defined as follows: a laser beam, having a defined power distribution, strikes the surface of an opaque substrate of infinite length but finite width and depth moving with a uniform velocity in the positive x-direction. The incident radiation is partly reflected and partly absorbed according to the value of the reflectivity. The reflectivity is considered to be zero at any surface point where the temperature exceeds the boiling point. This is because a "keyhole" is considered to have formed which will act as a black body. Some of the absorbed energy is lost by re-radiation and convection from both the upper and lower surfaces while the rest is conducted into the substrate. The part of the incident radiant power which falls on a keyhole is considered to pass into the keyhole losing some power by absorption and reflection from the plasma within the keyhole as described by a Beer–Lambert absorption coefficient. Matrix points within the keyhole are considered as part of the solid conduction network, but operating at fictiously high temperatures. The convective heat transfer coefficient is enhanced to allow for a concentric gas jet on the upper surface as used for shielding in welding. The system is considered to be in a quasi-steady state condition so that the thermal profile is considered steady relative to the position of the laser-beam.

The mathematical model could be used for the following purposes:
(1) to predict the temperature profile;
(2) to predict maximum welding speeds;
(3) to predict the heat affected zone;
(4) to predict the thermal cycle at any location or speed;
(5) to predict the effect of thickness or any other parameter (e.g. reflectivity); and

(6) to predict effect of supplementary heating or cooling.

Predictions of the fusion zone have been successfully made for Ti welds (Mazumder 1978, Mazumder and Steen 1979) and welds in mild steel (Eboo 1979). The comparison of predicted and experimental fusion zones for a partial penetration bead-on-plate weld in mild steel is shown in fig. 3.54 which also shows the effect of varying the absorption coefficient – note the beam diameter is not constant in the figure. Using an electric arc and a laser heat source together (Eboo 1979) – the arc having an assumed parabolic power distribution – a fit with the observed fusion zones and a comparison with the expected fusion zone from a more powerful laser is shown in fig. 3.55. By resetting the traverse speed during the relaxation iteration, the maximum speed at which the melting point is just achieved on the lower surface can be calculated. This would be expected to be similar to or slightly higher than the maximum welding speed. It compares well with experiments in the welding of 2 mm Ti (fig. 3.56).

It is also possible to calculate the thermal cycle experienced by titanium under these welding conditions (fig. 3.57). This datum is almost impossible to measure. However, the cooling rate has a marked effect on the resulting microstructure and is thus worth knowing.

Further it is also possible to integrate the thermal fluxes through different zones and so find the overall utilization of energy within the welding system. For example, welding at 1500 W and 7.5 mm/s and considering only a 10 cm length of weld, the heat in the slab is 245 W, the heat reradiated is 1175 W, and the heat convected and reflected is 80 W. Finally, by comparing

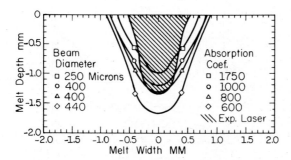

Fig. 3.54. Variation in laser melt widths with beam diameter and absorption coefficient at a laser power of 1570 W and reflectivity of 0.8. Welding velocity 33.5 mm/s (Mazumder and Steen 1980a).

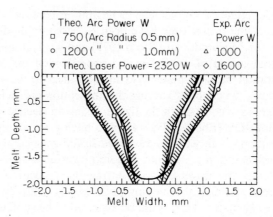

Fig. 3.55. A comparison of the predicted and experimental fusion zones associated with the 1.0 and 1.6 kW laser/arc welding runs. The theoretical arc power transferred: 75%, laser incident power: 1.57 kW, welding velocity: 33.5 mm/s, and absorption coefficient: 800 m^{-1} (Mazumder and Steen (1980a).

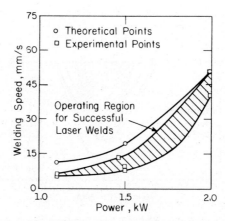

Fig. 3.56. Comparison of model predictions and experiment showing the variation of the welding speed for 2 mm titanium (6Al–4V) with laser incident power (Mazumder and Steen 1980a).

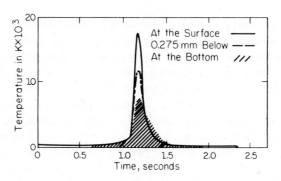

Fig. 3.57. Mathematically predicted cooling rate for Ti weld (6Al–4V); 1500 W, 7.5 mm/s, 2 mm thick (Mazumder and Steen 1980a).

experiments and calculations a quantitative estimate of the rate of grain growth or diffusion can be made.

The best way of using any mathematical model is to represent the data in dimensionless graphs. Graphs plotted from such data may be used for various other materials without having to rerun the computer program for each new parameter. All the numerical models discussed above provide such dimensionless plots for generalized use.

6. Summary

Numerous experiments since the development of multikilowatt CW CO_2 laser around 1970, have shown that the laser permits precision weld joints of a high quality only rivalled by an electron beam. Various commercial CO_2 laser systems having powers from a few watts to 15 kW are available for a multitude of laser processing applications. Today high power (up to 400 W average CW power) pulsed Nd–YAG lasers are also available for overlapping spot welding applications. Due to its shorter wave length (1.06 μm) the Nd–YAG laser can produce higher power density compared to a CO_2 laser. Otherwise welding principles for Nd–YAG lasers are the same as CO_2 lasers. Details of solid-state laser welding are discussed in a separate chapter of this book. The availability of high power lasers, the limitations of current welding technology and the rapid technological developments in the past decade have stimulated considerable interest in laser welding. It is beginning to be accepted as an industrial production technique.

Research carried out in laser welding has already demonstrated that high quality welding can be performed for many ferrous alloys, nickel alloys and titanium alloys. This process has also been found effective for some iridium alloys and many dissimilar metals which are very difficult to weld by alternative processes. Some partial success is also achieved for highly reflective materials such as aluminum alloys. But more research is necessary to improve the process for aluminum alloys.

Understanding about the laser welding variables and shielding gas requirements has already been achieved to a great extent. But the plasma formation and the keyhole mechanism during the laser material interaction is still a little understood process. Surface tension driven fluid flow inside the molten pool due to the characteristically high temperature gradient observed during laser welding also deserves more attention. Compared to the conventional welding processes more precise mathematical modelling is possible for laser welding due to the well defined characteristics of a laser beam. However, improvement to the predictive capabilities of mathematical models requires better understanding of plasma formation and fluid flow inside the pool.

References

Adams, Jr., C.M., 1958, Weld. J. **37** 210S.

Adams, H.J., 1970, Met. Constr. Br. Weld. J. **2** (1), 1.

Adams, M.J., 1973, CO_2 Laser Welding of Aero-Engine Materials, Rep. 3335/3/73 (British Welding Institute, England).

Alexander, J., 1981, unpublished work, University of Illinois, Urbana, USA.

Alexander, J. and Steen, W.M., 1980, The Effect of Process Variables on Arc Augmented Laser Welding, SYMPOSIUM OPTIKA' 80, Nov. 1980 (Budapest).

Alwang, W.G., Cavanaugh, L.A. and Sammartino, E. 1969, Weld. J. March 1969 **43** PTO 3, 105.

Anderson, J.E. and Jackson, J.E., 1965a, in: Proc. Electron Laser Beam Symp. Penn-State Univ., ed. A.B. EL-Kareh, p. 17.

Anderson, J.E. and Jackson, J.E., 1965b, Weld. J. **44** 1018.

Andrews, J.G. and Atthey, D.R., 1975, J. Inst. Maths. Appl. **15**, 59.

Andrews, J.G. and Atthey, D.R., 1976, J. Phys. D: Appl. Phys. U.K. **9**.

Arata, Y. and Inoue, K., 1973, Trans. JWRI **2**, No. (1), 41.

Arata, Y. and Miyamoto, I., 1972a, Trans. JWR1 **1** No. (4), 11.

Arata, Y. and Miyamoto, I., 1972b, Some Fundmental Properties of High Power Laser Beam as a Heat Source, vol. 3 (Japan Welding Society).

Arata, Y. and Miyamoto, I., 1973, Trans. JWRI **2** No. (2).

Arata, Y. and Miyamoto, I., 1977, Laser Focus No. (3).

Arata, Y. and Miyamoto, I., 1978, Technocrat May 1978, **11** No. (5), 33.

Arata, Y., Maruo, H., Miyamoto, I. and Inoue, Y., 1977, Dynamic Behavior of Laser Welding, IIW DOC. IV/222/77.

Baardsen, E.L., Schmatz, D.J. and Bisaro, R.E., 1973, Weld. J. April 1973 **52**, 227.

Banas, C.M., 1972a, High Power Laser Welding, ASM/NASA/GWV Symp. on Welding, Bending and Fastening, Virginia, May 30–June 1 1972.

Banas, C.M., 1972b, Laser Welding Developments, Proc. CEGB International Conf. on Welding Research Related to Power Plants, Southampton, England, Sept. 17–21, 1972.

Banas, C.M., 1975a, Electron Beam, Laser Beam and Plasma Arc Welding Studies, NASA Contractor Report No.: NASA CR – 132386, March 1975.

Banas, C.M., 1975b, United Technologies Research Center Report No. R75-412260-1, July 1975.

Banas, C.M., 1977, Laser Welding to 100 kW, UTRC Report No. R76-912260-2 (under Navy Contract N00173-76-M-0107) Feb. 1977.

Banas, C.M., 1979, Electron Beam, Laser Beam and Plasma Arc Welding Studies, Contract NAS 1-12565, NASA, March 1979.

Banas, C.M. and Peters, G.T., 1974, Study of the Feasibility of Laser Welding in Merchant Ship Construction, Contract 2-36214, U.S. Dept. of Commerce, Final Report to Bethlehem Steel Corp, August 1974.

Baranov, M.S., Metashop, L.A. and Geinrikhs, I.N., 1968a, Weld. Prod. G.B. **15** pt. 3.

Baranov, M.S., Metashop, L.A. and Geinrikhs, I.N., 1968b, Svar, Proizvod No. (3), 13.

Baranov, M.S., Kondratev, V.A. and Uglov, A.A. 1972, Pizika Khim. Obrabot Mat. Sept.–Oct. 1972 pt. 5, 11.

Baranov, M.S., Vershok, B.A., Geinrikhs, I.N. and Privezenster, V.I. 1976, Weld. Prod. May 1976 **23** No. (5), 19.

Baranov, M.S. et al., 1978, Weld Prod. **25** No. (5), 43.

Benson, L., 1970, Laser Weld. Simplified, Am. Mach. **114** No (8), 81.

Borland, J.C. and Jordan, M.F. 1972, The Institution of Metallurgist – Review Course, Oct. 1972, Series 2, No. 9, p. 95.

Bramson, M.A., 1968, Infrared Radiation: A Handbook for Applications (Plenum Press, New York).

Breinan, E.M. and Banas, C.M., 1971, WRC Bull. Dec. 1971, 201.

Breinan, E.M. Banas, C.M. and Greenfield, M.A., 1975, Laser Welding – The Present State of the Art. 11th Annual Meeting, DOC IV-181-75, (Tel Aviv, July 6–12, 1975) pp. 1–53.

7 No. (9).

)eep Penetration Laser Welding, AWS Annual Meeting

ectric Discharge Convection Lasers, paper presented at eeting (Washington, D.C.).

Conduction of Heat in Solids, 2nd ed. (Oxford Univ.

asers in Metallurgy, eds., K. Mukherji and J. Mazumder

ijermundsen, K., 1965, Br. Weld. J. Feb. 1965, 52.

J. Appl. Phys. **48** No. (9), 3895.

Cohen, M.I. Mainwaring, F.J. and Melone, T.G., 1969, Weld. J. **48**, 191.

Courtney, C. and Steen, W.M., 1978a J. Appl. Phys. 303.

Courtney, C., and Steen, W.M., 1978b, The Surface Heat Treatment of Eng Steel Using a 2 KW CW CO_2 Laser, Paper No. 2, Advances in Surface Coating Technology, Int. Conf., London, 13–15 Feb. 1978.

Crafer, R.C., 1976a, Weld. Inst. Res. Bull. U.K. April 1976, **17**.

Crafer, R.C., 1976b, Weld. Inst. Res. Bull. U.K. Feb. 1976 **17**.

Crafer, R.C., 1978, Advances in Welding Processes: Proc. 4th Int. Conf. Harrogate, Yorks., 9–11 May 1978, paper No. 46, pp. 267–278.

Danismevskii, A.M., Kastalskii, A.A., Ryvkin, S.M. and Yarsmetskii, I.D., 1970, Sov. Phys. JETP **31** 292.

David, S.A. and Liu, C.T., 1980, High Power Laser and Arc Welding of Thorium Doped Iridium Alloys, Report No. ORNL/TM-7258 (Oak Ridge, Tenn.) May 1980.

Dennery, G. and Guenot, R., 1962, Pubn. Tech. Ministere de l'Air, NT 112, Paris.

Duddenhagen, D.A., 1963, Lasers and Their Metallurgical Applications, A.S.T.M.E. Technical Paper Sp. 63–212.

Duley, W.W., 1976, CO_2 Lasers, Effects and Applications (Academic Press, New York).

Eboo, M., 1979, Ph.D. Thesis (London Univ.)

Eboo, M., Steen, W.M. and Clarke, J., 1978, Paper 17, Advances in Welding Processes, 4th Int. Conf. Harrogate, UK, May 1978 (British Welding Institute).

Engel, S.L., 1976, Laser Focus Feb. 1976, 44.

Engel, S.L., 1978, Weld. Des. Fabr. Jan. 1978, 62.

Frick, R.J., 1966, Met. Prog. **90** No. (3), 91.

Gagliano, F.P., 1969, I.E.E.E Wescon Tech. Paper, vol. 13, pt. 7, Session 8/1, Aug. 20–24, 1969.

Garashchuk, V.P. and Molchan, I.V. 1969, Avtomet Svarka **9** 12.

Garashchuk, V.P., Voravsky, V.E. and Velichko, O.A. 1972, Automal. Weld. **25** No. (3).

Gibson, A.F., Kimmit, M.F. and Walker, A.C., 1970, Appl. Phys. Lett. **17** 75.

Gibson, A.M. and Walker, A.C., 1971, J. Phys. C. Solid State Phys. **4** 2209.

Glasstone, S.L., 1961, Thermonuclear Reactions (Van Nostrand).

Guenot, R. and Racinet, J., 1970, Br. Weld. J., 427.

Guenot, R., Racinet, J. and Bousquet, A., 1966, Soud. Tech. Conn. **23**, 105.

Harry, J.E., 1974, Industrial Application of Lasers (McGraw-Hill, U.K.).

Hice, J.H., 1973, Lasers – An Evaluation of Their Performance as Energy Sources for Industrial Applications, A.S.T.M.E. Technical Paper Sp. 64–91.

Hochied, B., Klima, R., Beauvais, C. and Roux C., 1970, Mem. Sci. Rev. Metall. **LXVII** No. (9).

Jackson, J.E., 1965, Weld. Eng. **50** (2), 61.

Jhaveri, P., Moffatt, W.G., and Adams, Jr., C.M., 1962, Weld. J. **41**, 12S.

Jørgensen, M., 1980, Met. Constr. Feb. 1980 **12** No. (2), 88.

Klemens, P.G., 1976, J. Appl. Phys. **47**, 2165.

Kloepper, D., 1970, Laser Beam Welding Process Development, Grumman Interim Engineering Progress Reports IR622-9-(11) and (111), Jan. and April 1970.

Ledbedev, V.K. and Granista, V.T., 1972, Automat. Weld. **25** 63.

Locke, E.V. and Hella, R.A., 1974, I.E.E.E. J. Quantum Electron. **QE-10** (2), 179.

Locke, E.V., Hoag, E. and Hella, R., 1972a, Weld. J. **51** 245S.

Locke, E.V., Hoag, E. and Hella, R., 1972b, I.E.E.E. J. Quantum Electron. **QE8**, 132.

Mazumder, J., 1978, Ph.D. Thesis (London Univ).

Mazumder, J., 1978, unpublished, Laser Welding of Aluminum Alloys, research carried out at the Center for Laser Studies, Univ. of So. California, under contract from ALCOA Labs. Joining Div., ALCOA Center, Penn.

Mazumder, J. and Steen, W.M., 1979, Structure and Properties of a Laser Welded Titanium Alloy, TMS-AIME Fall Meeting, Paper No. F79-17, Sept. 1979.

Mazumder, J. and Steen, W.M., 1980a J. Appl. Phys. Feb. 1980, 941.

Mazumder, J. and Steen, W.M., 1980b, Met. Constr. **12** No. (9), 423.

Mazumder, J. and Steen, W.M., 1981, Weld. J. **60**, No.(6), 19.

Mazumder, J. and Steen, W.M., 1982, Met. Trans. A, **13A**(5), 865.

Meleka, A.H., ed., 1971, Electron Beam Welding Principles and Practice (McGraw-Hill, London) pp. 95, 96, published for the Welding Institute.

Metzbower, E.A. and Moon, D.W., 1979, in: Mechanical properties, Fracture Toughness and Microstructures of Laser Welds of High Strength Alloys, Proc. of the Conf. on Applications of Lasers in Materials Processing, 18–20 April, Washington D.C., E.A. Metzbower, ed. (American Society of Metals, Ohio) pp. 83–100.

Miller, K.J., 1966, Weld, Eng. **51** 46.

Miller, K.J. and Nunnikhoven, J.D., 1965, Weld. J. **44**, 480.

Moorhead, A.J., 1971, Weld, J. **50** 97.

Morse, P.M. and Feshback, H., 1953, Methods of Theoretical Physics, Part 1 (McGraw-Hill, New York).

Myers, P.S., Uyehara, O.A. and Borman, G.L., 1967, Weld. Res. Counc. Bull. No. (123), 1.

Nagler, H., (1976), Feasibility, Applicability, and Cost Effectiveness of LBW of Navy Ships, Structural Components and Assemblies, Contract N00G00-76-C-1370, vols. 1 and 2, Dec. 22, 1976.

Nichols, K.G., Dec. 1969, Proc. Inst. Electr. Eng. **116** pt-12, 2093.

Orrok, N.E., 1967, Met. Prog. **91** No. (2), 150.

Paley, Z. and Hibbert, P.D., 1975, Weld. J. Nov. 1975 **54** 385S.

Paley, Z., Lynch, J.N. and Adams, Jr., C.M., 1964, Weld. J. **43** 71S.

Patel, C.K.N., 1964, Phys. Rev. **136** A1187.

Pfluger, A.R. and Mass, P.M., 1965, Weld. J. **44** 1018.

Platte, W.M. and Smith, J.F., 1963, Laser Techn. Met. Joining Weld. J. Res. Suppl **42** No. (11), 481S.

Ready, J.F., 1971, Effect of High Power Laser Radiation (Academic Press, New York).

Rein, R.M. et al., Pat. 1448740, The U.K. Patent Office (1972).

Rosenthal, D., 1941, Weld. J. **20**, 2205S.

Rosenthal, D., 1946, Trans. ASME **48** 848.

Rykalin, N.N., 1951a, Calculation of Heat Flow in Welding (Moscow).

Rykalin, N.N., 1951b, The Calculation of Thermal Processes in Welding (Marshgiz), Z. Paley and C. M. Adams, Jr., 1963, transl. into English.

Rykalin, N.N. and Uglov, A.A., 1969, Fizika Khim, Obrobot Mat. Sept–Oct 1969 Pt. 5, 13.

Schwartz, M.M., 1971, WRC Bull. Nov. 1971 167.

Schwartz, M.M., 1975, WRC Bull. Oct. 1975, No. (210).

Schwartz, M.M. 1979 Met. Joining Manual (McGraw Hill, New York).

Schmidt, A.O., Ham, I. and Hoshi, T., 1965, Weld, J. Suppl., Nov. 1965, p. 481.

Seaman, F.D., 1977, The Role of Shielding Gas in High Power CO_2(CW) Laser Welding, SME technical paper no. MR77-982 (Society of Manufacturing Engineers, Dearborn, MICH.).

Seaman, F.D., 1978, Establishment of a CW/CO_2 Laser Welding Process, USAF Technical Report AFML-Tr-76-158, Sept. 1978.

Seaman, F.D. and Hella, R.A., 1976, Establishment of a Continuous Wave Laser Welding Process, IR-809-3 (1 through 10) AFML Contr. F336 15-73-C5004, Oct. 1976.

Seretsky, J. and Ryba, E.R., 1976, Weld, J. Suppl., July 1976 208S.

Shewell, J.R., 1977, Weld Des. Fabr. (June) 106.

Sickman, J.G. and Morijn, R., 1968a Phillips Res. Rep. **23** 367.

Sickman, J.G. and Morijn, R., 1968b, Phillips Res. Rep. **23**, 375.

Snow, D.B. and Breinan, E.M. 1978, Evaluation of Basic Laser Welding Capabilities, prepared by United Technologies Research Center, East Hartford, Conn., for ONR, Dept. of Navy, Report No. R78-911989-14, July 1978.

Spalding, I.J., 1971, Phys. Bull. U.K. July 1971, P402.

Steen, W.M. and Eboo, M., 1979, Metal Construction **11**(7) 332.

Steenbeck, M., 1932, Phys. Z. **33**, 80.
Swifthook, D.T. and Gick, E.E.F., 1973, Weld. J., 492S.
Tung-Po Lin, 1967, IBM. J. Sept. 1967 **11**, 527.
U.K. Patent 12681/75, 1975; 6008/78, 1978.
Velichco, O.A., Gasashchuk, U.P. and Moravskii, V.E., 1972a, Auto. Weld. **25** pt-4, 75.
Velichko, O.A., Garashchuk, V.P. and Moravskii, V.E. 1972b, Avtomet. Svarka **25**, No. (3), 71.
Velichko, O.A., Garashchuk, V.P. and Moravskii, V.E., 1972c, Avtomet. Svarka **25** No. (8), 48.
Westby, O., 1968, Temperature Distribution in the Workpiece by Welding (Dept. of Metallurgy
   and Metals Working, The Technical Univ. of Norway).
Wilgoss, R.A., Megaw, J.H.P.C. and Clark, J.N., 1979, Weld. Met. Fabr. March 1979, 117.
Williams, N.T., Thomas, D.E. and Wood, K., Met. Constr. **9**, Part 4, 157.
Yessik, M. and Schmatz, D.J., 1974, Laser Processing in the Automotive Industry, SME paper
   MR74-962.
Zaidi, M.A., 1978, M.Sc. Thesis (London Univ., Imperial College).
Zhukov, V.V., 1970, Automat. Weld. **23**, pt. 1, Jan. 1970 42.

CHAPTER 4

LASER HEAT TREATMENT

VICTOR G. GREGSON

Coherent, Inc.
Palo Alto, California 94303, USA

Laser Materials Processing, edited by M. Bass
© *North-Holland Publishing Company, 1983*

Contents

1. Early development and industrial use

The early references to the use of lasers for heat treatments are only brief. These references are difficult to find, but appear in German laser texts and papers from the early 1960s (De Michelis 1970). In the United States, researchers at US Steel used a ruby laser to harden steel (Speich et al. 1966, Speich and Fisher 1965, Speich and Szirmae 1969). They were the first US group to use laser heating in a controlled manner for metallurgical research, to use non-reflective coatings and to analyze heat flow characteristics. Other descriptions of laser heat treatment appeared a few years later in the Russian scientific literature (Veiko et al. 1969, Mirkin and Philipetskii 1968, Mirkin 1970, 1973, Moisa 1974, Barchukov and Mirkin 1969, Annenkov et al. 1974, Mirkin and Philipetskii 1967). However, the development of laser heat treatment in manufacturing practice started at the General Motors Corporation. Dr Jon Miller at the Saginaw Steering Gear Division, General Motors Corporation, began to use a laser at the Manufacturing Development Staff of the General Motors Technical Center. This work led to the first manufacturing laser heat treatment process in the Saginaw Steering Gear Division as well as the formation of an industrial laser group at the GM Technical Center to develop laser heat treatment.

This work grew rapidly and within five years had spread to other major corporations and laser manufacturers. Many corporations had already begun to pursue the industrial use of lasers, but for applications other than laser heat treatment. The work at the Ford Motor Company with a United Technology Research Laboratories multikilowatt laser to weld automotive body components is an example (Baardsen et al. 1973, Yessik and Schmatz 1975). With high power lasers at hand, the development of laser heat treatment moved quickly. The cooperation between major industrial corporations and laser companies helped promote development and caused a wider dissemination of the process as well as an active cross-fertilization of ideas and techniques.

A rather extensive list of potential applications has been reported for CO_2 lasers. This list will not be repeated in this review, but can be found in the literature marked by * in the list of references. Ruby, Nd:YAG and Nd:glass lasers have also been used for heat treating metals (Speich et al.

1966, Tice 1977). At first, the goal of laser heat treatment was only selective surface hardening for wear reduction. We now find that laser hardening is also used to change metallurgical and mechanical properties. The practical uses of laser heat treatment include:

(1) hardness increase;
(2) strength increase;
(3) facilitate lubrication;
(4) wear reduction;
(5) reharden martensitic stainless steel;
(6) temper metals;
(7) increase fatigue life;
(8) surface Carbide creation; and
(9) creation of unique geometrical wear patterns.

From the above list of potential industrial applications, it can be seen that the number of parts being laser hardened on a daily basis under full manufacturing controls is growing.

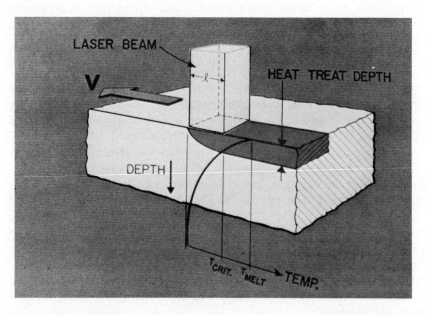

Fig. 4.1. The concept of transformation hardening of ferrous materials by a laser beam. (After Avco Everett Metalworking Lasers.) V is relative part velocity. L is beam width or diameter, T crit. is the minimum transformation temperature, T melt is the melt temperature of the ferrous material.

2. General description

Laser heat treatment is primarily used on steels and cast irons with sufficient carbon content to allow hardening. The metal surface is first prepared with an absorbing coating. After the coating is applied, the laser beam is directed to the surface. Fig. 4.1 illustrates the effects of the incident laser beam. As the beam moves over an area of the metal surface, the temperature starts to rise and thermal energy is conducted into the metal part. Temperatures must rise to values that are more than the critical transformation temperature but less than the melt temperatures. After the beam passes over the area, cooling occurs by mass quenching.

The laser beam is defocused or oscillated to cover an area such that the average power density has a value of 10^5 to 10^6 W/cm². A defocused laser beam can easily be visualized but the energy distribution of the oscillated beam may not be. Fig. 4.2 shows this energy distribution. At times segmented mirrors are used to redistribute or integrate the laser beam energy. The energy distribution of the integrating optics is also illustrated in fig. 4.2.

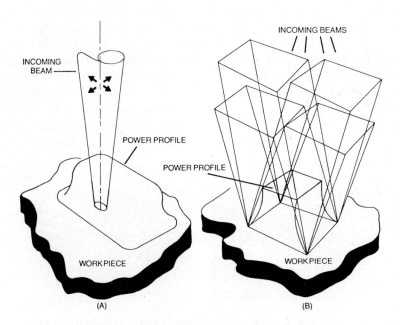

Fig. 4.2. Representation of laser power distributions that result from (A) oscillator optics and (B) integrating optics. (After Bonello and Howes 1980.)

In addition, fig. 4.3 shows examples of using the particular energy distribution of certain laser beams together with special optics to distribute the beam energy effectively in cylindrical geometries.

Using the power densities indicated above a relative motion between the workpiece and the beam of 0.5 and 5 cm/s will result in surface hardening. If surface melt occurs and this is not desired, relative motion should be increased. A decrease in power density will produce the same effect. If no hardening, or shallow hardening, occurs and is not desired, relative motion should be decreased. In this case an increase in power density will produce the same effect.

For production heat treatment, the coverage rate determines if laser hardening is a realistic process and this rate is highly dependent upon geometrical and laser parameters. However, most steels do respond within a factor of two to a similar set of values. Fig. 4.4 illustrates these idealized coverage rates for typical steels.

2.1. Coatings

Metals reflect most of the incident infrared energy as described in the Drude Free-Electron Theory (Wooten 1972, Wieting and Schriempf 1972). For power densities less than 10^5 W/cm^2, 90–98% of the incident laser power will be reflected away from most polished metal surfaces. In the early development of the carbon dioxide laser it was considered pointless by some to ever develop high power for metal processing. However, at higher power densities, other physical processes and effects dominate in the manner illustrated by Fig. 4.5 (Roessler and Gregson 1978). This seemingly anomalous absorption has been studied by many (Roessler and Gregson 1978,

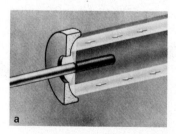

Fig. 4.3. Use of unstable resonator optics and toric mirrors for transformation hardening of (A) outer surface of a cylinder or (B) inner surface of a hollow cylinder (Sandven 1980). (Photos reprinted courtesy Avco Everett.)

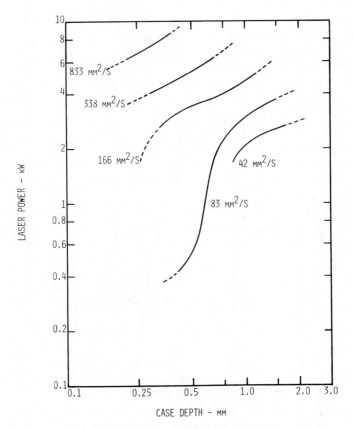

Fig. 4.4. Idealized coverage rates for transformation hardening of typical steels.

Basov et al. 1969, Bonch-Bruevich et al. 1968, Zavecz et al. 1975, Koo and Slusher 1976, Sanders and Gregson 1973, Ready 1976), but seems to be the result of minor impurities in the metals forming plasmas at the surface rather than a true decrease in reflectivity. These effects are important because they lead to the creation of "key-holes" in molten metals and laser power absorption of 90% through internal and trapped reflections within the "key-hole".

All laser heat treatment is not performed in the regime of absorption effects, but at power densities less than 10^5 to 10^6 W/cm^2. Attempts have been made to use the anomalous absorption effect in heat treatment, but the

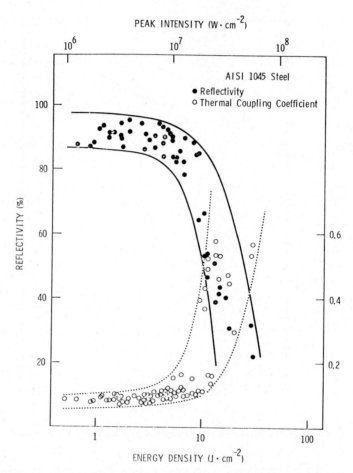

Fig. 4.5. The reflectivity and thermal coupling coefficient of AISI 1045 steel at 10.6 μm. The peak intensity is estimated on the basis of a 0.5 μs pulse duration (Roessler and Gregson 1978).

required degrees of in-line detection, feedback, and control have not been practical. This places an upper limit of 10^5 to 10^6 W/cm^2 on the power density used for heat treatments. As a result the applicable metal reflectivities are high and the Drude Free-Electron Theory holds. As such, coatings must be applied to the surface to increase the use of the available power. If coatings are not used the absorbed power is reflected and conducted away at a rate that does not permit the surface temperatures to reach the desired values. Table 4.1 lists coatings which have been used in laser heat treatment and is arranged such that the more efficient coatings appear near the top.

Table 4.1
List of coatings used in laser heat treatment

Phosphate (zinc, iron, lead or maganese)
Iron oxide
High temperature paint
Graphite (spray)
Molybdenum disulfide

Phosphate coatings are the result of a chemical conversion process in which the steel or cast iron surface is replaced by a metal phosphate. This metal can be zinc, iron, lead, or maganese. All phosphate coatings work, but maganese phosphate is used predominantly. It is a very efficient and very reproducible coating in the sense that the same power, speed, and spot size will always create the same depth of hardening.

The important variables in laser heat treatment are power, spot size, speed, and coating thickness. If the coating is too thin, it will be destroyed quickly in the laser processing and the metal will reflect power. If the coating is too thick, excessive power is required to burn through the coating and the resulting "plume or plasma" will absorb too much power. Coating thicknesses between 0.065 mm and 0.10 mm seem to be in the optimum range based on experience.

Paints are often used for coatings because of the ease with which they can be applied. However, coating thickness must be monitored constantly to ensure uniform depths of hardening. The best paints for laser heat treatment can be found by comparing labels which list contents. Paints which have high pigment and low organics are superior. The pigments and fillers should be titanium oxide, silicon dioxide, and carbon black which are all good absorbers of the carbon dioxide laser wave length. The organics will only burn away, thus creating a plasma which must be controlled and removed.

Most other coatings prove less satisfactory than paints and chemical conversion coatings. However, it should be mentioned that few systematic studies can be found comparing coatings (Megaw 1950, Carley 1977). Spray graphite coatings or molybdenum disulfide coatings used in die lubrication or as a parting compound have worked but results have been inconsistent. Metal oxide coatings also work but give inconsistent results. It has not been determined if coating variations are the cause or if these coatings are less efficient than others. Chemical etching has proved promising but again systematic studies are not available.

Coating thickness is an important parameter if any control or correlation of heat treated depths is to occur. Instruments to measure coating thickness

can be found in the coating or paint departments of most large corporations. In order to choose from among the many coatings which have been
tested it is instructive to review the coating chosen by the manufacturing
groups which are using lasers in production laser heat treatment. Three are
using the maganese phosphate conversion coating, three are using special
paints and one is using a black oxide coating.

2.2. Metal heating

The process of metal heating is a balance between the absorbed laser energy
and thermal energy conduction into neighboring regions. Absorption of
laser energy is a very localized surface phenomenon. From the
Hagen–Rubens formulation (Wooten 1972) of the Drude Free-Electron
Theory, the intensity (I) of the laser falls exponentially as $I \sim A e^{(-Z/Z_0)}$
where:

Z = distance into the material;
$Z_0 = \lambda / 4\pi K$;
λ = laser wavelength, 1.06 μm or 10.6 μm;
K = optical constant, absorption coefficient (values of 2 to 4); and
A = surface absorption.

We will assume that suitable coatings have been used to maximize
absorption ($A = 1$) and so we find that the first few micrometers of the
metal surface is the region where laser photon energy is converted into
thermal energy. Thus, temperatures in this region are the result of an energy
balance between the amount of incoming laser energy and the amount of
thermal energy being conducted away from the metal surface. If the rate of
incoming laser energy is high relative to the rate at which thermal energy is
conducted away then temperatures are high and concentrated near the
surface. If the rate of incoming energy is nearly equal to the rate at which
thermal energy is conducted away, the temperatures are lower, but a much
deeper zone is heated than in the first example.

Temperature distributions within metals can be calculated by conventional formulations of the thermal response of materials. A few of these
formulations are listed in Duley (1976), Ready (1971), Carslaw and Jaeger
(1959), Bechtel (1975), Sparks (1976), Veiko et al. (1968), Arata and
Miyamoto (1972). Elaborate computer code formulations also exist, but
these can be time consuming and costly if three-dimensional modeling is
attempted.

A few references exist in which the gaussian beam distribution is used as
the initial source function (Duley 1976, Speich and Fisher 1965). Arata et al.

(1978) have used the specific energy distribution of the Spectra Physics Model 971 laser in their analysis of laser heat treatment. Sandven (1980) has used the unique laser beam distribution of an AVCO HPL-10 in a specific geometry to calculate the laser heat treatment of cylinders.

For purposes of illustration, we will consider only the one-dimensional response of a uniform energy distribution onto a metal surface.

2.3. Metal cooling

As soon as the beam moves to another area of the metal surface, cooling occurs by thermal heat conduction into the surrounding metallic region. The equations which describe this process are discussed in the following sections (Duley 1976). For factories using lasers, this mass quenching by thermal heat conduction occurs within a few seconds after heat up. No water or oil is used on the surface to aid the cooling process.

In the laboratory, there are successful results of laser heat treatment which require a water quench. Generally, these are small parts which do not have enough metal to permit mass quenching to occur. Parts such as knife edges, saw blade teeth, or thin metal cross sections are examples which usually require a water quench. The choice of steel alloy, of course, determines the need to use a water quench. The pure iron–carbon alloys have only a few tenths of a second quench time while the alloyed steels can be cooled at a slower rate and still create the maximum hardness allowed by the carbon content of the steel.

2.4. Post processing

After heat treatment, manufacturing operations may still be required, but these are at the discretion of the manufacturer. About one-half of the laser heat treaters will use the part as it is after laser hardening. The laser hardening is the final operation in the part's manufacture. Generally, these groups use phosphate as the initial coating because it has additional advantages besides being a good absorber of laser power. The remaining factories will either perform a final hone or a paint removal operation to generate the required metal finish before the part is packaged or assembled.

2.5. One dimensional thermal heating and cooling

Laser energy distributions on the metal surface which are produced by:
  (1) high power multimode beams (Tophat Mode);
  (2) one and two axis scanning beams (dithered in a zig zag mode);

(3) beam integrators;
(4) segmented mirrors; and
(5) defocused beams;

are commonly used in laser heat treatment. All of these generate a reasonably uniform distribution of power over the central region of the beam path. The temperature distribution with depth during the temporal duration of the irradiation can be represented by equations derived from simple, but idealized models of one-dimensional heat transfer. A simple test to determine if this representation can be used is to examine the cross-section of a heat treated sample. If the bottom of the hardened zone is flat and parallel to the surface under the central part of the cross-section, then this analysis will accurately predict the temperatures in heated material. The edges of the cross section are regions where the problem is two-dimensional and the simple picture will not accurately predict the induced temperatures. One must use a more complex description to predict edge effects. The simple one-dimensional temperature prediction is separated into two parts, one equation for heat-up and one equation for cool-down (Duley 1976).

2.5.1. Heat-up

During the heat-up period light is incident on the sample with the following power density waveform:

$$F(t) = \begin{cases} F_0 & \text{for } t > 0, \\ 0 & \text{for } t < 0. \end{cases}$$

The surface source is idealized to be instantly applied, constant in time and uniform in the (x, y) plane. The temperature distribution as a function of time and depth in the material is then:

$$T(z, t) = \frac{\varepsilon 2 F_0}{K} (kt)^{1/2} \text{ierfc}\left(\frac{z}{2(kt)^{1/2}} \right).$$

2.5.2. Cool-down

When the light is turned off or after the beam has passed the material cools according to:

$$T(z, t) = \frac{2 F_0 k^{1/2}}{K} \left[t^{1/2} \text{ierfc}\left(\frac{z}{2(kt)^{1/2}} \right) \right.$$
$$\left. - (t_0 - t_1)^{1/2} \text{ierfc}\left(\frac{z}{2[k(t - t_1)]^{1/2}} \right) \right],$$

where:

> T = temperature (°C);
> z = depth from surface (cm);
> t = time (s);
> ε = emissivity, ~ 1;
> F_0 = average power density (W/cm^2);
> K = thermal conductivity (W/cm °C);
> k = thermal diffusivity (cm^2/s);
> t_0 = time start for power on (s);
> t_1 = time for power off (s); and
> ierfc = integral of the complimentary error function.

These equations are strictly valid with depth only on the central axis but in practice they have meaning over the region of a metallurgical cross section which shows a flat hardened zone. These equations are also valid only if the metal thickness is greater than $(4kt)^{1/2}$, otherwise there will be boundary effects from the rear side of the sample.

The surface temperature can easily be derived from these equations by setting $z = 0$. During heat-up:

$$T(0, t) = \frac{2\varepsilon F_0}{K} \left(\frac{kt}{\pi} \right)^{1/2}.$$

Two changes allow these equations to be easily handled by programmable hand calculators.

(A) A normalized coordinate can be established so that a single set of calculations for given thermal properties suffice. This coordinate is $T(z, t)/\varepsilon F_0$, which has units of °C/W/cm^2.

(B) The integral of the complimentary error function is re-written in terms of the error function and an approximation is used to compute the error function. The two relationships between ierfc and erf are:

(i) $\quad \mathrm{ierfc}(\alpha) = \dfrac{e^{-\alpha^2}}{\pi^{1/2}} - \alpha \mathrm{erfc}(\alpha),$

where erfc = complimentary error function,

(ii) $\quad \mathrm{erfc}(\alpha) = 1 - \mathrm{erf}(\alpha).$

As a result

$$\mathrm{ierfc}(\alpha) = \frac{e^{-\alpha^2}}{\pi^{1/2}} - \alpha[1 - \mathrm{erf}(\alpha)].$$

The following equation is an approximation to the error function which is accurate to within one part in 2.5×10^{-5} (Abramowitz and Stegun 1964):

$$\text{erf}(\alpha) = 1 - \left(a_1 b + a_2 b^2 + a_3 b^3\right)e^{-\alpha^2},$$

where:

$$b = (1 + P\alpha)^{-1},$$
$$a_1 = 0.3480242,$$
$$a_2 = -0.0958798,$$
$$a_3 = 0.7478556,$$
$$P = 0.47047.$$

Although the one-dimensional distribution is very useful and instructive, the actual energy distribution of the beam is required if more exact thermal distributions are desired. This is particularly true if the edge effects and hardened width is needed. It is also true if the part treated is small and mechanical edges and the back surface are heated.

2.6. Metallurgy

2.6.1. Hardening mechanism in steel

The process of laser heat treatment in terms of a grain by grain metallurgical description is illustrated in fig. 4.6. Fig. 4.6a shows the lamellar segregation of carbon in a grain of pearlite. This segregation occurs as sheets of ferrite, or pure iron, bounded by sheets of cementite, or iron carbide (FeC_3). Dimensions of individual grains depend upon the thermal history of the ferrous alloy, but can range from 0.25 mm to 0.025 mm. There can often be four to fourteen "sheets" or layers of ferrite and cementite in each grain of pearlite.

Fig. 4.6b illustrates the end result of a pearlite grain which would undergo three different thermal cycles. Thermal cycle A illustrates a pearlite grain which is heated to temperatures less than the critical phase temperature and then cooled to ambient temperatures. In this temperature regime there is little diffusion of carbon from the cementite layers into the ferrite layers, thus the grain is heated and cooled with little change in the physical properties. Pearlite remains as pearlite.

In cycle B, the pearlite grain is heated to temperatures above the phase transition. This regime has a very rapid diffusion of carbon in the cementite into the surrounding ferrite. This proceeds rapidly and the grain becomes a homogeneous distribution of iron and carbon called austenite. Upon cool-

A. Local Equilibrium within a Pearlite Grain

—Cementite
—Ferrite

B. Time-Temperature Constraint to Produce Martensite

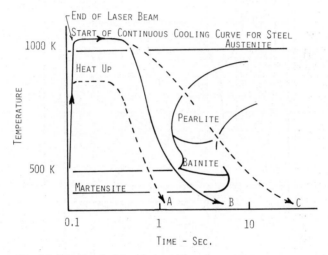

Fig. 4.6. Metallurgical description of laser heat treatment.

ing, the austenite will drop to a temperature less than several hundred degrees celsius and enter a region in which martensite forms. To enter this region the heated material must quench fast enough to "miss" the "nose" of the continuous cooling curve.

In cycle C, this pearlite grain is heated above the phase transition temperature as in cycle B. Austenite is formed, but the cooling rate is much slower. This cooling rate allows the austenite to reform pearlite by having the carbon diffuse back into layers of cementite and ferrite. Thus, pearlite undergoes transformation into austenite and then returns to pearlite for this thermal cycle. The grain boundaries and cementite/ferrite spacings may be different from the original pearlite because of a different thermal history.

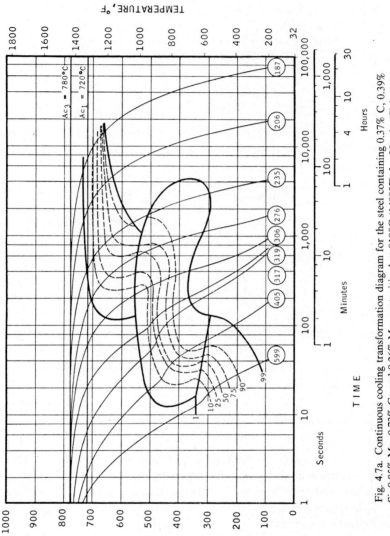

Fig. 4.7a. Continuous cooling transformation diagram for the steel containing 0.37% C, 0.39% Si, 0.85% Mn, 0.73% Cr and 0.26% Mo, austenitized at 810°C (1490°F) for 20 min (Siebert et al. 1977).

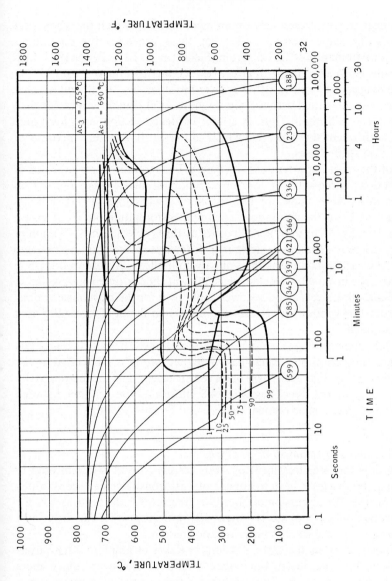

Fig. 4.7b. Continuous cooling transformation diagram for the steel containing 0.37% C, 0.36% Si, 0.82% Mn, 0.73% Cr and 0.76% Mo, austenitized at 795°C (1465 °F) for 20 min (Siebert et al. 1977).

Most heat treat processes rely on an equilibrium in which the whole part undergoes the same thermal cycle. For these processes the continuous cooling transformation (CCT) diagram (Siebert et al. 1977, Boyer 1977, Cias and Doane 1973) as in fig. 4.6b effectively summarizes the results.

These diagrams do not easily carry over into laser heat treatment because the process is concerned with a grain by grain local equilibrium rather than a whole part equilibrium. Thus, in a laser hardened section, each local area has had a different thermal cycle which can result in different mixtures of the common steel phases.

One of the most important uses of the CCT diagram is as an indicator of the cool-down time-window for the creation of martensite. Fig. 4.7 illustrates this time window for two different steels (Boyer 1977). For fig. 4.7a, the "nose" of the continuous cooling transformation occurs at twelve seconds. For fig. 4.7b, the "nose" occurs at thirty-two seconds. The major difference between these two steels is a difference in the alloy percentages whereas the carbon content is the same in both alloys. A pure iron–carbon steel with little or no alloy elements has an even shorter time window (0.5 to 1.0 s). Thus, alloy steels allow a slower cool down for the marteniste regime. The greater this window the greater the maximum depth of hardening that can be attained.

Another property of steels under the heat treatment process is that steels with a greater carbon percentage have a higher hardness. This effect is the exact relationship that has been observed in steels that are hardened by other heat treatment processes. Fig. 4.8 illustrates the relation between hardness and steel carbon content (Hodge and Orehoski 1946).

2.6.2. Hardening mechanisms in cast irons

The concepts for laser treatment of steels can be carried over and used in discussions of the laser heat treatment of cast-irons. For cast-irons which contain pearlite the discussion of laser heat treatment is the same as that for the previous discussion in steels (Angus 1976). Such pearlite cast-irons heat treat easily under the laser-induced transformation.

For some cast-irons the carbon occurs as flakes of graphite. These irons can also be laser hardened, but under conditions which are much more restricted than the above. These more limited conditions occur because much longer time-at-temperature is required to allow carbon to be diffused outward from each graphite flake. Thus, hardening occurs both as transformed grains of pearlite and as a rind of martensite around each graphite flake. A long soak time will allow a uniform case depth. This long time-at-

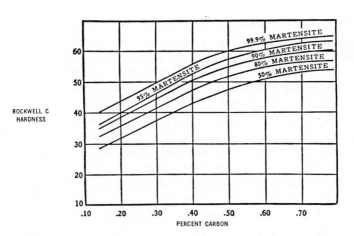

Fig. 4.8. Relationship between carbon content, hardness and martensite percentage (Hodge and Orehoski 1946).

temperature will permit the diffusion of carbon, but it will restrict the range over which the processing parameters can be varied.

Pearlite cast-irons are also a major grouping of the cast-iron alloys. These cast-irons may also have graphite flakes, but the dominance of the associated pearlite controls the hardening. The shorter time-at-temperature required to uniformly diffuse carbon in each pearlite grain is the same as the pearlite hardening dynamics in steel. Thus, the range over which the processing parameters can be varied will be similar to the steels.

2.7. Computational model

In a computer model of the laser heat treatment process, the calculation of temperature increase and subsequent cooling gives provides an insight into some of the process parameters. The time–temperature calculation can answer such questions as:

(i)   has the temperature exceeded the critical phase temperature such that the material is in the austenite regime.

(ii)   Has the temperature exceeded the melt temperature.

(iii)   Is the cooling rate fast enough to miss the "nose" of the continuous cooling curve diagram.

The calculation cannot answer:

(iv)   has time-at-temperature been sufficient to allow adequate carbon diffusion.

For this, we must develop a metallurgical model which incorporates the temperature information. Fig. 4.9 illustrates the simple one-dimensional metallurgical model which is used to determine the carbon diffusion of pearlite. Fig. 4.9a illustrates a pearlite grain which has alternating layers of ferrite and cementite. The layers will have an average spacing which can be determined by a metallographic examination. This spacing is used to fix the geometry for a one-dimensional calculation of carbon diffusion from a layer of cementite into an adjacent layer of ferrite. The geometry of this calculation is illustrated in fig. 4.9b. This calculation adopts the simplest diffusion model to illustrate the overall effect. This model is Fick's second law of diffusion (Shewmom 1968, Reed-Hill 1974) and can be found in most textbooks on diffusion or metallurgy. Obviously, more complex diffusion models can be used (Brown 1976), but we wish to illustrate only the main

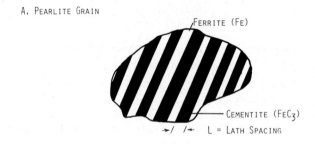

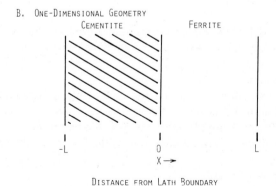

Fig. 4.9 One-dimensional model of carbon diffusion in pearlite.

effect. Fick's Laws permit the concentration of a diffusion substance (carbon in this case) to be determined as a function of distance and time. The differential equation to be solved is:

$$\frac{dC(x,t)}{dt} = D\frac{d^2C(x,t)}{dx^2},$$

where D is the coefficient of diffusion. If we define a dimensionless function F which is the ratio of the initial carbon concentration and the carbon concentration at the interface, then the error-function solution is:

$$F(x,t) = \frac{C_{-L}(x,t)}{C(0,t)} = 1 - \mathrm{erf}\left(\frac{x}{2(Dt)^{1/2}}\right).$$

The previous equations allow a simple calculator to be used in the solution. Fig. 4.10 illustrates the type of solution that will be obtained. The carbon ratio permits the calculation to avoid having to find the exact carbon percentages which will differ for different alloys. At the start, the car-

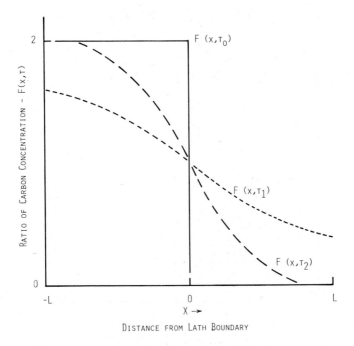

Fig. 4.10. Expected solution to the one-dimensional diffusion equation.

V.G. Gregson

bon ratio in the cementite will have a constant relative value of 2. In the ferrite, the carbon ratio will have a value of zero. As time progresses, carbon will diffuse across the boundary until the ratio eventually reaches a value of 1. As a rule, if the carbon ratio (F) at $x = L/2$, the midpoint of the ferrite lath, has a value of 0.25 then sufficient diffusion has occurred to produce martensite under adequate cooling rates.

An important part of this model and calculation is the value of D, the diffusion coefficient. The diffusion coefficient of carbon in iron is temperature dependent. Fig. 4.11 shows the change in diffusion coefficient as the temperature varies. This graph is a compilation of many studies (Holgrook et al. 1932, Grace and Derge 1958, Hilbert and Lange 1965, Heisterkamp and Lohberg 1966, Kaibicher and Leninskih 1970, Goldberg and Belton 1974, Parris and McLellan 1976) but adequately represents the value of diffusion coefficient for the purposes of this simplified model. The computational procedure is to determine the temperature as a function of position

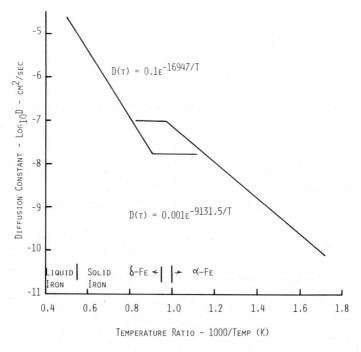

Fig. 4.11. Temperature dependent diffusion coefficient of carbon in iron.

and time through the temperature calculation. This value of temperature is then used to determine the diffusion coefficient. The values of the diffusion coefficient and time are then used to compute the value of *F*. If this value exceeds the above rule of thumb, then the steel or cast-iron is in a state which will produce martensite if cooled within the required time interval.

3. Experimental model

Many laser users prefer to develop their heat treat process parameters by a trial-and-error approach of changing processing parameters and measuring the resulting hardened depth. Results are quickly apparent and optimization occurs within the first twenty-five trials. Others have struggled with algorithms to logically and systematically summarize the results for insight and future use. The parameters which affect the process are:

 (a) power;
 (b) speed;
 (c) spot size;
 (d) beam energy distribution;
 (e) steel or cast-iron alloy; and
 (f) absorbing coating.

A simple algorithm has evolved which allows most of the parameters to be systematically summarized. This experimental data can then be compared with similar results which can be summarized with the help of the preceding calculations. Reasonable agreement has occurred when such results were compared for one or two steels.

The formalism for this approach has been derived from similar studies in weld parameter systematics. (Harth and Leslie 1975) This formalism was first tried on laser heat treatment by Harth et al. (1976) and has proved to be an effective, but simple way to characterize the laser heat treatment process.

Using the model of the thermal equation of state illustrated by fig. 4.12 (Charschan 1977, p. 5.5), the energy required to move from point A to point B is well defined. For this, an isolated volume of metal is heated from temperature A to temperature B by introducing a given amount of energy into the volume. The relation between temperature and energy is controlled by the specific heat of the metal. Thus:

$$E/V = C\rho(T_B - T_A),$$

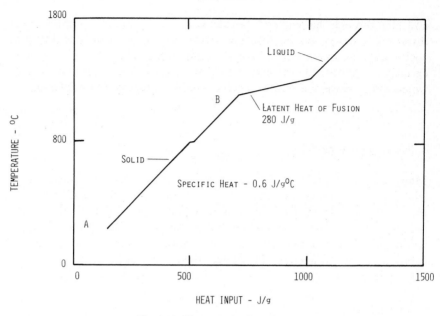

Fig. 4.12. Thermal equation of state.

where:

$$E/V = \text{energy per unit volume (J/cm)};$$
$$C = \text{specific heat of metal (J/g K)};$$
$$\rho = \text{specific density of metal (g/cm}^3\text{); and}$$
$$(T_B - T_A) = \text{temperature interval (K)}.$$

Since

$$\frac{\text{energy}}{\text{unit volume}} = \frac{\text{energy}}{\text{unit area}} \frac{1}{\text{hardening depth}}$$

and

$$\text{energy} = \text{power} \times \text{time} \ (J = W \ s),$$

we may write

$$\frac{J}{\text{unit area}} = \text{depth} \left[C\rho (T_B - T_A) \right] = \frac{W \ s}{\text{unit area}}.$$

Joules/unit area is called energy density in the graphs and illustrations which accompany this discussion. Please be aware that there are many combinations of parameters which have been used by other writers in representing hardening data. All are correct and none have shown such a superiority of data representation as to have won universal acceptance. My preference is to plot Joules/unit area versus speed for a desired hardening depth. These are calculated from the following measured quantities:

(1) relative speed of translation between metal and beam;

(2) laser beam diameter on metal surface. Beam width and length for segmented mirror optics or dithered optics;

(3) laser power at metal surface; and

(4) measured depth of hardening by optical analysis.

The value of Joules/unit area is calculated by using the measured values of power and area together with the time duration for which the laser beam is directed on a specific spot on the metal surface. This time duration is calculated from:

$$\frac{\text{laser beam diameter}}{\text{speed}} = \frac{\text{cm}}{\text{cm/s}} = \text{s.}$$

Thus:

$$\frac{\text{W}}{\text{area}} \frac{\text{diameter}}{\text{speed}} = \frac{\text{W s}}{\text{area}} = \frac{\text{J}}{\text{cm}^2}.$$

Sometimes this data is plotted in log–log coordinates in order to linearize data and extend the representations. Fig. 4.13 illustrates the representation of such data in addition to that offered by Harth et al. (1976). Such data representations delineate regions where hardening does not occur and regions where surface melt begins to occur. Obviously, the region between these extremes is the regime for laser heat treatment. This data representation differs from alloy to alloy. Of the six parameters for laser heat treatment, four are characterized by the plot (power, speed, spot size and alloy) and two are fixed (coating and beam energy distribution).

Fig. 4.13 is confusing to casual observation. The data seems to trend in the wrong direction. One must remember that time is used in both coordinates. An example of how data is used in such representations may help to illustrate the trends which occur. Consider laser power of 500 W which is focused to a spot diameter of one millimeter. This spot is moved at two speeds which are one cm/s and two cm/s. Power and spot diameter are used to calculate a power density of 63 662 W/cm². We must now determine the time interval in which each element of surface metal is exposed to the beam. This is 0.1 cm/(1 cm/s) or 0.1 s and 0.1 cm/(2 cm/s) or 0.05 s, respectively. Using the relationship between power and energy

(1 W × 1 s = 1 J) we have coordinates of:
  (A) 6366 J/cm² and 1 cm/s,
  (B) 3183 J/cm² and 2 cm/s.
For each of these two parameter sets, the depth of metal hardening is measured. One can surmize that data set B has less total energy and less total hardening. Thus the data does trend in the correct manner in fig. 4.13.

Fig. 4.13 has one use in addition to being a simple representation of many processing parameter. It can be "deconvolved" into representations of power, spot size and speed which correctly interpolates these parameters between and beyond the original data set. This process is described in fig.

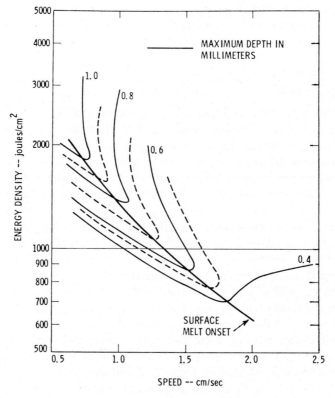

Fig. 4.13. Energy density plot for AISI 1060 steel.

4.14. Most manufacturing processes in laser heat treatment are defined around a given depth of hardening. From fig. 4.13, a depth of hardening is selected and coordinates of energy density and speed are selected. This information is used in the following algorithm

$$(\text{energy density}) \times (\text{speed}) \times (\text{spot size}) = \text{power}.$$

By selecting a range of spot sizes, a matrix of process parameters can be calculated (speed, spot size and power) which would produce the required depth of hardening. Such a matrix of values has been generated from fig. 4.13 for 0.4 mm depth of hardening. This data set is represented as a data plot in fig. 4.15. Two additional pieces of information are included in fig. 4.15 to further define the available region for martensite hardening without surface melt. Fig. 4.13 indicates a speed of 1.9 cm/s or faster will show surface melt if the indicated energy density is applied to the steel surface. Thus, fig. 4.15 shows surface melt at speeds greater than 1.8 cm/s. The continuous cooling curve for 1060 steel is used to obtain the time interval to the nose of the curve for bainite/pearlite. This time interval is used with the

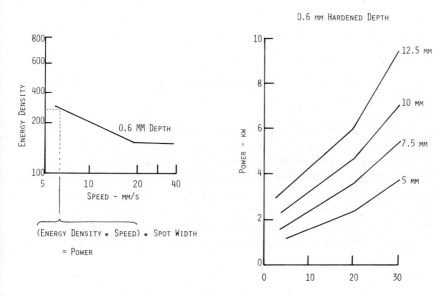

Fig. 4.14. Use of energy density plot to generate matrix of power, speed and spot size parameter.

LASER HARDENING OF AISI 1060 STEEL

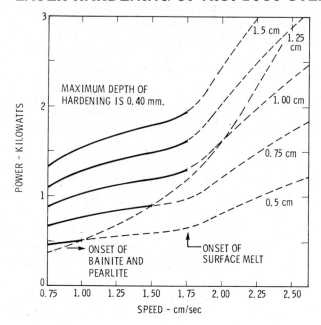

Fig. 4.15. Graph of process parameters derived from fig. 4.13.

spot size to estimate an effective speed (spot size/time interval) which is also plotted in fig. 4.15. These two sets of information constrain the original matrix to values which are represented by the solid line. This procedure can be repeated for each of the lines of equal hardened depth.

A verification of the energy density plot (fig. 4.13) can be made by creating a similar plot using the information calculated by the equations for heat-up and cool-down together with the equations for diffusion. Such a comparison has shown excellent agreement when it has been tried for one or two steel alloys. Similar comparisons by other investigators or extensions of the model to include greater detail should improve our understanding of the laser heat treatment process and our confidence in the validity of the model.

4. Summary

Between 1965 and 1970, many major US corporations, such as the US Steel and General Motors Corporation, were exploring laser heat treatment as a possible industrial process should economic indicators prove favorable. Of the several hundred technical successes, at least seven have reached the production floor as the preferred manufacturing process.

The technical process of matching the thermal response of the metal to the rate of delivered laser beam energy requires the use of absorptive coatings and defocused beams. Metal temperatures must exceed the critical transformation values but remain less than the melt temperature. The use of absorbing coatings is important because these coatings start the absorption and transfer of energy into the metal under conditions which would normally reflect most of the energy.

A simple thermal model of one-dimensional heating and cooling can be used to calculate the transient temperatures at the surface and at distances below the metal surface. The model can be extended by using a simple model of diffusion which adds details of the metallurgical response. More detailed three-dimensional thermal models can be used as well as more detailed models of diffusion.

The hardening mechanism in steel is the creation of martensite through rapid heating and cooling. The hardening mechanism in cast-iron also results from the creation of martensite. If pearlite is initially present, the creation of martensite is easily achieved. If graphite is present, martensite is created through the diffusion of carbon across greater distances and corresponding longer times at the required temperatures. The experimental conditions required to achieve hardening can be compared to the calculated parameters of power, speed and spot size with reasonable agreement.

References

References marked by * contain potential applications for CO_2 lasers.

Abramowitz M. and Stegun I.A., eds., 1964, Handbook of Mathematical Functions, National Bureau of Standards, Applied Mathematics Series No. 55 (US GPO, Washington, DC).

Angus H.T., 1976, Cast-Iron – Physical and Engineering Properties, Second Ed., (Butterworths, London) pp. 366 to 368.

Annenkov V.D., Barchukov A.P., Davydov Yu I., Zhukov A.A., Kokora A.N., Krishtal M.A. and Uglov A.A. 1974, Fiz. i Khim. Obrab. Materialov, No. 2, 38.

*Anon, 1978, Locking in on lasers, Heat Treating 10 No. 7, 22.

Arata Y. and Miyamoto I., 1972, Some fundamental properties of high power laser beam as a heat source (Report 3) – metal heating by laser beam, Trans. Japan Weld Society, e No. 1, 1.

*Arata Y., Maruo H. and Miyamoto I., 1978, Application of lasers for material processing – heat flow in laser hardening, International Institute of Welding, IIW Document IV-241-78.

Baardsen E.L., Schmatz D.J. and Bisaro R.E., 1973, High speed welding of sheet steel with a CO_2 laser, Welding J., 227.

*Banello L. and Howes M.A.H., 1980, How cost, technical factors affect laser heat treatment, Heat Treating, 32.

Barchukov A.I. and Mirkin L.I., 1969, Fiz. i Khim. Obrab. Materialov, No. 6, 126.

Basov N.G., Boiko V.A., Kronkhin O.N., Semenoy O.G. and Sklizkov G.V., 1969, Sov. Phys. Tech. Phys. **13** 1581.

*Bates W.F., Laser shock processing of aluminium alloys, Applications of Lasers in Materials Processing, Proc. Conf., Washington, DC, 18–20 April 1979, (American Society for Metals, Metals Park, OH 44703), p. 317.

Bechtel J.H., Heating of solids targets with laser pulses, 1975, J. Appl. Phys. **46** No. 4, 1585.

*Belforte D.A., 1977, High power laser surface treatment, Dearborn, USA, SME Paper No. 1077-373.

*Belforte D.A., 1980, Sep.–Oct., High power laser applications in manufacturing technology, The Carbide and Tool J., 16.

Bellis, J., ed., 1980, Lasers, Operation, Equipment, Application and Design, Engineering Staff of Coherent, Inc. (McGraw-Hill, New York).

Bonch-Bruevich A.M., Imas Ya. A., Romanov G.S., Libenson M.N. and Maltsev L.N., 1968, Sov. Phys. Tech. Phys. **13** 640.

Boyer, H.E., consulting ed., 1977, Atlas of Isothermal Transformation and Cooling Transformation Diagrams (American Society for Metals, Metals Park, Ohio 44073).

*Breinan E.M., Kear B.H. and Banas C.M., 1976, Processing materials with lasers, Phys. Today **44**.

Brown L.C., 1976, Diffusion-controlled dissolution of planar, cylindrical, and spherical precipitates, J. Appl. Phys. **47** No. 2, 449.

*Bryant A.L. and Koltuniak F.A., 1974, Heat treating with a high energy laser, Society of Manufacturing Eng. Tech. Paper MR74-964.

*Carley L.W., 1977, Feb., Laser heat treating, Heat treating, 16.

Carslaw H.S. and Jaeger J.C., 1959, Conduction of Heat in Solids, 2nd Ed. (Oxford University Press, London).

*Charschan S.S., et al., 1977, *Guide for material processing by lasers*, Laser Institute of America.

Cias W.W. and Doane D.V., 1973, Phase transformation kinetics and hardenability of medium-carbon alloy steels, Climax Molybdeum Company Publication, also found in Met. Trans. **4** Oct., 2257.

*Clauer A.H. and Fairand B.P., 1979, Interaction of laser induced stress waves with metals, Applications of Lasers in Materials Processing, Proc. Conf., Washington DC, 18–20 April 1979 (American Society for Metals; Metals Park, Ohio 44073) p. 291.

*Clauer A.H., Fairand B.P. and Wilcox B.A., 1977a, Pulsed laser induced deformation in an Fe-3wt pct Si alloy, Metallurgical Transactions **8A** No. 1, 119.

*Clauer A.H., Fairand B.P. and Wilcox B.A., 1977b Laser shock hardening of weld zones in aluminium alloys, Metallurgical Transactions **8A** No. 12, 1871.

*Cline H.E. and Anthony T.R., 1977, Heat treating and melting with a scanning laser or electron-beam, J. Appl. Phys. **48** No. 9, 3895.

De Michelis C., 1970, Laser interaction with solids – a bibliographical review, IEEE J. Quantum Electron. **QE-6** No. 10, 630.

Duley W.W., 1976, CO_2 Lasers, Effects and Applications (Academic, New York).

*Engel S.L. 1976a, Basics of laser heat treating, Dearborn USA, SME Paper No. MR76-857.

*Engel, S.L., 1976b, Heat treating with lasers, Am. Mach. 107.

*Gnanamuthu D.S., 1979, Laser surface treatment, in: Application of Lasers in Materials Processing, Proc. Conf., Washington DC, (American Society for Metals, Metals Park, Ohio 44073, 1979) p. 177.

*Gnanamuthu D.S., Shaw C.B., Jr., Lawrence W.E. and Mitchell M.R., Laser transformation hardening, Conf. Applications of Lasers in Material Processing, Washington, DC, April

1979.

Goldberg D. and Belton G.R., 1974, The diffusion of carbon in iron–carbon alloys at 1560°C, Met. Trans. **5** 1643.

Grace R.E. and Derge G., 1958, Trans. AIME **212** 331.

Harth G.H. and Leslie W.C., 1975, A new diagram for the application of welding theories, Welding J., Res. Supplement, **54**.

*Harth G.H., Leslie W.C., Gregson V.G. and Sanders B.A., 1976, Laser heat treating of steels, J. Metals **28** 5.

Heisterkamp F. and Lohberg K., 1966, Arch. Eisenhuttew **37** 813.

*Hella R.A., 1978, Materials processing with high power lasers, Optical Engineering **17** No. 3, 198.

Hilbert M. and Lange N., 1965, J. Iron and Steel Institute **203** 273.

*Hill J.W., Lee M.J. and Spaulding I.J., 1974, Surface treatments by laser, Opt. and Laser Technol. **6** No. 6 276.

Hodge J.M. and Orehoski M.A., 1946, Relationship between hardenability and percentage of martensite in some low-alloy steels, Transformation of AIME **167** 627.

Holgrook W.F., Furnas C. and Joseph T.L., 1932, Ind. Eng. Chem. **24** 993.

*Hsu T.R., 1973, Application of the laser beam technique to the improvement of metal strength, J. Testing and Evaluation **1** No. 6, 457.

Kaibicher A.B. and Leninskih G.M., 1970, Ixv. Akad. Nauk. SSSR, Metallurgy, No. 6, 72.

*Kawasumi H., 1978, Metal surface hardening CO_2 laser, Technocrat **11** No. 6, 11.

Koo J.C. and Slusher R.E., 1976, Appl. Phys. Lett. **28** 28.

*Locke E.V. and Hella R.A., 1974, Metal processing with a high power CO_2 laser, IEEE J. Quantum Electron **QE-10** 179.

*Locke E.V., Gnanamuthu D. and Hella R.A., 1974a, March, High power lasers for metal working, Avco Everett Research Laboratory, Inc., Research Report 398.

*Locke E.V., Gnanamuthu D. and Hella R.A., 1974b, High power lasers for metal working, Dearborn, USA, SME Paper No. MR74-706.

*Megaw J.H.P.C., 1980, Laser surface treatments, Surfacing J. **11** No. 1, 6.

*Megaw J.H.P.C. and Kaye A.S., 1978, Materials processing and heat treatment – surface modification and welding by laser, Laser 78, Proc. Conf., London, 9–10 March 1978 (Engineers' Digest).

*Miller J.F. and Wineman J.A., 1977, Laser hardening at Saginaw steering gear, Metal Progress **111** No. 5, 38.

Mirkin L.I., 1970, Plastic deformation of metal caused by a 10^{-8} s laser pulse, Sov. Phys. – Dokl. **14** No. 11, 1128.

Mirkin L.I., 1973, Formation of oriental structures by action of a laser beam on metals, Sov. Phys. – Dokl. **17** No. 10, 1026.

Mirkin L.I. and Philipetskii N.F., 1967, Physical nature of the hardening of steels under the influence of light pulses, Sov. Phys – Dokl. **12** No. 1, 89.

Mirkin L.I. and Pilipetskii N.F., 1968, Strengthening high-speed steel by the action of a light ray (laser beam), Izv. VUZ-Chern. Met. **11** No. 11, 124.

Moisa M.I., 1974, Corrosion resistance of AISI 5140 type steel after a laser treatment, Fiz. Khim. Mekh. Mat. **10** No. 1, 94.

*Oakley P.J., 1981, Laser heat treatment and surfacing techniques – a review, The Welding Institute Research Bulletin **22** No. 1, 4.

Parris D.C. and McLellan R.B., 1976, The diffusivity of carbon in austentite, Acta Metallurgica **24** 523.

Ready J.F., 1971, Effects of High Power Laser Radiation, (Academic, New York).

Ready J.R., 1976, IEEE J. Quantum Electron. **QE-12** 137.

*Ready J.F., 1978, Industrial applications of lasers (Academic, New York 1978).

Reed-Hill R.E., 1974, Physical Metallurgy Principles (D. Van Nostrand, New York).

Roessler D.M. and Gregson V.G., Jr., Reflectivity of steel at 10.6 μm wave-length," Appl. Opt. **17** No. 7, 992.

Sanders B.A. and Gregson V.G., Jr., 1973, in: Proc. Electro-Optical Systems Design Conf., N.Y., Sept. 1973, p. 24.

Sandven O.A., 1980, Heat flow in cylindrical bodies during laser surface transformation hardening", Avco-Everett Metal Working Lasers, Sommerville, MA.

*Seaman F.D. and Gnanamuthu D.S., 1975, Using the industrial laser to surface harden and alloy, Metal Progress **108** No. 3, 67.

Shewmom P.G., 1968, Diffusion in Solids (McGraw-Hill, New York).

Siebert C.A., Doane D.V. and Breen D.H, 1977, The Hardenability of Steels – Concepts, Metallurgical Influences, and Industrial Applications, (American Society for Metals, Metals Park, Ohio 44073).

Sparks M., 1976, Theory of laser heating of solids: metals, J. Appl. Phys. **47** No. 3, 837.

Speich G.R. and Fisher R.M., 1965, Recrystallization of a rapidly heated $3\frac{1}{4}\%$ Silicon Steel, Recrystallization, Grain Growth and Textures, (American Society for Metals, Metals Park, Ohio) ch. 13; Note the Appendix pp. 586 to 598.

Speich G.R. and Szirmae A., 1969 (Appendix by Speich G.R. and Richards M.J.) Formation of austinite from ferrite and ferrite-carbide aggregates, Trans. of Metallurgical Soc. of AIME **245** May, 1063; Note Appendix: Diffusion equations for pearlite dissolution, p. 1073.

*Speich G.R., Szirmae A. and Fisher R.W., 1966, Rapid heating by laser techniques, ASTM STP 396, Advances in Electron Metallography, Ch. 6, p. 335.

*Stanford K., 1980, Lasers in metal surface modification, Metallurgia, **47** No. 3, 109.

*Steen W.M. and Courtney C., 1979, Surface heat treatment of EN8 steel using a 2 kW continuous wave CO_2 laser, Metals Technology **6** No. 12, 456.

Tice E.S., 1977, Laser annealing of copper alloy 510, IEEE/OSA Conf. on Laser Engineering and Applications, IEEE J. Quantum Electron. **QE-13** No. 9, 10.

*Trafford D.N.H., Bell T., Megaw J.H.P.C. and Bransden A.S., 1979, Paper 9, Heat Treatment 1979, Proc. Conf., Birmingham, May 1979 (The Metals Society, American Society for Metals).

*Van Cleave D.A., 1977, Lasers permit precision surface treatments, Iron Age **219** No. 5, 25.

Veiko V.P., Kokora A.N. and Libenson M.P., 1968, Experimental verification of the temperature distribution in the area of laser-radiation action on a metal, Sov. Phys. – Dokl. **13** No. 3, 231.

*Wick C., 1976, June, Laser hardening, Manufacturing Engineering.

Wieting T.J. and Schriempf J.T., 1972, Report of NRL Progress (US Govt. Printing office, Washington, DC) p. 1.

Wooten F., 1972, Optical Properties of Solids (Academic, New York).

*Yessik M. and Scherer R.P., 1975, Practical guidelines for laser surface hardening, Dearborn, USA, SME Paper No. MR75-570.

*Yessik M. and Schmatz D.J., 1974, Laser processing in the automotive industry, Soc. of Manufacturing Eng. Tech. Paper MR74-962.

Yessik M. and Schmatz D.J., 1975, Laser processing at Ford, Metal Progr. **107** No. 5, 61.

Zavecz T.E., Saifi M. and Notis M., 1975 Appl. Phys. Lett. **26** 165.

List of laser heat treatment patents

(1) US Patent No. 4015100, 29 March 1977. Surface modification, D.S. Gnanamuthu, E.V. Locke/Avco Everett Res. Lab. Inc.

(2) US Patent No. 4093842, 6 June 1978. Ported engine cylinder with selectively hardened bore, D.I. Scott/General Motor Corp.

(3) US Patent No. 4151014, 24 April 1979. Laser annealing. Controlled tempering of non-ferrous metal workpiece, S.S. Charschan, S.E. Tice/Western Electric Company, Inc.

(4) US Patent No. 4157932, 12 June 1979. Surface alloying and heat treating processes. Treatment with laser before introducing alloying material, V.I. Chang, C.M. Yen/Ford Motor Co.

CHAPTER 5

RAPID SOLIDIFICATION LASER PROCESSING AT HIGH POWER DENSITY

E.M. BREINAN and B.H. KEAR

Terminal Drive Corporation
Tulsa, Oklahoma 74115, USA
and
Exxon Research and Engineering Co.
Linden, New Jersey 07086, USA

Laser Materials Processing, edited by M. Bass
© *North-Holland Publishing Company, 1983*

Contents

1. Introduction

The interaction between a laser beam and a metal or alloy surface is controlled by a number of variables including the wavelength of the radiation, the incident power density, and the available interaction time. For the case of a continuous, convectively cooled, high-power CO_2 laser with a wavelength of 10.6 μm the laser–material interaction as a function of incident power density and available interaction time is as shown in fig. 5.1. A particular combination of power density and interaction time defines a specific operational regime within the "interaction spectrum", and each operational regime results in the occurrence of a unique materials-processing effect. In order of increasing power density, the various processing effects include transformation hardening (Hella and Gnanamathu 1976, Breinan et al. 1976a), bulk surface alloying and cladding (Gnanamuthu and Locke 1976), deep penetration welding (Brown and Banas 1971, Breinan et al. 1975) and the LASERGLAZE™ effect (Breinan et al. 1976b,c), drilling of holes and metal removal effects (Voorhis 1964), and laser shock-hardening (Fairand et al. 1972, 1974). In fig. 5.1, it may be noted that the materials-processing effects, or operational regimes, are clustered along a diagonal running from "high power density–short interaction time" to "low power density–long interaction time". This is a consequence of several factors. First, the very high power densities necessary for explosive material vaporization effects, such as those producing shock hardening and drilling, are obtained with pulsed lasers and thus are available only for short times. Repeated pulses generally are not additive. Also, the quantities of energy necessary to generate the various thermal effects ranging from heating to "red heat" up to vaporization do not differ in energy consumed by more than two orders of magnitude. Thus, it is not primarily the quantity of energy applied, but the rate and power density at which it is applied, which gives rise to the specific materials-processing effect desired. Since both power density and interaction time span six or seven orders of magnitude, the diagonal is the primary area of interest. A study of fig. 5.1 should make this clear, along with the fact that, since the range of power densities extends far above that produced by most common thermal sources, even the

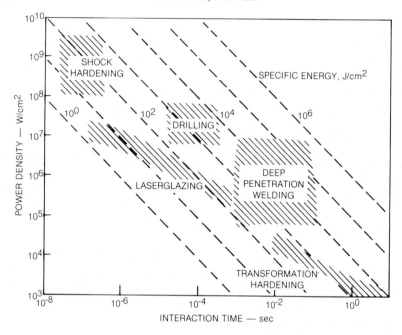

Fig. 5.1. Operational regimes for various laser materials processing techniques.

long interaction time laser processes are in fact quite rapid, with interaction times rarely approaching 0.1 s.

The LASERGLAZE[TM] process was conceived to utilize the high power densities available from focused laser beams in conjunction with short interaction times to effectively limit thermal effects to a shallow surface layer (Breinan et al. 1976b,c), with specific energy inputs ranging from 10^{-2} to 1 J mm^{-2}. Rapid surface melting thus occurs in a time in which an almost negligible amount of thermal energy can be conducted into the base metal, producing extremely sharp temperature gradients between the solid and the liquid. Since there is always intimate solid/liquid contact, very rapid quenching of the melt to the cold, solid substrate material results from these high thermal gradients. As indicated below, the quench rate is ultimately dependent on melt layer thickness, with cooling rates of $10^{4}-10^{8}$ K s^{-1} attainable in thin melt layers. As a result of this rapid chilling, a variety of interesting metallurgical structures are produced.

Initial development of the laserglaze process thus involved use of lasers to melt only thin surface layers of metals and alloys at high melting efficiencies (Breinan et al. 1976b, Kear and Breinan 1977). When such melting was achieved, rapid solidification and subsequent solid state cooling occurred in a predictable manner, as described below. The variety of interesting micro-structures which were produced gave the impression that rapid solidification laser processing could be used to control the structure of alloys to a greater degree than had previously been possible. Due to the nature of the laser/material interaction, however, and to the heat flow considerations, the laserglaze process was found to be effective in producing high cooling rates only in relatively thin layers.

In order to develop the technology for structural control of thicker sections, a process that combines continuous material feed with laserglazing has been developed. This LAYERGLAZE™ process is capable of producing rapidly solidified bulk structures by sequential buildup of thin, rapidly solidified layers (Breinan and Kear 1978). Initial development of the process involved addition of feedstock in the form of filler wire. In addition to some physical difficulties which were encountered with wire feeding, there was also the practical problem of manufacturing wire from a number of experimental compositions, many of which were substantially ductile only *after* rapid solidification processing. As a result, the use of powder as a feedstock was considered, and a system for layerglaze processing using a powder feed source was developed. This development was incorporated into the primary application being investigated; namely, the formation of a scale model gas turbine disk for spin testing. In addition to the fabrication of bulk shapes entirely from layerglaze deposits, the process, especially when powder feeding is utilized, has demonstrated significant flexibility in its ability to apply coatings and to build appendages or sections onto a number of components using a variety of addition materials. Although the specific applications of these systems are at present proprietary, some of the systems themselves are described, in addition to a review of the turbine disk fabrication programme. A significant portion of the disk fabrication and test programme is the identification of an alloy which is compatible with layerglaze processing and, at the same time, exhibits mechanical properties which represent an advance in the state-of-the-art of disk technology. Significant strides have been made in alloy design and development, and detailed structural analysis of candidate alloys has been described by Snow et al. (1980).

2. Theory

2.1. Heat transfer considerations

The local melting and resolidification of a small portion of the surface of an alloy with an intense heat source (such as a laser beam) presents a unique opportunity for controlled crystal growth under high temperature gradient, high cooling rate, and possibly plane front conditions. Absorbed power densities on the order of 10^5 to 10^6 W/cm^2, at the appropriate scan rate, produce layers of melt about 0.025 mm deep in which most of the absorbed energy is concentrated in the melt layer, resulting in very steep temperature gradients (about 10^5 °C/cm). This in turn causes rapid cooling of the melt layer (at rates of about 10^6 °C/s) and high solidification rates (250 cm/s) as the melt interface moves toward the surface of the material. It is known that when the ratio (G/R) of the temperature gradient, G, to the solidification rate, R, at the melt interface is high, the resulting crystal structure exhibits desirable metallurgical properties. It is of interest, therefore, to determine the quantitative effect of power density and scan time on G/R for a given melt depth and, further, to examine the transient behavior of G/R at the melt interface during the entire solidification process.

If the laser surface melting process is considered to be one-dimensional, the equations of heat conduction are far simpler than for the cases of two or three dimensions, but are still nonlinear because of the latent heat term at the interface. This is true even if the properties of the material, such as specific heat and thermal conductivity, are constant. Normally, this type of problem is solved by numerical solution of the conduction equations or by a finite-element, heat-balance technique (Murray and Landis 1959, Ruhl 1967a). In a previous study of the cooling rates in metals (Greenwald 1975), a finite-element technique was used to calculate cooling rates based on temperature distributions in the melt layer and solid substrate. The initial temperature distributions were determined by analog computer solution of the nonlinear (with latent heat) melting problem (Cohen 1967). Such methods, while closely approximating the exact solution of the equations, are tedious and time-consuming to set up and usually costly in terms of computer time per case. Hence, an entirely analytical, approximate method for obtaining information on the G/R ratio at the melt interface was conceived and used to avoid the complexities of a numerical solution and to provide results rapidly and economically.

The major assumption made to justify the approximate method of solution was to neglect the latent heat of melting. If the properties of the material are essentially the same in the solid and liquid states, the problem becomes one of just surface heating with (in effect) no change of phase. In making this assumption it was reasoned that the fundamental character of the rapid-heating/rapid-cooling phenomenon would be preserved because of the high temperature gradients, even if the absolute answers were only of order-of-magnitude accuracy. The other assumptions were those used in

previous analytical studies of laser surface heating (Greenwald 1975, Greenwald 1976a,b). These assumptions are as follows:
 – the material is a one-dimensional semi-infinite solid. This assumption is valid providing the width of the melted zone is large compared to the melt depth.
 – The material is at a constant initial temperature of 21°C throughout.
 – Heat input is constant during the application of power. Cooling is by conduction only; radiation and convection are neglected.
 – The material is homogeneous with constant properties.
 – For the purposes of this analysis the material was assumed to be pure nickel with the properties indicated in table 5.1.

The one-dimensional conduction heat transfer equation applying to this problem is given in eq. (5.1), and the boundary and initial conditions are defined by eqs. (5.2) and (5.3), respectively (a glossary of terms is given in table 5.2).

$$\frac{\partial^2 T(x,t)}{\partial x^2} + \frac{g(x,t)}{k} = \frac{1}{\alpha}\frac{\partial T(x,t)}{\partial t} \qquad \text{for } 0 \leqslant x < \infty, \quad t < 0, \quad (5.1)$$

$$\partial T(x,t)/\partial x = 0 \qquad \text{at } x = 0, \quad t > 0, \tag{5.2}$$

$$T(x,0) = T_0 = \text{constant for } 0 \leqslant x < \infty, \quad t = 0. \tag{5.3}$$

The heat input to the material is defined as an internal heat generation term $g(x,t)$ in eq. (5.1) where

$$g(x,t) = q_0 \delta(x-0)\eta(\tau-t). \tag{5.4}$$

The parameters δ and η are the delta and Heaviside functions, respectively. Eq. (5.4) defines a surface heat input of magnitude q_0 for a time τ and of magnitude zero thereafter.

Table 5.1
Properties of pure Ni

Melting temperature	1455°C
Vaporization temperature	2732°C
Conductivity	0.915 W/cm°C
Density	8.89 g/cm^3
Specific heat	0.105 cal/g°C

Table 5.2
Glossary of terms

erfc = complimentary error function
k = conductivity (W/cm°C)
g = internal heat generation (W/cm^3)
q_0 = absorbed power density (W/cm^2)
t = time (s)
T = temperature (°C)
T_0 = initial temperature (°C)
x = depth beneath the surface (cm)
α = diffusivity = conductivity/(density × specific heat) (cm^2/s)
δ = delta function
$\gamma = t - \tau$ s
η = Heaviside function
τ = heating time (duration of laser power) (s)

An integral transform technique (Ozisik 1968) was used to solve eqs.
(5.1)–(5.4), yielding the following closed-form solutions for the transient
temperature field during heating and cooling.
During heating:

$$T(x,t) = \frac{q_0}{k}\left[\left(\frac{4\alpha t}{\pi}\right)^{1/2} e^{-[x/(4\alpha t)^{1/2}]^2} - x\,\mathrm{erfc}\left(\frac{x}{(4\alpha t)^{1/2}}\right)\right] + T_0. \quad (5.5)$$

During cooling:

$$T(x,t) = \frac{q_0}{k}\left\{\left(\frac{4\alpha t}{\pi}\right)^{1/2} e^{-[x/(4\alpha t)^{1/2}]^2} - \left(\frac{4\alpha\gamma}{\pi}\right)^{1/2} e^{-[x/(4\alpha\gamma)^{1/2}]^2}\right.$$

$$\left. - x\left[\mathrm{erfc}\left(\frac{x}{(4\alpha t)^{1/2}}\right) - \mathrm{erfc}\left(\frac{x}{(4\alpha\gamma)^{1/2}}\right)\right]\right\} + T_0. \quad (5.6)$$

At the surface of the material ($x = 0$) eq. (5.5) reduces to

$$T(t) = \frac{q_0}{k}\left(\frac{4\alpha t}{\pi}\right)^{1/2} + T_0 \quad (5.7)$$

and eq. (5.6) becomes

$$T(t) = \frac{q_0}{k}\left[\left(\frac{4\alpha t}{\pi}\right)^{1/2} - \left(\frac{4\alpha\gamma}{\pi}\right)^{1/2}\right] + T_0. \quad (5.8)$$

During cooling, the temperature gradient, $G(x,t)$, and the cooling rate,

$\dot{T}(x, t)$ are also of interest. These quantities are given, respectively, by eqs. (5.9) and (5.10).

$$G(x, t) = \partial T / \partial x = -\frac{q_0}{k} \left[\text{erfc}\left(\frac{x}{(4\alpha t)^{1/2}} \right) - \text{erfc}\left(\frac{x}{(4\alpha \gamma)^{1/2}} \right) \right], \quad (5.9)$$

$$T(x, t) = \partial T / \partial t = \frac{q_0}{k} \left[\left(\frac{\alpha}{\pi t} \right)^{1/2} e^{-[x/(4\alpha t)^{1/2}]^2} - \left(\frac{\alpha}{\pi \gamma} \right)^{1/2} e^{-[x/(4\alpha \gamma)^{1/2}]^2} \right].$$

$$(5.10)$$

Finally, the solidification rate, R, is related to G and $\dot{T}$ by the relation

$$\dot{T} = GR, \quad (5.11)$$

where G and R are evaluated *at* the melt interface.

To better understand the heat transfer during rapid solidification, the cooling process was examined for a constant depth of melt with absorbed power density as the primary variable. Of course, the approximate analysis does not include an actual change of phase; the material is assumed to be "melted" at a given depth if the temperature is above the melting point. The algorithm used to calculate the transient value of G/R at the melt interface during solidification is as follows.

For a given depth of melt, d, an absorbed power density, q_0, is chosen. Since the condition of melting implies a melt temperature, T_{melt}, at depth d, eq. (5.5) can be solved iteratively for the time required to produce this condition. Let this time be τ. Thus, the material has been surface-heated from time-zero to time τ with a power density q_0 and melted to depth d, at which depth the temperature is T_{melt}. The surface temperature, T_{max}, at this time (τ) can be found from eq. (5.7). This is the highest temperature that occurs anywhere in the material at any time.

With the material melted to depth d at time τ, the energy input ceases and the material begins to cool. As solidification proceeds, the depth at which the temperature equals T_{melt} moves toward the surface. The surface temperature drops and when it reaches T_{melt} the solidification is complete.* The time at which this occurs, τ, can be determined from eq. (5.8). The temperature field in the material during the cooling process (between times τ and τ') is defined by eq. (5.6).

To determine the temperature gradient and cooling rate at the melt interface as solidification proceeds toward the surface of the material, it is

---

*Although the mathematics of this analysis do not include actual melting, the above description is still accurate in terms of the temperature history of the melting process.

first necessary to establish the position of the interface as a function of time. This was accomplished by selecting, in the time interval $\tau' - \tau$ (during which solidification occurs), a number of discrete points, t_i, and at each point solving eq. (5.6) iteratively for the depth, s_i, at which the temperature equals T_{melt}. This procedure is illustrated schematically in fig. 5.2.

Fig. 5.2 shows the temperature history at different depths in the material. The upper curve shows how the temperature at the surface starts at the initial temperature T_0, begins to melt at time t_a, and reaches a temperature of T_{max} at time τ when energy input ceases. The surface temperature then falls until it reaches T_{melt} at time τ' at which time solidification is complete. The lower curve of fig. 5.2 shows the temperature history at depth d. The time τ at which the temperature at this depth reaches T_{melt} defines the required heating time. The middle curve shows the temperature history at an intermediate depth s_i which melts at time t_b and solidifies at time t_i. It is seen that solidification occurs during the time interval $\tau' - \tau$ over which the surface temperature varies from T_{max} to T_{melt}.

Fig. 5.3 shows temperature profiles in the material at selected times, in effect, a cross-plot of fig. 5.2. The upper curve illustrates the temperature profile at time τ when the material at the surface has reached a temperature of T_{max} and the material at depth d has just melted. The middle curve shows the temperature profile at a subsequent time t_i when the material has

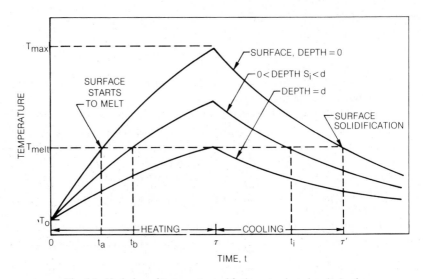

Fig. 5.2. Variation of temperature with time at selected melt depths.

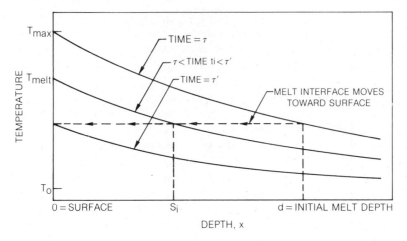

Fig. 5.3. Temperature profiles at selected times.

solidified to depth s_i. The lower curve illustrates the temperature profile at time τ' when the surface of the material has just solidified. The locus of points s_i at times t_i when the temperature equals the melt temperature, defines the position of the melt interface as a function of time during solidification.

Knowing the depth of the melt interface as a function of time, eqs. (5.9) and (5.10) can be evaluated to yield the temperature gradient G and cooling rate $\dot{T}$ at the interface, and eq. (5.11) can then be solved for the instantaneous freezing rate R. The ratio G/R at the melt interface is thus determined both as a function of time and as a function of depth.

2.2. Validation of analytical technique

To establish the accuracy with which the analytical method described above predicts the thermal behavior of a melting and solidifying solid, a comparison was made with results from a previous study (Greenwald 1975) in which the temperature history of rapidly cooling pure nickel was calculated by a finite-element method (latent heat was included during heating in the latter results).

Fig. 5.4 shows a comparison of the melt depth histories predicted by the two techniques. The melt depth is nondimensionalized by the initial melt depth, d, which is the depth of melt at the cessation of energy input. A dimensionless time parameter is used as the independent variable, defined

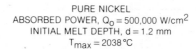

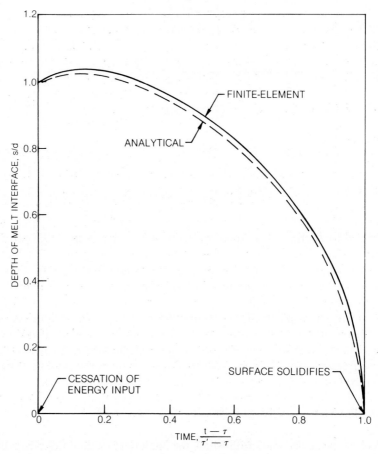

Fig. 5.4. Comparison of melt depth history.

as the time measured from the cessation of power, nondimensionalized by the total time required for the material to freeze to the surface. The time parameter thus varies from zero to 1.0.

In both the analytical and finite-element cases a maximum surface temperature of about 2038°C was predicted. Note that the material continues melting after the energy input has ceased (as indicated by the initial

rise in the curve) even though the surface has begun to cool. Solidification is seen to begin at a dimensionless time of about 0.15.

Similar agreement between the two methods was obtained for the solidification rate R, the temperature gradient G, and cooling rate $\dot{T}$.

The heating time, τ, predicted by the analytical method was 0.0046 s while the heating time for the finite-element case was 0.006 s. The difference probably reflects the additional energy input for the latent heat of melting accounted for in the finite-element case but neglected in the analytical case. The total cooling time, $\tau' - \tau$, was 0.00057 s for the finite-element calculation and 0.00055 for the analytical approximation.

2.3. Parametric study

The effect of power density and melt depth on heat transfer parameters at the melt interface was examined for a wide range of parameters. Fig. 5.5 shows the transient behavior of the melt interface as a function of time, with absorbed power density as a parameter. The curves are for a constant initial melt depth, d, of 0.025 mm. The word "initial" refers to the depth at the cessation of power and not the maximum depth.

Note that as power density increases the material melts to a greater fraction of the initial melt depth before solidification begins. Even for very low power densities there is an initial advance of the melt interface prior to solidification. Also, for each absorbed power density there is a characteristic maximum surface temperature which occurs at the cessation of power. An absorbed power density of 550 000 W/cm^2 was assumed to be the maximum value practical for surface-melting a depth of 0.025 mm since it results in a surface temperature equal to the vaporization of nickel.

The curves of s/d versus $(t - \tau)/(\tau' - \tau)$ are independent of melt depth if, when the depth is increased by a factor M, the power, q_0, is reduced by the same factor M.

The variation of the temperature gradient, G, at the melt interface with melt depth is shown in fig. 5.6 for various absorbed power densities. The gradients have a maximum value at the start of solidification and approach zero as the surface solidifies. The gradient is a strong function of the absorbed power density, and hence, so is cooling rate. The values for the gradients shown in fig. 5.6 are actually negative since temperature decreases with increasing depth in the material. The curves of fig. 5.6 were obtained by evaluating eq. (5.9) at the appropriate values of depth and time.

It was found that the temperature gradient at the melt interface scales with initial depth such that if the depth d is increased by a factor M and the

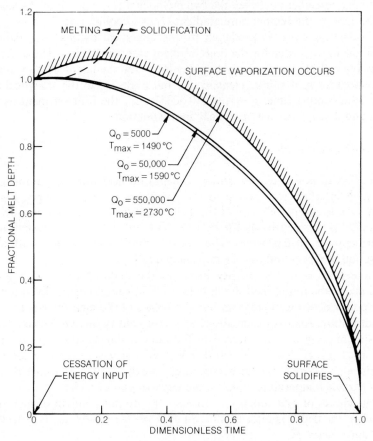

Fig. 5.5. Transient behavior of melt interface.

power q_0 is reduced by the factor M, the resulting gradients will have the same shape as those of fig. 5.6 except that they will be lower by a factor of M if plotted versus s/d. The same would be true if the time variable $(t - \tau)/(\tau' - \tau)$ were used instead of s/d.

For example, using fig. 5.6, if it is desired to obtain the temperature gradient at a value of $s/d = 0.6$ for a 0.25 mm melt depth ($M = 10$) at a power density of 5000 W/cm^2, one would read from fig. 5.6 the value

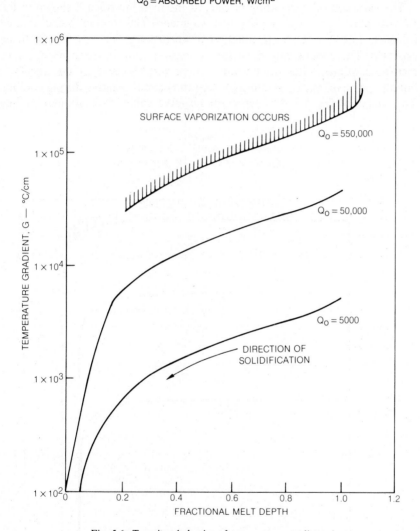

Fig. 5.6. Transient behavior of temperature gradient.

20 000°C/cm at a power density of 50 000 W/cm² and divide by 10 to obtain 2000°C/cm.

The variation of the cooling rate, $\dot{T}$, at the melt interface is shown in fig. 5.7 with absorbed power density as a parameter. The cooling rate, evaluated using eq. (5.10), is a strong function of absorbed power, q_0, as would be expected. The cooling rate must start at zero as solidification begins since melting continues after the cessation of power. Hence $\dot{T}$ at the interface, initially positive, must pass through zero to become negative during cooling. The curves of fig. 5.7 thus represent negative values even though cooling

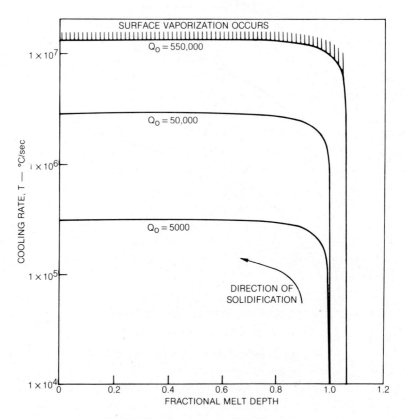

Fig. 5.7. Transient behavior of cooling rate.

rates are reckoned as positive. As solidification approaches the surface of the material, the cooling rates approach a constant value.

The cooling rate at the melt interface scales with initial melt depth, d, such that if d is increased by a factor M and the power is reduced by the factor M the resulting cooling rates will have the same shape as those of fig. 5.7 except that they will be lower by a factor of M^2 if plotted as a function of s/d. The same scaling law would obtain if the dimensionless time $(t - \tau)/(\tau' - \tau)$ were used as the independent variable.

As mentioned earlier, the slope of the melt-interface versus time curve is proportional to the freezing rate, R. This quantity, evaluated by means of eq. (5.11), is plotted in fig. 5.8 as a function of melt depth, with absorbed power density as a parameter and for a constant initial melt depth of

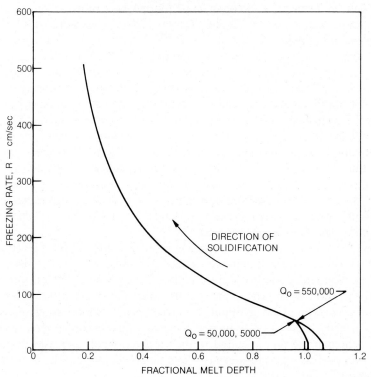

PURE NICKEL
INITIAL MELT DEPTH = 0.025 mm
Q_0 = ABSORBED POWER, W/cm^2

Fig. 5.8. Transient behavior of freezing rate.

0.025 mm. Note that the freezing rates shown are negative if one considers that $R = ds/dt$ is negative during freezing and positive during melting. The curves exhibit several interesting characteristics. These are:

(1) the freezing rate curves all begin at zero because, if the initial tendency is a continuation of melting after the cessation of energy input, $R = ds/dt$ must change sign and therefore must pass through zero.

(2) The freezing rate tends to infinity as solidification of the surface is approached.

(3) Except for the initial transient (which begins at different depths) the freezing rate curves are practically indistinguishable for a given melt depth, regardless of power density.

(4) There is a definite scaling law which relates freezing rate to initial depth: if the depth, d, is increased by a factor, M, and the power, q_0, is reduced by the factor M (from the curves of fig. 5.8) the resulting freezing rates will have the same shape as those of fig. 5.8 except that the freezing rates will be lower by a factor of M if plotted as a function of s/d. The same result would be true if the time $(t - \tau)/(\tau' - \tau)$ were used as the independent variable.

The parametric dependence of the parameter G/R at the melt interface is shown in fig. 5.9 with absorbed power as a parameter for an initial melt depth of 0.025 mm. Characteristically, the curves begin at infinity at the onset of solidification and end at zero as solidification approaches the surface. This behavior can be inferred from the curves of the gradient G (fig. 5.6) and the freezing rate R (fig. 5.8) for a constant absorbed power. As solidification begins, the gradient is large but finite, while the freezing rate is zero; thus the ratio G/R is infinite as solidification begins. As solidification approaches the surface, the gradient, which is continuously decreasing, tends to zero while the freezing rate, which is continuously increasing, tends to infinity. The ratio G/R, therefore, tends to zero as the melt interface approaches the surface of the material.

Power density is seen to have a strong influence on G/R for a given melt depth as well as on the maximum surface temperature, T_{max}. Interestingly, the time required to solidify a given depth of melt is nearly independent of absorbed power as shown by the dashed lines of constant time, measured from the cessation of energy input. This trend becomes more pronounced as power density decreases; i.e. lines of constant time become more vertical. Thus, although the time required to melt a given depth increases dramatically with decreasing power density, the time to solidify that depth tends to a constant.

Referring back to fig. 5.9, it is apparent that while the ratio G/R is infinite at the onset of solidification, and zero when solidification is com-

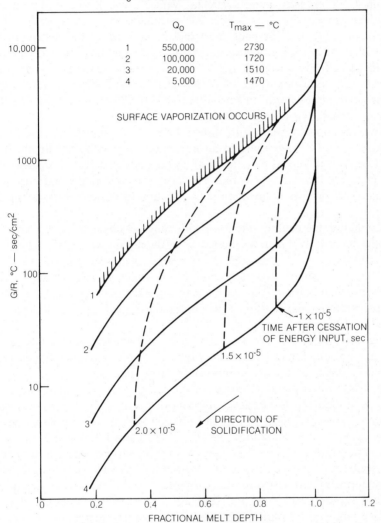

PURE NICKEL
INITIAL MELT DEPTH = 0.025 mm
Q_0 = ABSORBED POWER, W/cm^2

	Q_0	T_{max} — °C
1	550,000	2730
2	100,000	1720
3	20,000	1510
4	5,000	1470

SURFACE VAPORIZATION OCCURS

1×10^{-5}
TIME AFTER CESSATION
OF ENERGY INPUT, sec

1.5×10^{-5}

2.0×10^{-5}

DIRECTION OF
SOLIDIFICATION

G/R, °C — sec/cm^2

FRACTIONAL MELT DEPTH

Fig. 5.9. Variation of G/R at solidification interface.

plete, the effect of power density on G/R appears to be the *time* spent at or above a certain G/R value, or, alternatively, the depth of solidification over which G/R exceeds some particular value. Thus, if the initiation of a physical process during solidification is known to be enhanced by high values of G/R, it is perhaps the temporal and/or spatial distribution of G/R that is the real influence, rather than just the magnitude of the ratio itself. Indeed, the results of fig. 5.9 indicate that, regardless of the power density, any arbitrary value of G/R will occur at some time (and depth) during the solidification; this must be true if G/R varies continuously from infinity to zero during solidification.

The variation of G/R with dimensionless melt depth, s/d, or cooling time, $(t - \tau)/(\tau' - \tau)$, scales in an interesting way with melt depth. If the melt depth is increased by a factor M and the power density is decreased by the factor M the ratio G/R does not vary. Thus, the shape and magnitude of the G/R profiles is preserved as a function of dimensionless depth and time.

Examining the cooling rate for a wide range of absorbed power and melt depths, more fully illustrates the nature of the rapid surface melting process. Referring to fig. 5.7, it is seen that for a given absorbed power there is a characteristic, almost constant cooling rate which is attained as the final 50% of melt solidifies. Thus, as the melt interface moves toward the surface of the material the cooling rate curves of fig. 5.7 appear to approach horizontal tangents.

Fig. 5.10 shows the effect on the final (or characteristic) cooling rate of absorbed power for different values of melt depth. The line of melt depth equal to 0.025 mm corresponds to the case illustrated in fig. 5.7. For comparison, the three dark circles on the 0.025 mm melt depth line of fig. 5.10 correspond to the final cooling rates shown on fig. 5.7 for values of absorbed power of 5000, 50 000 and 550 000 W/cm². The cooling rates for these cases are seen to be about 3×10^5, 3×10^6, and 1.3×10^7 °C/s.

It is interesting to note that lines of constant surface temperature appear as parallel straight lines on the log–log plot of fig. 5.10. Following a line of constant melt depth leftward (decreasing absorbed power) from the point of surface vaporization, it is seen from the values of the isotherms that the surface temperature rapidly approaches the melting temperature while the cooling rate drops by more than two orders of magnitude. There is, of course, a large increase in the melting time as power density is decreased. However, the cooling time from cessation of energy input to when the surface solidifies remains fairly constant.

Table 5.3 gives the heating (melting) and cooling times as a function of power for a melt depth of 0.025 mm.

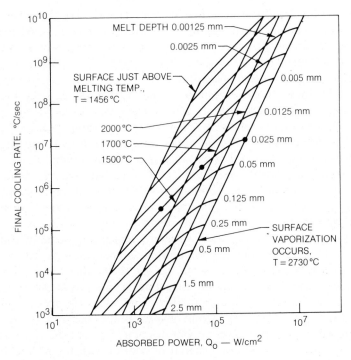

Fig. 5.10. Effect of absorbed power and melt depth on final cooling rate.

Table 5.3
Heating and cooling times as a function of power for 0.025 mm melt depth

Power density (W/cm²)	Surface temp. (°C)	Melting time (s)	Cooling time (s)
550 000	2730	6.71×10^{-5}	3.02×10^{-5}
200 000	1960	2.64×10^{-4}	2.47×10^{-5}
50 000	1590	2.77×10^{-3}	2.24×10^{-5}
5000	1469	0.236	2.17×10^{-5}
500	1456	23.1	2.10×10^{-5}

3. Microstructural considerations

The major effects of varying thermal conditions on solidification micro-structure are well known. Increasing the gradient-rate (G/R) ratio causes a progressive change in the solidification characteristics, ranging from fully dendritic → cellular dendritic → planar front growth. On the other hand, increasing the cooling rate, or gradient-rate product, since $\dot{T} = GR$, gives rise to shorter diffusion paths and finer structures. In other words, the ratio G/R controls the character of the microstructure, whereas the product GR determines the scale of the microstructure.

Fig. 5.11 shows some examples of microstructural changes in $\langle 100 \rangle$ oriented superalloy single crystals, after various laser surface melting treatments. As indicated, a relatively deep penetration "homogenizing" pass is applied to the surface prior to the application of one or more superimposed laserglazing passes. Without such an homogenizing treatment the glazed layers exhibit incomplete dissolution of MC carbide particles, and other refractory constituents in the initial microstructure. In all cases, laserglazing produces epitaxial growth in the resolidified layers, accompanied by a marked refinement in the scale of the microstructure. Within a given recast layer, the scale of the dendritic structure remains reasonably constant, although it undergoes obvious discontinuous changes with varying melt depth. This is to be expected, since fig. 5.7 shows that the cooling rate, which controls the scale of the microstructure, is relatively constant, after an initial transient stage. Close examination of the glazed layers under higher magnification also reveals significant changes in the character of the microstructure within a given melt zone. For example, fig. 5.12 shows a very thin layer of plane-front solidified material in contact with the melt/substrate interface. Moreover, with increasing distance from this interface, the structure becomes cellular (one-dimensional dendritic), cellular–dendritic, and finally fully dendritic. Such changes are clearly in accord with the calculated high initial G/R ratio, and its rapid fall-off as solidification proceeds towards the free surface, fig. 5.9. The second pass in fig. 5.12 shows the cellular growth extends to the free surface, which is indicative of a persistently high G/R ratio in this high cooling rate regime of solidification.

Fig. 5.12 shows that the region encompassing the second pass, including a thin heat-affected zone (HAZ) in the underlying first pass does not etch-up as well as the rest of the material. The reason for this is that the cooling rate is fast enough to prevent precipitation of γ' particles only in the second pass and its associated HAZ. Apparently, in the HAZ, the elimination of the γ' particles is a consequence of much faster kinetics for the solution of the γ' phase than for its precipitation. The region including the HAZ and the second pass exhibits a striking difference in size, distribution and morphology of MC carbides. In the second pass, the carbides occur as small platelets, unidirectionally aligned within the cellular boundaries. On the other hand, in the first pass, the more massive carbide platelets are distrib-

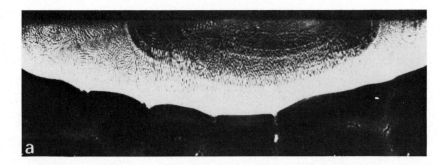

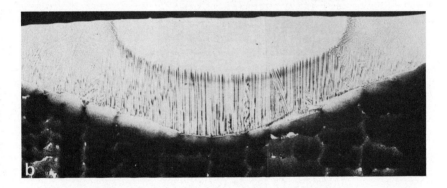

Fig. 5.11. Superimposed laser melting passes in single crystals of (a) Udimet 700, (b) Mar-M200 and (c) B-1900 superalloys.

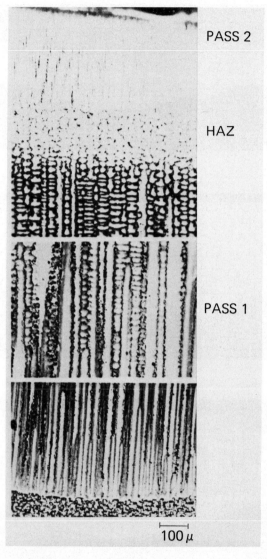

PASS 2

HAZ

PASS 1

100 μ

Fig. 5.12. Enlargement of fig. 5.11(c) showing evidence for epitaxial growth in laser melted regions.

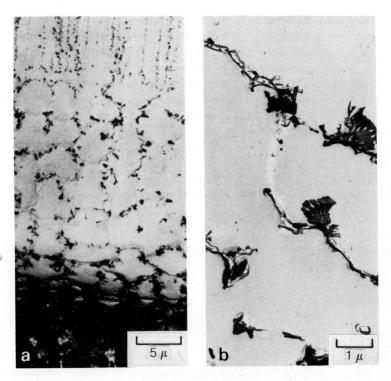

Fig. 5.13. Carbide and γ' extraction replica of HAZ in fig. 5.12. The dark band below the HAZ represents extracted γ' particles.

uted in an irregular manner, following the tortuous paths defined by the interstices of a fully-developed dendritic structure (fig. 5.13). The step change in the scale of the carbides is going from the first pass to the second pass is to be expected, since their formation is governed by the degree of segregation, and scale of the interdendritic interstices, both of which diminish with increasing cooling rate.

In laserglazing, the crystalline substrate necessarily is in intimate contact with the melt. This raises the question as to whether the presence of favorable nucleation sites for crystallization would prevent the undercooling necessary to achieve an amorphous structure. Test experiments on a low melting point eutectic alloy (Pd–4.2Cu–5.1Si) demonstrated that, under appropriate conditions of fast cooling, this alloy can be made amorphous by laserglazing. In this instance, the melted layer (~ 0.18 mm maximum in thickness (was obtained with an incident power density of 6×10^6 W/cm^2 and an interaction time of 3×10^{-5} s. As shown in fig. 5.14, within the melted region, there are no discernible microstructural features, in contrast to the eutectic structure in the substrate. One manifestation of the non-crystalline amorphous nature of the material is the symmetrical pattern of curved shear bands formed around hardness indentations, fig. 5.14d. Another manifestation is the vein-like character of the fracture surface, fig. 5.14c. Confirmatory evidence was obtained by TEM; the thin foils had a speckled appearance and the electron diffraction pattern showed diffuse rings, fig. 5.14b.

Similar tests carried out on low melting point TLP-type eutectic alloys gave no clear indication of an amorphous transition by laserglazing, even in

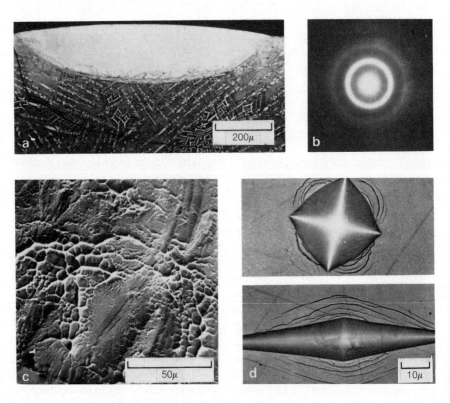

Fig. 5.14. Micrographs of Laserglaze™ processed Pd–4.2Cu–5.1Si alloy; (a) cross section of melt pass, (b) electron diffraction pattern, (c) fracture surface, and (d) hardness indentations.

thin sections (~ 0.025 mm) that had cooled at high rates. More typically, the result of laserglazing these alloys was the formation of either an ultra-fine dendritic structure, with varying degrees of phase decomposition, or an extremely fine eutectic structure. Examples of these two types of behavior were found in the three boron-rich nickel-base alloys shown in fig. 5.15. Alloy (b) has the preferred composition for TLP-21, an interlayer composition used for diffusion bonding of Udimet 700. Alloys (a) and (c) are, respectively, low and high boron modifications of this multicomponent alloy. As indicated in fig. 5.15, increasing the boron in this series of alloys increases the volume fraction of $\gamma + Ni_3B$ eutectic, at the expense of primary γ. Alloy (c) contains a mixture of two primary phases ($\gamma + M_3B_2$), whereas alloys (a) and (b) contain only one primary phase (γ). All three alloys responded to laserglazing by forming featureless regions, at least under the optical microscope. Fig. 5.16 shows a series of overlapping passes in alloy (a), whereas fig. 5.17 shows a series of parallel passes with varying interaction time in alloy (c). The most sharply distinguishing feature of these two examples is the extensive cracking in alloy (c), and the absence of cracking in (a). This has nothing to do with variations in laser melting parameters, but merely reflects the different strengths and ductilities of the substrates and laser melted regions. Both the substrate and laser melted regions are very hard in alloy (c), whereas they are much softer in alloy (a). Thus, cracking is avoided, or at least reduced, in (a) because of the ability of the material to undergo plastic deformation in response to the thermal

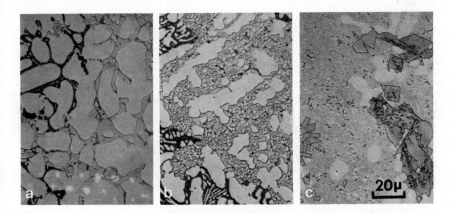

Fig. 5.15. Micrographs of a series of B-rich Ni–15Co–15Cr–5Mo alloys. (a) 1.5 w/o B, (b) 2.75 w/o B, (c) 4.0 w/o B.

Fig. 5.16. Overlapping Laserglaze [TM] passes at 3.0 kW, 25 cm/s, in the 1.5 w/o B alloy from fig. 5.15(a) (VHN of melted region = 600 kg/mm^2).

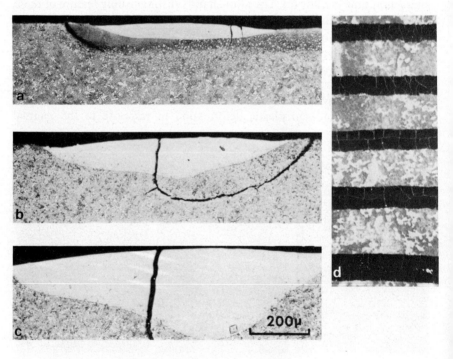

Fig. 5.17. Series of parallel Laserglaze[TM] passes with varying interaction time in the 4.0 w/o B alloy of fig. 5.15(c). VHN = 1300 kg/mm^2; (a) 3 kW, 50 cm/s; (b) 3 kW, 25 cm/s; (c) 3 kW, 10 cm/s. (d) Top surface showing cracking.

strains developed during quenching. In (c), cracking typically took the form of a single longitudinal mid-rib crack, with many short transverse cracks, fig. 5.17d. Alloy (b) also showed extensive cracking, which is consistent with the presence in this alloy of a large volume fraction of the hard, brittle $\gamma + Ni_3B$ eutectic, comparable with that in alloy (c). Although the microstructures of the laser melted regions in these alloys appeared to have no structure, close examination by SEM and TEM techniques showed that this was not the case. On the contrary, alloys (a) and (b) were composed of an ultra-fine dendritic structure, fig. 5.18a, whereas alloy (c) possessed a remarkable ultra-fine filamentary eutectic structure, fig. 5.18b. A general refinement in these structures was noted with decreasing thickness of the melt zone, as would be expected in view of the higher cooling rates. Under polarized light, with the material in the as-polished condition, a columnar grain structure became visible in alloy (c), fig. 5.19. Using this technique, it was shown that the superposition of two or more melting passes resulted in epitaxial growth, albeit with a thin transition layer of what appeared to be a spheroidized structure. A similar transition microstructure also occurs quite naturally at the interfaces between the melt zone and its substrate. In alloys (a) and (b), the transition zone was quite wide, reflecting the wide melting range of TLP-21 (1060°C solidus–1200°C liquidus). Selective melting of the eutectic, leaving the dendrites unmelted, can be seen in the transition zone of alloy (a) in fig. 5.16.

Laserglazing experiments performed on eutectic superalloys yield a broad spectrum of microstructures, which differ mainly in degree, or scale, rather

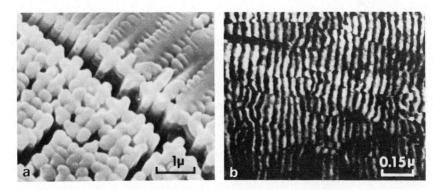

Fig. 5.18. Micrographs of Laserglaze^{TM} processed B-rich alloys: (a) ultrafine dendrites in 1.5 w/o B alloy, (scanning electron micrograph), (b) Ultrafine, enriched eutectic in 4.0 w/o B alloy (transmission electron micrograph).

Fig. 5.19. Optical micrograph of 4.0 w/o B alloy under polarized light. Columnar grain
structure in the laserglazed region is shown.

than in kind. Under high cooling rates, the microstructure is typically
composed of a fairly uniform distribution of MC carbide particles embedded in a solid solution matrix phase, similar to that found in splat
quenched material. Under more moderate cooling rates, the structure is
distinctly dendritic, with the MC carbide phase concentrated exclusively in
the interdendritic interstices. These two representative types of microstructures are shown for a CoTaC-type eutectic alloy in fig. 5.20. In the laser
melted pass at 3 kW–51 cm/s, fig. 5.20a, the MC carbide phase occurs as
discrete particles having a platelet morphology, whereas in the laser melted
pass at the lower cooling rate (3 kW–12.7 cm/s) fig. 5.20b, the carbide
phase has a filamentary morphology. The very fine scale of the filamentary
eutectic structure can be appreciated by reference to fig. 5.20c, which shows
the transition zone between the melted layer and its substrate; the latter was
solidified under carefully controlled plane front conditions. This abrupt
change in microstructure is illustrative of the sensitivity of the microstructure in this alloy to solidification conditions. It is well known that a high
ratio of temperature gradient to solidification rate, (G/R), is a prerequisite
for coupled growth in this multicomponent, monovariant eutectic alloy.
Another noteworthy feature with respect to the response of these alloys to
laserglazing was the absence of cracking, both in the melts and in the
substrate heat affected zones. Apparently, in these alloys, the solid solution
matrix phase is sufficiently ductile as to ensure the necessary plastic
accommodation in both surface melt layer and substrate during cooling
from the melt.

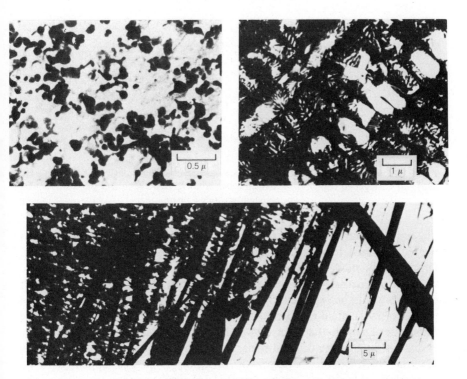

Fig. 5.20. Extraction replicas of Laserglaze™ processes CoTaC-3 alloy. Morphology and scale of TaC carbides varies with local solidification conditions.

In contrast to the behavior of the carbide eutectic alloys, laser melting of the γ/γ'-δ eutectic did not give a well organized, fine-scale lamellar eutectic structure. In this case, the typical result was the development of a fine dendritic structure, fig. 5.21a, with some precipitation of γ' and δ, but particularly the δ phase. Moreover, cracking was observed in the laser melted regions, especially in the deeper melts, but not in the substrate. Evidently in this case, plastic deformation occurs quite freely in the substrate during fast cooling, and evidence for this is shown in fig. 5.21b. On the other hand, deformation is inhibited in the laser melted regions because of precipitation of γ' and δ, which sharply increases the strength of the material.

Laserglazed γ' precipitation hardened superalloys invariably exhibit dendritic growth. Direct evidence has been obtained by SEM observations on

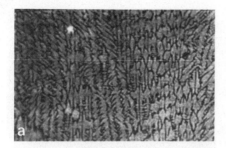 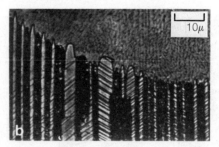

Fig. 5.21. Micrographs of Laserglaze™ processed γ, γ'− δ directionally solidified eutectic alloy. (a) Laserglazed region. (b) Interface between laserglazed region and substrate showing twinning in substrate δ-phase.

the external surface along the center line of the melt, where, due to inadequate melt feeding, the dendritic structure stood out in sharp relief. Observations made in this way showed that the scale of the dendritic structure decreased with increasing cooling rate, or decreasing laser melt depth, as would be expected, fig. 5.22. Comparable structures were observed in transverse metallographic sections; moreover, strong indications were found for regrowth of certain, if not all, of the partially-consumed substrate grains residing at the original melt/substrate interface. In monocrystalline material, such oriented overgrowth, or epitaxial growth, appeared to be particularly favored on a cube oriented substrate, apparently because the normal to the substrate coincides with a fast-growing dendrite direction; namely $\langle 001 \rangle$. This is shown for PWA 1418L in fig. 5.11. The rather dramatic refinement in the scale of the dendritic structure following the first melting pass at 4 kW–3.5 cm/s is quite obvious in this micrograph. A further refinement in structure occurs in the two additional overlapping surface melting passes just visible at the top of the micrograph. When misoriented grains are present in the epitaxial zone, fig. 5.11 indicates that these tend to be eliminated by faster growth of the more favorably oriented cube-oriented grains. Misoriented grains are frequently nucleated in the vicinity of γ/γ' eutectic located at the melt/substrate interface. Examination of the various interfaces in the laser melted material, fig. 5.11b, shows that in the first pass the structure develops by cellular growth, and as the gradient falls off this degenerates into cellular dendritic and finally dendritic growth. In the second and third passes, the cellular dendritic mode of solidification is predominant. The relationship of these structures to the heat flow considerations developed in section 1, has been explained in detail above.

Fig. 5.22. SEM micrographs of dendrites emerging from external surface of P&WA 1418L superalloy. (a) 3 kW, 30 cm/s; (b) 3 kW, 96 cm/s.

More compelling evidence for epitaxial growth was obtained from observations on the external surface of the laser melted material. As shown in fig. 5.22, the surface is covered with a network of intersecting slip bands, which are continuous despite extraordinary variations in the general appearance of the dendritic structure, figs 5.23b and 23c. Slip bands in the same orientations were also visible in the adjacent substrate. Clearly, this would not be possible unless the entire laser melted material had the same orientation as the monocrystalline substrate. Obviously, this deformation represents the plastic accommodation that must occur during freezing if cracking is to be avoided. No cracking, in fact, was found in the melted PWA 1418L. Oriented overgrowth without cracking has also been seen in other orientations in this material, notably in [110] and [112] substrate orientations.

An interesting case of dendritic growth during laserglazing occurs in the {110} surface orientation, where, owing to the fact that two equivalent ⟨100⟩ growth directions are favored, the dendritic colonies impinge on one another along the center line of solidification to form a dendritic boundary, with no change in crystallographic orientation. Such dendrite boundaries tend to become sharply delineated as the interface between the melt zone and substrate assumes some curvature. Thus, in a deep-penetration homogenizing pass, the dendritic boundary is sharply defined (fig. 5.24).

Most alloy steels in the normal heat-treated condition contain a fine dispersion of one or more carbides. The high-speed steels have a relatively large volume fraction of strengthening carbide phases. In vanadium steels, the predominent carbide is MC. In steels with high concentrations of the refractory elements W and Mo, the principal carbide phase is M_6C. Nearly all the materials contain Cr-rich $M_{23}C_6$ carbide. Alloys selected for laser-

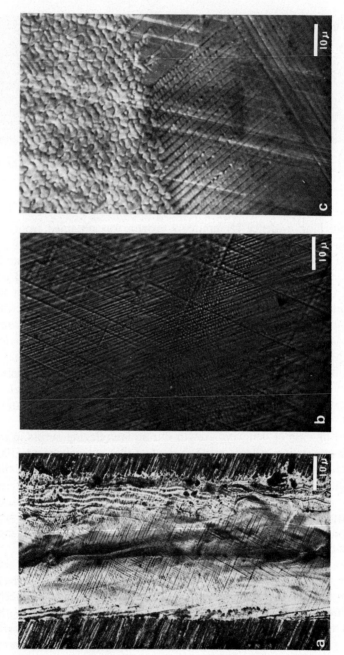

Fig. 5.23. Evidence of slip band on surface of Laserglazed P&WA 1418L superalloy. Note that slip bands are continuous across changes in dendrite morphology, indicating the single crystal character of the glazed material.

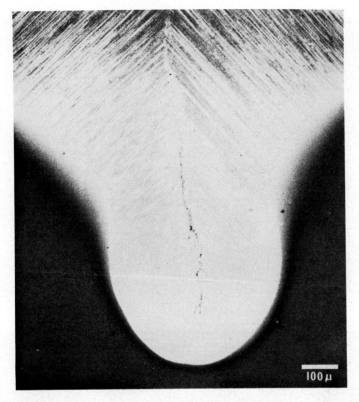

Fig. 5.24. Dendrite boundary in {110} surface orientation of rapidly solidified laser melt pass.

glazing experiments included 440C, M-50, M-2, and 4350. The primary goal was to determine the influence of cooling rate on measured Vickers hardness in the glazed material, including both melt and heat affected zones. In the glazed condition, alloy 440C exhibited a dendritic structure of Cr-stabilized ferrite, interspersed with thin sheets of carbide in the interdendritic region (fig. 5.25). The melt zone was associated with a broad heat-affected zone (HAZ) in the substrate. The dark-etching band in the substrate appeared to define the lower limit of microstructural changes in the HAZ. The measured hardness in the melt zone (dendritic structure) was HV ~ 470 kg/mm^2. This is to be compared with HV ~ 825 kg/mm^2 for the carbide-strengthened martensitic substrate. So in this case the effect of glazing was to produce softening of the material. On the other hand, the region of the HAZ just below the melt interface was somewhat harder (HV ~ 900

Fig. 5.25. Microstructure of Laserglaze™ Processed alloy 440C; (a) dendritic structure, (b) extracted carbides in interdendritic regions.

kg/mm^2) than the substrate. It appears, therefore, that laser heat treatment, rather than melting, is the approach to take to further harden this particular material.

Alloy M-50 also showed evidence of dendritic growth in the melt zone (fig. 5.26), albeit in a form more difficult to detect than in alloy 440C. Moreover, in marked contrast to the behavior of alloy 440C, M-2 shows some improvement in hardness in the melt zone compared with the substrate: HV ~ 850 versus 800 kg/mm^2, respectively. The heat affected zone also showed a small increase in hardness. Since the original carbide particles in the substrate are still visible in the HAZ (fig. 5.26), it seems likely that the high hardness of this zone is caused by the formation of fine martensite, not resolvable under the optical microscope. A similar situation exists in the melt zone, except that the fine martensite forms within the dendritic structure, probably interspersed with fine carbides. It is noteworthy that the original coarse carbide particles readily dissolve in the melt during glazing. In double or multiple passes, a clear indication was obtained of the formation of martensite (or bainite) on a coarse scale, accompanied by some reduction in hardness.

In complete contrast to 440C and M-50, alloy M-2 showed no indications of a dendritic structure. On the contrary, the melt zone appeared to be composed of a fine-grained, two-phase structure (fig. 5.27). Hardness measurements showed this two-phase structure to be softer (HV ~ 600 kg/mm^2) than the substrate (HV ~ 1000 kg/mm^2), at least in the region of a single pass. In the HAZ associated with a double pass, hardness values as high as HV ~ 900 kg/mm^2 were recorded, demonstrating that the strength in the

Fig. 5.26. Laserglaze^{TM} processed M-50 steel showing undissolved carbides in HAZ. These carbides are dissolved in the melt zone.

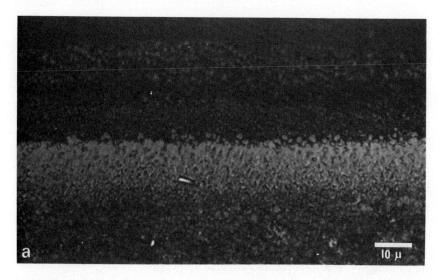

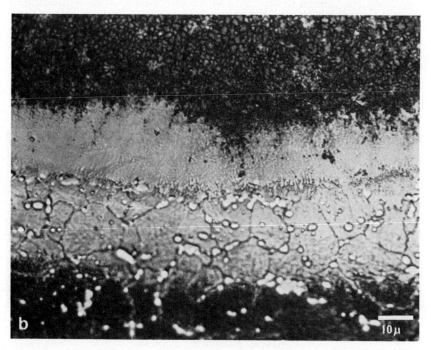

Fig. 5.27. Double glazing pass in alloy M-2 showing duplex $\alpha + \gamma$ structure.

melt zone can be recovered by a post-glazing heat treatment. Possibly, this occurs as a result of the formation of tempered martensite. Recently it has been shown that the as-glazed, two-phase structure is composed of ferrite and austenite, but not martensite. The absence of martensite explains the low hardness of the glazed material.

The preceding results show the varied response of optimally hardened alloy steels to glazing. It is of interest also to consider what happens to a material that is initially in the softened condition, and in fact normally used as such. Illustrative of the significant changes that can be induced by glazing of such materials is the behavior of spheroidized annealed alloy 4350 (HV ~ 450 kg/mm^2). Laserglazing of the material produces a dramatic harden-

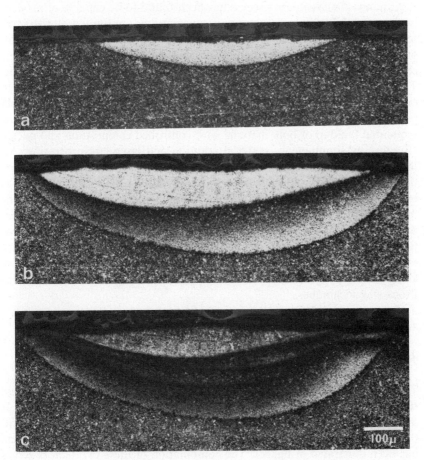

Fig. 5.28. Superimposed glazing passes in alloy 4350.

ing, with the maximum hardness corresponding to the thinnest glazing passes or, in other words, the fastest cooling rates. Typical hardness values range from 800 to 850 kg/mm^2, depending on cooling rate. In the HAZs associated with double or multiple passes (fig. 5.28) the hardness is much reduced compared with that in the as-glazed material, but still higher than that of the annealed substrate.

There is thus significant evidence that rapid chilling of alloys by laserglazing gives rise to a variety of interesting and potentially useful metallurgical microstructures. In addition, consideration of these structures and the means by which they are produced has highlighted the laser as a potentially important tool for future materials processing applications.

Although the laserglaze process has been found to be capable of producing rapidly chilled microstructures, its applications are limited by the small section thickness required to achieve high cooling rates. Typically, an order of magnitude decrease in the melted and resolidified section thickness results in an increase in the average cooling rate of two orders of magnitude. The laserglaze process is thus suited only for treatment of thin layers. In order to achieve the fabrication of rapidly-solidified thick sections (bulk parts), the layerglaze concept was evolved. Layerglaze is essentially the sequential buildup of bulk material with controlled composition and microstructure by simultaneous material addition and laserglaze processing. Although the deposition of each subsequent layer produces a high-temperature thermal transient in the layer beneath, the time duration of this transient is extremely brief. As a result, although the process will not permit the formation of bulk amorphous structures, it does allow fabrication of relatively homogeneous crystalline structures with little or no phase transformation between layers. In the section to follow, the experimental details of rapid solidification processing by laserglaze and layerglaze will be discussed, along with a prognosis about future application of these Directed Energy Processing techniques.

4. Processing

4.1. Surface melting and alloying (laserglaze processing)

The nature of the interaction of a high power laser beam with a material surface has been discussed extensively in previous sections of this book. It depends primarily on the absorbed power density and interaction time, and the variety of interactions obtainable as a function of these two parameters has been illustrated in fig. 5.1. The combination of these two factors serves to define operational regimes for various materials processing techniques. At very high power densities of $\sim 10^9$ W/cm^2, attainable currently with only pulsed laser equipment, nearly instantaneous surface vaporization occurs on interaction of the beam with the material. If pulse duration is kept to $\sim 10^{-7}$ s, typically involving an energy input of the order of 10^2 J/cm^2, interaction is limited to the surface and the rapid expansion of the vaporized metal produces an effect similar to a blast wave (Fairand et al. 1972, 1974, Clauer and Fairand 1975). As a consequence, a shock wave propagates and reflects within the material causing work hardening. The advantage here is that the laser provides for precise control of the input energy. In the deep penetration welding regime (Brown and Banas 1971, Locke et al. 1972, Breinan et al. 1975), the incident power density must be of $\sim 10^6$ W/cm^2 in order to establish the deep penetration mode. Interaction time is relatively longer than for shock hardening since substantial energy must be deposited within the material to induce melting and vaporization to a reasonable depth. In contrast to the low energy input per unit area for shock hardening, therefore, welding requires energy inputs of $\sim 10^4$ J/cm^2. For transformation hardening (Hella and Gnanamuthu 1976, Breinan et al. 1976a), a somewhat lower power density is required in order that some in-depth heating takes place without causing melting to occur at the top surface. The required power density is thus governed by the rate at which thermal energy diffuses through the workpiece material; the value must be high enough to effectively localize the thermal input to a thin region at the surface but low enough to avoid surface melting. Since the thermal interaction must proceed to a specified depth, thermal energy input per unit area required is intermediate between that for shock hardening and deep penetration welding and may typically be of the order of 10^3 J/cm^2. Slightly higher values will induce melting and are therefore suitable for surface alloying (Hella and Gnanamuthu 1976). The laserglaze region, first explored in the work reported in Cohen (1967) and Greenwald (1975, 1976a, b) involves high incident power density of the order of that utilized for welding, but substantially shorter interaction times. With this combination, specific energy inputs may range from 10 to 100 J/cm^2 and the thermal effect is concentrated in a very thin region at the material surface. Localized melting occurs very rapidly in a time period during which little thermal energy penetrates into the base material. This leads to the establishment of extremely sharp temperature gradients which facilitate rapid cooling of the melt following the interaction as detailed in the previous sections. Cooling rates approaching 10^6 °C/s have been estimated to be readily attainable with rates to the range of 10^8 °C/s possible. As a consequence of the high cooling rate, ultra-microcrystalline or amorphous microstructures have been obtained, as documented above.

Typical apparatus for laserglaze processing has involved use of continuous, multikolowatt CO_2 lasers, although the effect is dependent on power density and can thus be achieved at power levels below 1 kW. Both Gaussian and unstable resonator output beams have been used. A typical experimental setup is schematically illustrated in fig. 5.29, and pictured in action in fig. 5.30. In fig. 5.29, a nominal 7.5 cm diameter beam from the laser is directed toward and focused upon the workpiece by reflective optics. In a typical test, a 46 cm focal length mirror would be used to provide an effective minimum spot diameter of 0.05 cm at the workpiece. At 3.0 kW, these optics would provide a maximum incident power density of approximately 1.5×10^6 W/cm^2, a power density equivalent to that provided by a black body thermal radiative source at 22 800°C. This high power density is essential for localizing the energy input at the material surface, and further promotes effective coupling of the laser energy with the material, despite the initially high reflectivity of metallic surfaces to the 10.6 μm wavelength of carbon dioxide laser radiation. The 3 kW power level is a convenient one, in that it promotes effective beam coupling, but does not create significant plasma generation problems. As noted in fig. 5.29, plasma suppression is accomplished by means of an inert gas shield, which further prevents atmospheric contamination of the melt. Cooling due to the inert gas flow was estimated to be negligible, in comparison with the heat-sinking effects of the unheated substrate material.

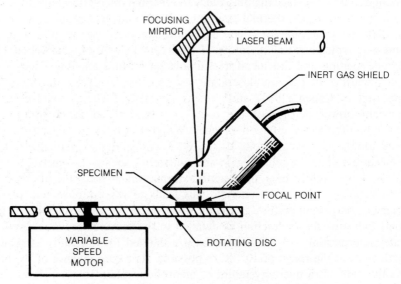

Fig. 5.29. Schematic drawing of the Laserglaze^{TM} process.

Fig. 5.30. Photograph of laserglazing being performed (gas shield removed for visibility).

A range of laserglaze melt depths may be achieved by varying the translational speed of the workpiece under the focused beam. Linear speeds of from 150 to greater than 6000 cm/min are attained by using a variable speed rotating disk with specimens located at a fixed radius on the disk surface. Under the above conditions, such a speed range would yield energies/unit length of melt ranging from 1200 down to 30 J/cm. Higher energy inputs can be used as initial homogenizing passes, with the lower energy inputs resulting in rapid solidification effects. The necessity for using homogenizing passes stems from the fact that inhomogeneous materials, when subjected to high speed passes, often do not remain in the molten state for sufficiently long to promote full homogenization of the various phases.

Rotating disk work-handling equipment of the type described above, facilitates irradiation of the multiple samples in a single pass. Other types of equipment can, of course, be used, including linear traverse equipment, and motion of the beam instead of the workpiece; however, due to the high traverse speeds involved, rotating equipment has been the most convenient for laboratory work. Samples are generally arranged adjacent to each other with their top surfaces located at the focal plane of the beam focusing optics. A smooth machined surface is presented to the focused laser beam. Coatings, such as those employed for some laser surface hardening work, are typically not used. In general, such coatings are not effective at the high power densities characteristic of the laserglaze regime and, further, are to be avoided since they might lead to contamination of the melt zone. Bulk specimens of ~ 0.5 cm thick should be utilized in order to ensure that essentially no heating of the bulk substrate would occur. This thickness is well above the analytically predicted requirement that the specimen thickness be at least four times the melt depth in order to facilitate rapid self-quenching. Most tests are conducted with the specimens initially at room temperature; however, tests have been conducted with the specimens initially chilled to liquid nitrogen temperature. Although the results may not have general applicability, it was noted that this prior cooling did not significantly alter the results obtained. This is probably due to the fact that the additional increment in temperature afforded by cooling from room temperature to liquid nitrogen temperature is small compared with the temperature difference between that of the liquid at the melting point and that of the solid at room temperature. Clearly, the precooling of the material becomes more important the lower its melting temperature.

A condensed, or "summarized" version of the thermal calculations on melt depth and cooling rate, which were discussed above is presented in figs. 5.31 and 5.32. These two figures represent a set of convenient "working figures" which can be used to estimate first, melt depth from absorbed power and interaction time, and then cooling rate from melt depth and power density, as follows:

Fig. 5.31 illustrates the transient surface melting characteristics for nickel. The solid lines indicate melt depth versus interaction time for various power densities. Interaction time is defined as the time required for the incident laser beam to traverse one spot diameter, and is computed as the ratio of spot diameter to beam sweep speed, D/V. The period is initiated from a uniform initial temperature of 21°C and thus includes the time during which the material is heated up to and through the melting point. Subsequent to the onset of melting, the advance of the melt interface was found to be approximately linear with time. The dashed lines in fig. 5.31 indicate constant specific energy inputs to the material and are computed as the products of the absorbed specific power and the interaction time. It can be seen that the specific energy required to obtain a given melt depth decreases as power density is increased and interaction time is decreased. The reason for this, of course, is that at higher power densities less time is available for heat conduction into the solid, i.e., the energy is more concentrated in the melt. Thus, increased melting efficiency is not the only benefit of higher power densities; the steeper temperature gradients in the material also enhance more rapid cooling.

Fig. 5.32 shows the effect of melt depth on average cooling rate for selected values of power density. It can be seen that there is a maximum cooling rate for a given absorbed power density corresponding to the

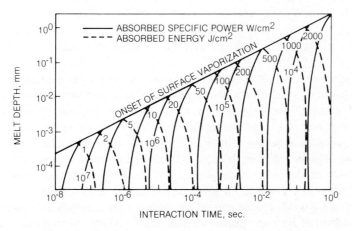

Fig. 5.31. Transient surface melting characteristics of nickel.

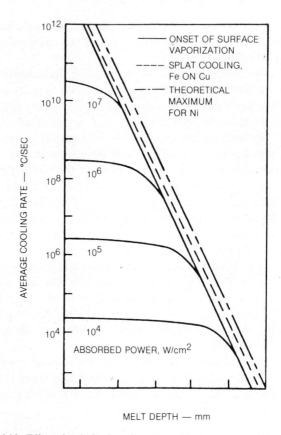

Fig. 5.32. Effect of melt depth and power density on average cooling rate.

limiting case in which melt depth approaches zero. This maximum cooling rate was shown analytically to increase with the square of the power density. For a constant melt depth, the increase in cooling rate with absorbed specific power is a consequence of the increased steepness of the temperature gradients. Roughly, doubling the power density doubles the cooling rate for a given melt depth. Fig. 5.32 also shows that a minimum cooling rate exists at the point corresponding to the onset of surface vaporization (maximum melt depth for a given power density). For reference, fig. 5.32 also shows the theoretical maximum cooling rate in nickel as a function of melt depth as well as results for the splat cooling of iron on a copper substrate (Ruhl 1967). The former curve is obtained by considering that all of the absorbed energy is initially concentrated in the melt, such that a temperature discontinuity equivalent to the melt temperature exists at the solid–liquid interface. It is noteworthy that the theoretical maximum cooling rate, the splat cooling rate, and the average cooling rate at the onset of surface vaporization all follow an inverse square law with melt depth. The effects of convection and radiation heat losses on cooling rate were examined for a wide range of parameters. Even at a radiative emissivity of 1.0 and a high convective heat transfer film coefficient of 0.57 W/cm^2 °C, the average cooling rate was not changed by any more than a few per cent for absorbed power densities of 10^5 W/cm^2 and greater. The obvious conclusion is that when high power densities are utilized for laserglaze processing, the subsequent rapid cooling phenomenon is controlled by conduction heat transfer.

Laser surface melting followed by rapid self-quenching (laserglaze) has thus been shown to be capable of reproducibiy producing cooling rates of greater than 10^6 °C/s in appropriately thin sections. The range of microstructures achieved by the laserglaze process includes amorphous metallic solids, extended solid solutions, ultrafine eutectic structures, and refined dendritic structures, as detailed in the previous section. Some work has been done to determine the utility of the laserglaze process as a means to improve surface affected properties such as the corrosion, erosion, and wear properties of metallic alloy surfaces. Some promising examples of such improvements have been obtained by different research groups. Very little work, however, has been reported on the laserglaze process combined with surface compositional modification, although this constitutes an area of obvious high potential. In some limited experiments, preplacement of alloying material on the workpiece prior to laser melting has been used, primarily to harden specific wear regions of a part. Of significantly more interest, however, has been a process which incorporates continuous delivery of alloying material to the workpiece surface prior to laser melting. By delivering material to the interaction zone on a continuous basis, surfaces may be coated with any desired composition, or bulk, rapidly-solidified material may be fabricated to "near-net shape" configurations. This process has been termed the LAYERGLAZE™ process for sequential buildup of bulk rapidly-solidified structures.

To date, the primary investigation of the layerglaze process has taken the form of a program to develop alloys suitable for rapid solidification laser processing and to develop the capability for fabricating parts. These studies will be summarized below:

4.2. Bulk rapid solidification (layerglaze processing)

4.2.1. Experimental processing of layerglazed materials

Apparatus. In order to facilitate the smooth addition of feedstock material, and to ensure adequate bonding with the substrate, power densities in the range of $0.15–1 \times 10^5$ W/cm^2 are commonly utilized. An environmental chamber allows the maintenance of a helium atmosphere around the interaction region. The process has been automated using an $x – y – z – \Omega$ numerically controlled vertical milling machine which is capable of linear speeds of up to 7.62 cm/s. The experimental setup is shown in fig. 5.33. The laser beam is focused by a spherical mirror (not shown) and enters the work chamber through a small aperture at the top. Interaction with the workpiece and feedstock takes place approximately 5 cm below the aperture. A positive pressure of helium gas within the box flows steadily outward through the aperture. The wire or powder feed nozzles are attached to the chamber, and direct the feed material to the interaction zone between the beam and the arbor. The most significant effort in layerglaze processing has been the development of capability to fabricate a scale-model gas turbine disk. This effort will be described herein, to illustrate the technology.

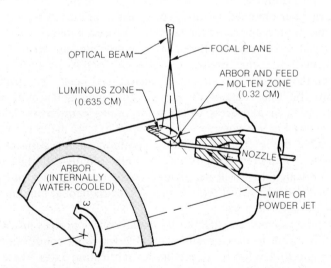

Fig. 5.33. Details of the Layerglaze^{TM} interaction zone.

For disk fabrication the arbor itself is a type 304 stainless steel tube, 5 cm long × 3.7 cm diameter and with a 3.2 mm wall thickness, and is internally water-cooled through a specially designed mandrel. In cases where other parts are to be coated, they are substituted for the mandrel. Since the automated system is capable of x, y, z axis and rotational motion, it is possible to program the traversing of irregular surfaces. To date, however, experimental difficulty has been minimized by working primarily with shapes of revolution.

The process has been developed using both wire and powder feed. The wire feed drive unit is a commercially available dual-drive unit which is capable of delivering wires from 0.051 to 0.127 cm diameter over a wide range of speeds. When powder is utilized as the feedstock, a simple mechanically-vibrated gravity-flow powder feed system has been employed. Fig. 5.34 shows the collimated stream of powder which is emitted from the nozzle under typical operating conditions. Since the nozzle-to-workpiece distance is 6.3 mm, the powder stream within this range approximates a dense, cylindrical column.

Process criticality. The goal of layerglaze deposition is the achievement of a uniform, homogeneous, flaw-free, and fully dense part. In initial experi-

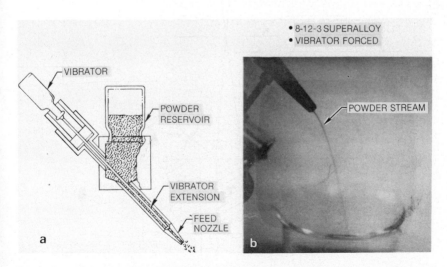

Fig. 5.34. (a) Schematic drawing of internally vibrated powder feeder and (b) photograph of collimated powder stream.

ments utilizing wire feed, a drop transfer phenomenon was encountered, whereby the liquid melted at the end of the feed wire did not flow smoothly and continuously onto the workpiece, but rather was transferred in discrete drops causing a bumpy, irregular deposit. The irregularities become self-per-petuating, due to the fact that the bumps, once established, tended to rake further amounts of liquid from the end of the wire. Under certain experi-mental conditions, it was noted that drop transfer was nearly nonexistent, and that material transfer was quite smooth. The final, most optimum relationship between the arbor and the wire feed is depicted in fig. 5.33. With the arbor rotating at a speed of w, a teardrop-shaped molten zone is obtained at steady state when the impingement spot of the optical beam is circular. The length of the molten zone is typically twice the beam spot diameter. The location of the feedstock, at this stage, turned out to be the most critical parameter. With the impingement angle of the wire feed at 30°, it was determined that most satisfactory transfer occurred when the wire contacted the arbor at the exact edge of the molten pool. If the wire is

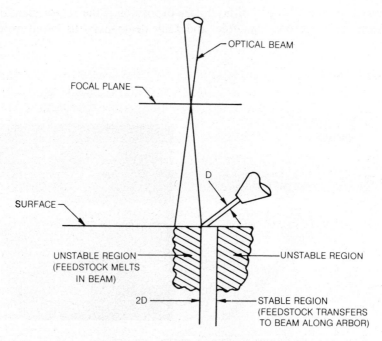

Fig. 5.35. Stability Criterion for Layerglaze™ process feedstock interaction zone.

displaced into the pool, significant drop transfer results, whereas if the wire is displaced away from the pool, incomplete melting occurs. The maximum tolerance on the wire position is approximately two wire diameters, as illustrated in fig. 5.35.

Structures produced using wire feed. Initial parametric studies were conducted using stainless steel wire feed and a stainless steel substrate as a model system for the wire feed process. Wire diameter was 0.089 cm and the goal of the tests was to establish the most uniform possible deposition for the wire feed process. A typical series of tests, in which the arbor advance rate was the variable under study is shown in fig. 5.36. The macroetched cross sections of the five-layer deposits showed that the slower rate resulted in the formation of the most uniform deposits.

The wire feed process was used to fabricate a larger deposit of 58 layers. Cross sections of this part are shown in fig. 5.37. Two major concerns can be noted when examining the cross-section in fig. 5.37(a). One is the deep-penetration type of spiking that occurs throughout the section, and the other is the numerous spherical voids that are associated with the deep-penetration spikes. These features of the structure are more clearly discernible in figs. 5.37(b) and (c). The deep-penetration spiking is unexpected, since it was known that deep-penetration requires power densities in excess of 10^6 W/cm^2. The conditions under which this sample was generated, however, involved a power density of only 0.62×10^5 W/cm^2, well below that normally required for deep-penetration. Similar melt cross sections obtained on flat surfaces under identical conditions did not show any tendency for spiking or deep penetration.

A variety of possible causes for the spiking were considered. The only plausible one stems from the roughened surface of the arbor which is present after a single layer of wire has been deposited. This roughening occurs as a result of the overlap of two adjacent crowned passes, which creates a sharp "furrow" in between, such as can be seen in fig. 5.36. The high freezing rates associated with the process, along with the surface tension in the liquid, combine so as not to allow the material to flow into a flat, uniform surface. As a consequence, a furrow is formed on the surface of the first and all subsequent layers to be deposited. As the direction of travel periodically reverses and the wire is deposited with an "opposite spiral", a cavity is formed at the intersection of the top layer and the layer immediately beneath it. This cavity becomes a steep-walled radiation trap for the 10.6 μm CO_2 laser radiation. Under these conditions, it is reasonable to expect the power density at the base of the cavity to exceed 10^6 W/cm^2,

Fig. 5.36. Layerglaze test arbors with five layer deposits.

thus causing the deep-penetration type of spiking and the associated shrinkage voids.

Although considerable effort has been made to eliminate spiking during wire feed buildup, the conclusion was that although it could be minimized by a variety of means, it could not be eliminated. Since porosity in the final part was highly undesirable, an alternative approach to material feed was taken. Powder feed was attempted, since powder in concept could allow a more continuously variable and controllable flow of feedstock material to the part.

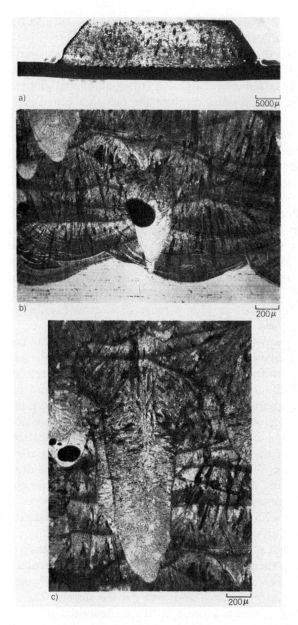

Fig. 5.37. Macroetched cross section of 58 layer deposit. Note porosity associated with undesirable deep penetration spiking effect.

Development of powder feed capability. In order to alleviate the spiking difficulty associated with wire feed, it was necessary to either greatly reduce the wire feed diameter or to substitute a form of material addition that divided the feedstock into substantially smaller increments, so as to reduce the scale of surface irregularities. A strong additional motivation for going to a powder feed process involved the fact that conversion of layerglaze alloys into powder was considerably less difficult than converting these alloys into wire from the nonrapidly solidified condition.

A powder delivery mechanism using variations in nozzle diameter and vibration for flow control, and calibrated for flow rate and uniformity was shown in fig. 5.34. Uniform powder flow was achieved by controlling the nozzle diameter and powder size. The typical nozzle diameter used for a − 170 mesh to + 500 mesh mixture was 0.089 cm, and produces a uniform flow rate of 0.14 g/s. This powder size range is considered optimum from the feeding standpoint. If only the finer powder size fractions are used, clogging of the feeder results. With only the larger size fractions, the feeder is able to achieve free flow, but the flow is nonuniform and cannot be adequately controlled. From the standpoint of powder utilization, it should be noted that use of as wide a range of sizes as possible is desirable to improve the yield factor.

Initial development of the powder feed process served to confirm that the criticality of alignment in the interaction zone between the beam, the arbor, and the feedstock is identical to that determined for wire, as depicted in fig. 5.35. The location of the stable region is the same, and in the unstable region the appearance of the deposit changed significantly, and is appreciably different from that of the wire deposit. In the unstable region, the powder injected directly into the beam would melt and spatter. Also, optical breakdown of the beam would result from the seeding effect of the misplaced powder. When operating in the stable region, the deposit surface is very smooth and uniform. A typical sample of effective deposition is shown in fig. 5.38, in which a deposit of a layerglaze alloy powder approximately 0.3 cm thick × 2.5 cm wide has been applied to a standard, water-cooled stainless steel arbor. This test sample shows that the quality of the deposit is superior to that achieved with wire feed. Fig. 5.38 contains a comparison of the surface in the as-deposited condition, and after a portion of it has been laserglazed. The glazing was achieved by turning off the powder and allowing the laser to remelt and even out the surface. The small spherical type "spatter" deposits located on the as-deposited section are remelted, leaving an almost mirror-smooth surface.

Fig. 5.38. Layerglaze™ processed arbor and microstructure of powder feed test sample. (Half of the sample surface was reglazed on the final pass.)

During powder feed studies it was noted that an additional benefit of the powder feed is that the laser coupling efficiency is much higher than with wire feed, leading to a much higher energy deposition efficiency, and, as a result of the facilitated coupling, enhanced process stability is achieved in the interaction zone.

The powder feed process has been optimized with respect to such parameters as energy density and mass flow rate. Typical optimized parameters at a laser power setting of 5 kW include a power density of 0.62×10^5 W/cm^2, surface speed of 15 cm/s, and deposited layer thickness of 0.25 mm/layer.

4.2.2. Fabrication experience

Bulk Sections. Preliminary, experimental fabrication of thin section deposits has been followed by the production of heavier sections for microstructural evaluation and mechanical test. Initially, a bulk sample 6.45 cm in diameter was deposited on the 3.7 cm diameter mandrel. This part was machined on the side surfaces, macroetched to indicate grain structure, and then sliced into pieces for metallographic and mechanical test purposes. The as-fabricated part is pictured in fig. 5.39.

When the sample was sectioned, polished, and macroetched, a uniform layered structure was revealed. The end view indicated a strong tendency for epitaxial growth across the built-up layers, giving the appearance of a

Fig. 5.39. 6.45 cm diameter test part fabricated by powder feed Layerglaze™ process. Power: 5.0 kW, surface speed: 8.5–14.2 cm/s; 8–12–3 powder (−100, +500 mesh).

"radially directionally solidified structure". The visible distortion of the mandrel was evidence for considerable compressive stresses which the deposit imposed upon it. Subsequent structural analysis and mechanical testing of this preliminary part revealed that the laser-deposited material was free from major flaws, and had good structural integrity along with reasonable strength and ductility.

An additional part was fabricated in order to demonstrate the capability to produce a significantly larger diameter part for a program which involved the spin testing of a scale model turbine disk. This part is shown in fig. 5.40 both in the as-fabricated condition, and following machining to a typical disk configuration. This part was fabricated from a combination of alloys, the bore section consisting of a Ni–8Al–12Mo–3Ta alloy (at%) (8–12–3) which was one of a family specifically developed for laserglaze fabrication, and the outer diameter consisting of a conventional superalloy, IN-718. The use of two different alloys served to demonstrate the feasibility of altering material composition within a single part.

Surface layers. The layerglaze process, by virtue of the fact that it continuously deposits material, is well-suited for coating of surfaces and buildup of surface layers. The process is in fact straightforward and results in dense, well-bonded coatings, provided that the coating and substrate are compati-

Fig. 5.40. Layerglaze™ fabricated model turbine disk (preform (a) and machined (b) part). Final diameter: 13.2 cm.

ble and that the glaze material is not extremely crack sensitive. Among the systems which have been applied by layerglaze to data are pure titanium on titanium alloy substrates, and copper on stainless steel. Also, pure aluminum is easily clad onto aluminum alloy substrates, and iron base hard-facing alloys have been clad onto ferrous alloys.

In some instances, where the substrates and surfacing material may not be compatible, layerglazing has shown some significant improvement over other methods of application. By keeping the interaction times very short, an absolute minimum of alloying occurs in the small interaction zone. This procedure allowed deposition of copper on a titanium substrate. When copper is applied to titanium under higher heat-input conditions, with considerable alloying, a brittle deposit is formed, which cracks and spalls off.

In the case of application of hardfacing alloys to steel substrates by layerglaze, thermal expansion mismatch often resulted in cracking of the hardfacing alloy. This problem was usually alleviated by preheating of the substrate material, and successful application of hardfacing alloys by the layerglaze process have been accomplished. Layer thickness may vary over a wide range, and is completely controllable.

4.3. Future perspectives for rapid solidification laser processing

Although to date the utilization of layerglaze processing has been concerned
with the fabrication of simple axisymmetric shapes, it should be pointed out
that this process is not limited to them, and in fact is capable of producing
almost any desired shape. Some of the possibilities envisioned are illustrated
in fig. 5.41. As indicated, the production of more complex shapes requires
the use of a numerically controlled work station, which is capable of
simultaneous motion about at least three axes. The simple case of translation back and forth on a straight edge has obvious implications for
hardfacing of cutting tools.

Experience with the basic layerglaze technique has also pointed the way
to a potential means for controlling structure and properties more closely
than has previously been possible. The technique has been termed Directed
Energy Processing and is illustrated schematically in fig. 5.42. The process-

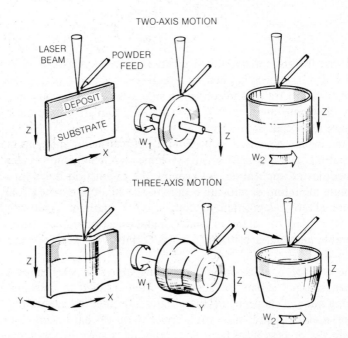

Fig. 5.41. Examples of the potential of the Layerglaze™ process for near-net shape fabrication.

ing starts at work station # 1 with material being added by the laser melting and deposition of feedstock. This step essentially creates the part. At additional work stations, other materials processing operations can be sequentially accomplished on each incremental layer. These include inspection, mechanical deformation, and/or heat treatment. Like deposition, which derives advantages from rapid solidification of a thin section, each of these subsequent operations also derives advantage from being applied to a small layer. Incremental inspection can thus be accomplished without destructively testing the disk. Mechanical processing can be applied to small increments without resulting in significant distortion, and thermal treatments or even remelting can be accomplished immediately by interrupting the material feed.

In addition to these operations, the sequential nature of deposition and the ability to change the feedstock continuously will permit the future production of parts more specifically tailored to their applications than is presently possible. The composition, structure, and even the residual stresses in a part can be modified and controlled for optimal properties. While there exists some practical limitations on the degree to which the structure can be tailored using incremental processing, the technique clearly holds the potential for more specific structure/property control than has previously been possible.

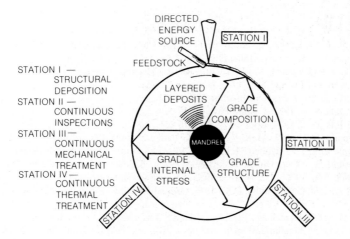

Fig. 5.42. Schematic representation of the generalized concept for directed energy in-situ processing.

Directed energy processing, as applied to part fabrication, appears to offer some advantages over layerglaze processing. In the first place, the ability to detect and to remove imperfections during the actual buildup of the part is clearly beneficial. This suggests that the process has the potential to make a virtually flaw-free part. This in turn would enable the full benefit to be derived from the high structural strength of the material, without being compromised by concern for defect-limited fatigue behavior. Another favorable aspect of directed energy processing is its ability to erase all traces of the initial columnar grained structure in the disk by inducing in-situ recrystallization. The resulting fine-grained structure should possess higher ultimate strength and better ductility than the columnar-grained material, which would be an additional advantage. However, this incremental recrystallization could be halted at any desired part diameter in order to retain a columnar-grained structure if desired. It thus appears that the combination of rapid solidification laser processing with the directed energy incremental processing concept offers a more flexible means of part fabrication than has previously been possible.

References

Breinan E.M. et al., 1975, IIW Document IV-181-75.
Breinan E.M. et al., 1976a, Proc. ASM Conf. on Laser surface treatment for automotive applications, Detroit, MI.
Breinan E.M. et al., 1976b, Superalloys – Metallurgy and Manufacture (Claitor's Publishing Division, Baton Rouge, La., USA).
Breinan E.M. et al., 1976c, Phys. Today 1 44.
Breinan E.M. and Kear B.H., 1978, Rapid solidification laser processing of materials for control of microstructures and properties, Proc. Conf. on Rapid solidification processing at reston, VA (Claitor's Publishing Div., Baton Rouge, LA) p. 87.
Brown C.O. and Banas C.M., 1971, 52nd Annual Meeting, American Society, San Francisco.
Clauer A.H. and Fairand B.P., 1975, J. Metals 27 A51.
Cohen M.I., 1967, J. Franklin Institute 283, 4, 271.
Fairand B.P. et al., 1972, J. Appl. Phys. 43 3875.
Fairand B.P. et al., 1974, Appl. Phys. Lett. 25 431.
Gnanamuthu D.S. and Locke E.V., 1976, US Patent No. 4015100.
Greenwald L.E., 1975, United Technologies Research Center Report R75-111321-1.
Greenwald L.E., 1976a, United Technologies Research Center Report UTRC76-115.
Greenwald L.E., 1976b, United Technologies Research Center Brochure.
Hella R.A. and Gnanamuthu D.S., 1976, Annual Meeting, American Welding Society, St. Louis.
Kear B.H. and Breinan E.M., 1977, Laserglazing, a new process for production and control of rapidly-chilled metallurgical microstructures, Proc. Sheffield Int. Conf. on Solidification and casting, Ranmoor House, Sheffield University.

Locke E.V. et al., 1972, IEEE J. Quantum Electron. **8** 132.
Murray W.D. and Landis F., 1959, Trans. of ASME **106**.
Ozisik M.N., 1968, Boundary value problems of heat conduction (Inter. Textbook Co.).
Ruhl R.C., 1967, Mat. Sci. Eng. **1** 313.
Snow D.B., Breinan E.M. and Kear B.H., 1980, Rapid solidification processing of superalloys using high power lasers", Proc. 4th Int. Symp. Superalloys, Seven Springs, PA.
Voorhis W.G., 1964, Proc. Ion and electron beam symp.

CHAPTER 6

SHAPING MATERIALS WITH LASERS

STEPHEN COPLEY

Department of Materials Science
University of Southern California
Los Angeles, California 90089-0241, USA

MICHAEL BASS, BRUNO JAU and RUSSELL WALLACE

Center for Laser Studies
University of Southern California
Los Angeles, California 90089-1112, USA

Laser Materials Processing, edited by M. Bass
© North-Holland Publishing Company, 1983

Contents

1. Introduction

The development of high power lasers has made possible a variety of material removal techniques. In chapter 2, Steen and Kamalu describe straight line cutting and hole drilling with lasers. In this chapter, we discuss laser turning and milling. Two approaches are considered: laser-assisted machining (LAM), in which a laser beam heats material to be machined by a single point cutting tool; and, laser machining (LM), in which the laser

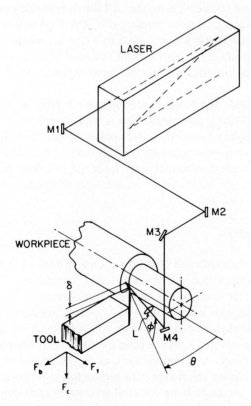

Fig. 6.1. Experimental arrangement for laser-assisted machining.

forms a groove in the material by vaporization. At this point, LAM has been applied only to metallic systems. Laser machining has been applied mainly to ceramics, however, the application of this approach to metallic systems is being investigated.

2. Laser-assisted machining

An arrangement for LAM is shown in fig. 6.1. A CW laser beam is directed by turning mirror (M1) along a path parallel to the turning axis of the lathe. The beam is then reflected from carriage mirror (M2), cross-slide mirror (M3) and workpiece mirror (M4), and finally focussed onto the workpiece a distance δ in front of the edge of the cutting tool. Because the direction of the carriage motion is parallel to the turning axis and that of the cross-slide motion is perpendicular to the turning axis, the beam retains its alignment during operation of the lathe.

The absorption of the beam is maximized if it is oriented perpendicular to the surface it is heating. Thus in turning, if it is desired to heat the shoulder of the workpiece, the angle θ is set equal to $\pi/2 - \kappa_r$, where κ_r is the major cutting edge angle. On the other hand, in facing, θ is set equal to κ_r. If it is desired to heat the cylindrical surface of the workpiece, then θ is set equal to $\pi/2$.

In our experiments, the beam is produced by a 1400 W, CW–CO_2 laser operating in the TEM_{00} mode (gaussian spatial distribution of intensity). The mirrors are highly polished, water-cooled copper flats and the lens is ZnSe. For very high intensity beams, the lens may be replaced by a focusing mirror. The three components of tool force (F_c, F_t and F_b), which are defined in fig. 6.1, are measured by a piezoelectric dynamometer mounted in the base of the tool holder.

2.1. Potential benefits

In LAM, the laser is employed as a heat source. It is used in two ways: (1) to alter the workpiece material prior to chip formation; (2) to heat the material on the shear plane as the chip is being formed.

In the first case, the laser is used to vaporize, melt and resolidify or solutionize the workpiece material with the objective of improving its machinability. The advantage of the laser for such treatments in comparison to other directed energy heat sources such as the gas flame or plasma arc is its capability of altering the material to be machined without changing the microstructure of the underlying material to a significant degree. This is a consequence of the heat flow associated with the small spot size and high intensity of the laser beam and the high speeds at which the beam sweeps over the surface in a normal machining operation.

In the second case, the advantage of the laser is its capability of heating material on most of the shear plane as the chip is being formed without heating significantly the material that contacts the edge or face of the cutting tool. This, of course, is also a consequence of the heat flow associated with laser heating. Heating the material on the shear plane may result in benefits such as decreased cutting forces, increased material removal rate, increased tool life and improved surface conditions such as smoothness, residual stress or flaw distribution. However, heating the material that contacts the tool is likely to decrease tool life.

In applications where the laser is used to alter the material prior to chip formation, the economics of the process depends critically on whether or not the laser induced changes can take place at a volume rate equal to the material removal rate for the machining process as it is normally performed. If, in fact, the anticipated benefit is an increase in material removal rate, then clearly the volume rate for the laser-induced change must even exceed the normal material removal rate. If the anticipated benefit relates to improved surface conditions or tool life, then the volume change rate should be approximately equal to the material removal rate. If the volume change rate for the laser-induced change is much less than the normal removal rate then a question arises as to whether the laser treatment and the machining operation should be done together or separately. In either case, the time for the operation would be dictated by the volume change rate for the laser induced process. Thus, to be attractive, the benefit due to the laser process would have to be very substantial, e.g. to make possible the shaping of a material that could not be shaped any other way.

Of course, the volume change rate for the laser-induced change depends on the power of the laser and the efficiency of the beam delivery system. The volume change rate can normally be increased if the power absorbed by the workpiece is increased. In general, the higher the power of the laser that is required, the greater the cost. In some cases, the need for higher absorbed power can be satisfied through the use of coatings to enhance absorption.

In applications where the laser is used to heat the material on the shear plane as the chip is being formed, the laser scan speed must equal the speed at which the cutting tool passes over the workpiece. Heating the material on the shear plane may change the mode of the chip formation from discontinuous to continuous or decrease the tendency to form a built-up-edge. Such changes affect the smoothness and flaw distribution of the machined surface. In section 2.2, we will show that both these effects take place in Inconel 718. Heating ferrous materials may result in a softening of the shear plane material due to austenization, and the development of a surface compressive stress due to the formation of layer of martensite at the machined surface. We will present evidence for this effect in section 2.2, for a M43 tool steel. One of the most important reasons for heating material on the shear plane is to produce a decrease in cutting force. Such a decrease is anticipated in precipitation hardened alloys, if they can be heated by combined laser heating and shear plane heating into the temperature range where the yield stress decreases markedly with increasing temperature. We will present evidence for this effect in Inconel 718 in section 2.2. Such a decrease in machining forces may result in a direct benefit such as making possible the machining of a workpiece that would deflect elastically too much to maintain tolerances or, perhaps, even plastically deform under conventional machining conditions. Such a decrease in machining forces might also produce an increase in tool life, if the concomitant small increase in temperature of the material contacting the tool due to laser heating does not offset this effect. Finally, such a decrease may make possible an indirect but very important benefit; namely an increase in material removal rate and thus a decrease in machining cost. A discussion of this possibility will now be given.

Consider the cost of turning a unit volume of material. If C_o is the labor and overhead rate, C_t is the tool cost per edge, T is the tool life and t_m is the machining time, then the cost is given by the equation

$$U = C_o t_m + C_t \frac{t_m}{T}. \tag{6.1}$$

Nonproductive time such as loading and unloading time and tool change time have been omitted in deriving this equation because they depend on the specific workpiece and the shop conditions. For machining a unit volume, the machining time is given by

$$t_m = 1/Z, \tag{6.2}$$

where Z is the material removal rate. Thus

$$U = \frac{1}{Z}\left(C_0 + \frac{C_t}{T}\right). \tag{6.3}$$

One approach to decreasing the cost is to increase tool life; however, even for an infinite tool life the cost of machining a unit volume of material is C_0/Z. A more effective approach is to increase the material removal rate.

In single point turning, the material removal rate is given by the equation

$$Z = v a_c a_p, \tag{6.4}$$

where v is the cutting speed; a_c is the undeformed chip thickness, which in single point turning is equal to the feed; and, a_p is the back engagement, which is also known as the depth of cut. The magnitude of the cutting force is approximately proportional to the undeformed chip cross-section ($a_c a_p$). For roughing and semi-roughing operations, commercial practice dictates selecting as large an undeformed chip cross-section as possible taking into consideration the rigidity of the tool–workpiece system and the power of the lathe. If, for such force-limited cutting, it is possible by laser heating to decrease cutting force, then it is possible to increase the undeformed chip cross-section and thus the material removal rate. The use of a laser would, of course, increase the magnitude of C_0 in eq. (6.3). Also it might result in a decrease in tool life, T. Ultimately, the use of the laser would have to be justified on the basis of decreased cost.

2.2. Results

At this time, LAM is under active investigation in several laboratories including our own. Although there is data available supporting many of the benefits mentioned in section 2.1, insufficient data is available to draw firm conclusions regarding the technological feasibility of the LAM process. In this section, we first discuss the possibility of melting and resolidifying the material to be machined. Then, we discuss a novel premachining treatment for Ti–6Al–4V workpieces involving hole drilling with a pulsed laser. Finally, we give results for an M43 tool steel and Inconel 718, where the shear plane was laser heated during chip formation.

Melting and resolidifying a workpiece material prior to machining may increase its machinability. Many materials contain large, hard precipitates that contribute to tool wear. Such precipitates would be dissolved during melting. Because of the rapid growth rates and cooling rates associated with the subsequent resolidification and cooling, reprecipitation is likely either to be suppressed or to occur on a scale too fine to damage the tool.

Although no data is available supporting the assertion that tool wear might be decreased by melting and resolidifying the surface of an alloy containing hard precipitates prior to machining, data is available that makes possible an estimate of the volume change rate for this type of laser induced process. Esquivel et al. (1980) have investigated the shape and size of melt zones in Udimet 700, a nickel-base superalloy, as a function of power and scan rate. Fig. 6.2 shows transverse sections of laser melted trails produced at speeds ranging from 3 to 111 cm s^{-1}. The incident power was 535 W. In fig. 6.3, the depth and width of these trails is plotted as a function of velocity. Also plotted in fig. 6.3 are melt depths and widths, observed by Hill and Rajagapal (1980), in laser melted nickel. In this case, the incident beam intensity was 12 kW.

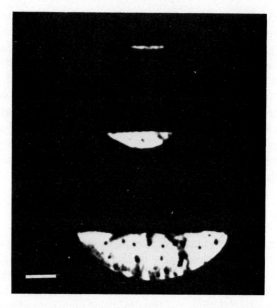

Fig. 6.2. Transverse sections of laser melted trails in Udimet 700 for an incident beam intensity of 535 W: (top) $v = 111$ cm s^{-1}; (middle) $v = 23$ cm s^{-1}; and (bottom) $v = 3$ cm s^{-1}. Marker represents 0.1 mm.

According to Gorsler (1980) the current machining limit for Inconel 718, an alloy similar to Udimet 700, for a semi-rough cut using a carbide tool is $v = 50$ cm s^{-1}, $a_c = 0.250$ mm, and $a_p = 0.152$ and 0.318 cm. At 565 W and 50 cm s^{-1}, the depth of the melt trail is much less than that of the undeformed chip, which equals a_c. Even at 12 kW and 50 cm s^{-1}, the depth of the melt trail is less than half that of the undeformed chip. Thus, one would conclude that melting and resolidification as a premachining treatment for Udimet 700 would be attractive only if the resulting benefit was very significant, because the volume change rate of the laser-induced change is much less than the material removal rate.

Jones (1980) is investigating a novel premachining treatment for Ti–6Al–4V alloy workpieces involving drilling holes in the material in front of the advancing cutting tool to decrease machining forces and to assist in chip breaking. The arrangement for this type of LAM is shown in fig. 6.4. A Nd–YAG laser with a pulse rate of 400 pps is used to drill holes in the workpiece. The average energy per pulse was 0.31 J/pulse resulting in holes 0.0375 cm to 0.050 cm deep. The cutting speed was 32 cm s^{-1} so that the spacing between holes along the circumference was 0.08 cm. Force reduc-

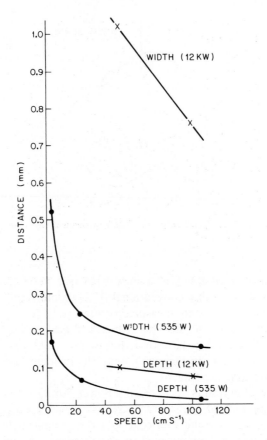

Fig. 6.3. The depth and width of laser melted trails as a function of velocity in Udimet 700 and nickel at 535 W and 12 kW, respectively.

tions of 30 to 45% were observed with laser treatment for a 0.05 cm depth of cut. A 20% force reduction was observed with laser treatment for 0.1 cm depth of cut. Two feeds were investigated, 0.00055 cm/rev and 0.0022 cm/rev. These results demonstrate the feasibility of reducing cutting forces by a hole drilling treatment. The volume treatment rate demonstrated, however, is much less than the material removal rate for semi-rough cutting with carbide tools. According to Gorsler (1980), the current machining limit for semi-rough cutting Ti–6Al–4V with a carbide tool is $v = 70$ cm s^{-1}, $a_c = 0.025$ cm and $a_p = 0.152$ to 0.318 cm.

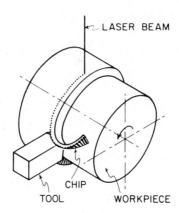

LASER BEAM

CHIP

TOOL   WORKPIECE

Fig. 6.4. Arrangement for laser assisted machining involving pulsed laser pretreatment.

We have investigated LAM of a fully hardened M-43 tool steel in the turning configuration with a conventional (C-2) carbide cutting tool. Three components of tool force were measured continuously during the turning of a 2.6 cm diameter M-43 tool steel bar, which had been through hardened to 62 R_c. The cutting conditions were as follows: $v = 22.5$ cm^{-1}, $a_c = 0.0025$ cm^{-1} and $a_p = 0.050$ cm. In the experiment, the incident laser beam power was varied during the experiment.

Fig. 6.5 shows the results of increasing power during LAM of M-43 tool steel. All components of tool force decreased with increasing power. The distance between the cutting tool edge and the impingement point of the laser beam on the workpiece was also varied ($\delta = 3$, 5 and 8 mm). The decrease in tool force with increasing power was most pronounced in the experiment where $\delta = 8$ mm.

Our results demonstrate the feasibility of turning a fully hardened tool steel with a laser assist. With laser beam heating the chip became continuous and a smooth surface finish was observed. These observations suggest that the laser-induced changes are due to the reversion of martensite to austenite. Of course, any residual austenite left behind by the tool would be quenched back to martensite and should produce a residual compressive layer. The material removal rates reported here are very low and no data was obtained regarding tool life. Such tool life data is essential to evaluate the process.

The greatest amount of research on the effect of heating the shear plane during chip formation has been carried out on Inconel 718. Fig. 6.6 shows

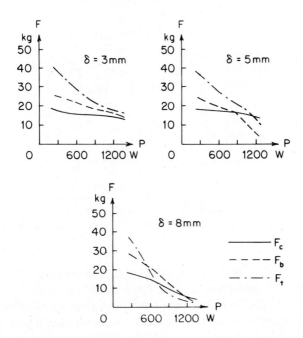

Fig. 6.5. Tool force versus incident power for LAM of M-43 tool steel for laser beam to tool cutting edge distances of 3, 5 and 8 mm. The force components, F_c, F_t and F_b are the cutting force, thrust force and back force, respectively.

the results of an experiment designed to illustrate the effect of velocity on force drop. A workpiece was machined in the facing configuration so that the cutting velocity continuously decreased as the cutting tool approached the turning axis. During a facing cut the laser was turned on and off several times and the force recorded continuously. Turning the laser on was observed to produce a force drop. The magnitude of the force drop increases with decreasing velocity. The experimental conditions for this experiment were 600 W incident beam power, beam diameter $(1/e) = 212$ μm, $\delta = 5$ mm, $a_c = 0.0075$ cm and $a_p = 0.0375$ cm. Thus the material removal rate was very low compared to the current machining limit for Inconel 718 ($v = 50$ cm s^{-1}, $a_c = 0.0250$ cm and $a_p = 0.152$–0.318 cm).

Fig. 6.7 shows the effect of varying the distance δ between the impingement point of the laser and the cutting tool edge on the reduction in cutting force for two cutting speeds, 10 and 20 cm s^{-1}. The other conditions for the experiment were: incident beam power = 600 W; beam diameter $(1/e) = 212$ μm; $a_c = 0.0075$ cm; and, $a_p = 0.050$ cm. At both cutting speeds, the

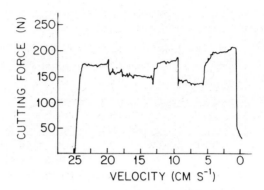

Fig. 6.6. The effect of cutting speed on force drop in the laser assisted machining of Inconel 718.

reduction in cutting force first increases and then decreases with increasing distance, δ. The maximum cutting force reduction occurs in the range $\delta = 3-4$ mm. At distances less than this range, chip interference is thought to account for the decrease in force reduction. At distances greater than this range, the material has time to cool so the average temperature of the shear plane with the laser turned on decreases. At the 20 cm s^{-1} cutting speed, there is less time for the absorbed power of the laser beam to heat the material on the shear plane than at 10 cm s^{-1}. Thus, the reduction in force observed at highest velocity is less than that observed at the lowest velocity.

From the preceding discussion, one might conclude that by increasing the incident power it might be possible (1) to increase the magnitude of the force reduction and (2) to obtain a significant force reduction at increased material removal rates. It was found, however, that when the beam power

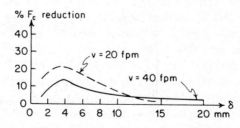

Fig. 6.7. The effect of varying the distance, δ, between the impingement point of the laser and the cutting tool edge on percent cutting force reduction.

was increased for cutting velocities in the 10–20 cm s^{-1} range, the observed force changes were not reproducible and, sometimes, when the laser was turned on the force actually increased. At 50 cm s^{-1}, it was not possible to obtain a significant force drop with the available incident beam power.

To obtain a better understanding of the changes in cutting force induced by heating the material on the shear plane with a laser, an investigation was carried out in which chip formation was videotaped and displayed on a split screen with simultaneously recorded tool forces. This investigation showed that at low cutting speeds ($v < 30$ cm s^{-1}), a stable built-up-edge forms on the cutting tool, see fig. 6.8. Laser heating at low power does not disturb this built-up-edge. If the beam power is increased, however, the built up edge becomes unstable. Loss of the built-up-edge changes the effective rake angle from ~ 20° to 8°, and should result in an increase in cutting force. The force increase associated with such a loss could outweigh the force decrease

Fig. 6.8. Stable built up edge formed on cutting tool in machining of Inconel 718 at $v =$ 15 cm s^{-1}.

due to heating the material on the shear plane. Thus, the loss of a built-up-edge provides a satisfactory explanation for the lack of reproducibility in the force change and the occasional increase in force induced by laser heating.

At cutting speeds greater than 30 cm s^{-1}, no built up edge was observed. The failure to obtain a significant force drop at 50 cm s^{-1} appears to be due to insufficient incident power with the 1400 W laser.

Recent results reported by Rajagapal et al. (1980) support the hypothesis that our failure to observe significant force drops at 50 cm s^{-1} was due to insufficient incident beam power. Their data is reproduced in fig. 6.9, which shows a plot of cutting force versus incident beam power. The data was obtained at: constant cutting speed, 50 cm s^{-1}; constant feed, 0.025 cm rev^{-1}; and, constant depth of cut, 0.050 cm. After recording the cutting force with no laser heating, the laser was turned on at 2 kW and then increased in 2 kW steps up to 14 kW. The cutting time at each power step was 5 s. Tool wear effects were taken into account by repeating the experiment at decreasing power levels, again changing the power in steps of 2 kW. The value plotted at each power setting was found be averaging the values obtained for increasing power and for decreasing power. In this experiment, the distance δ was 12 mm and the focused beam was a 0.1 cm × 0.4 cm ellipse. Clearly, sufficient shear plane heating occurred at the

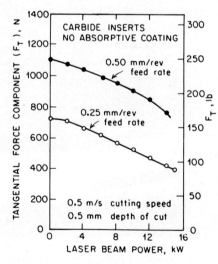

Fig. 6.9. Cutting force versus incident beam power in machining of Inconel 718 at 50 cm s^{-1}.

higher incident powers to produce a significant force drop. Unfortunately, with carbide tools excessive tool wear was observed suggesting that heating by the laser penetrated too deeply. In contrast, experiments with ceramic tools indicate that laser heating improves tool wear.

2.3. Analysis

The feasibility of obtaining significant force reductions by laser heating material on the shear plane has been demonstrated for conditions approaching the current machining limits for Inconel 718. A high incident beam power was required and considerable wear was observed in the case of carbide tools. In this section we analyze the heat flow resulting from laser heating and shear plane heating. Our objective is to estimate the incident beam power required for a specific increase in material removal rate at constant cutting force.

2.3.1. Laser heating

The experimental arrangement for LAM is shown schematically in fig. 6.10. The laser beam moves at constant velocity v in the positive x direction.

According to Rosenthal (1946), the temperature distribution due to a point heat source moving in the positive x direction at constant velocity is given by the equation

$$T = \frac{P}{C_p D 2\pi r} \exp\left(-\frac{v(r+x)}{2D}\right), \tag{5}$$

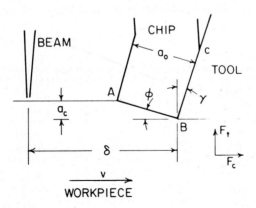

Fig. 6.10. Schematic diagram defining the parameters used in describing laser assisted machining.

where P is the absorbed beam power, C_p is the volumetric specific heat, D is the thermal diffusivity, v is the beam speed and $r = (x^2 + y^2 + z^2)^{1/2}$. Since the beam diameter in our experiments (0.02 cm) is much less than δ (0.40 cm), it is reasonable to expect the Rosenthal equation to give a good approximation to the temperature distribution.

Parameters employed in calculating the temperature distribution due to laser heating are given in table 6.1. The average value for the volumetric specific heat used in our calculations is based on the temperature range 38–1204°C and is 4.28 J cm^{-3} °C^{-1}. The average value for thermal diffusivity is based on the same temperature range and is 0.0466 cm^2 s^{-1}.

2.3.2. Shear plane heating

According to Weiner (1955), the temperature distribution on the shear plane due to shear plane heating can be calculated given: the undeformed chip thickness (a_c); the back engagement (a_p), the cutting velocity (v); the tool geometry; the machining forces $(F_c, F_t, $ and $F_b)$; and, the thermal properties. In applying the Weiner solution to turning, we assume orthogonal cutting equations can be applied. This assumption, which greatly simplifies the calculation, is reasonable because $a_p \gg a_c$.

Parameters used in our calculations are based on the following data: rake angle, $\gamma = 8°$; $v = 50$ cm s^{-1}; $a_c = 0.0150$ cm; $a_p = 0.0875$ cm, shear plane angle, $\phi = 20°$, $F_c = 380$ N; $F_t = 175$ N; and, thermal properties given in table 6.1.

As an example, we calculate the temperature distribution on the shear plane due to shear plane heating for Inconel 718 machined under conditions approaching the current machining limit: namely, $v = 50$ cm s^{-1}; $a_c = 0.025$

Table 6.1
Parameters employed in calculating temperature distributions

Temperature		C_p	D
(°F)	(°C)	(J cm^{-3} °C^{-1})	(cm^2 s^{-1})
100	38	2.92	0.0404
500	260	3.45	0.0425
1000	538	4.12	0.0455
1500	816	4.78	0.0487
2000	1093	5.40[a]	0.0517[a]
2200	1204	5.65[a]	0.0527[a]

[a]Extrapolated values based on 100–1500°F data.

cm; and, $a_p = 0.125$ cm. It is assumed that F_c and F_t are proportional to the undeformed chip cross-section. Thus, for $a_c a_p = 0.025$ cm $\times 0.125$ cm, $F_c = 905$ N and $F_t = 417$ N. Details of the calculation are as follows:

the machining power is

$$P_M = F_c v = 452 \ W; \tag{6.6}$$

the force parallel to the rake face of the tool, F_R, is

$$F_R = F_c \sin\gamma + F_t \cos\gamma = 539 \ N; \tag{6.7}$$

the velocity parallel to the rake face, v_0, is

$$v_0 = \frac{v \sin\phi}{\cos(\phi - \gamma)} = 17.5 \ \text{cm s}^{-1}; \tag{6.8}$$

thus, the power dissipated at the rake face, P_R, is

$$P_R = F_R v_0 = 94.3 \ W; \tag{6.9}$$

the power dissipated on the shear plane, P_s, is

$$P_s = P_M - P_R = 358 \ W; \tag{6.10}$$

the area of the shear plane, A_s, is

$$A_s = \frac{a_c a_p}{\sin\phi} = 9.14 \times 10^{-3} \ \text{cm}^2. \tag{6.11}$$

The heat liberated on the shear plane per unit area associated with the formation of the chip, q, is

$$q = P_s/A_s = 3.91 \times 10^4 \ \text{W cm}^{-2}. \tag{6.12}$$

Weiner (1955, fig. 3) gives a plot of $T\sin\phi$ as a function of $y^{1/2}$, where T is the dimensionless shear plane temperature and y is the dimensionless distance measured along the shear plane from the workpiece surface. The parameter y is given by the equation

$$y^{1/2} = \left(\frac{v\psi^2\cos\phi}{4D}\right)^{1/2} \xi^{1/2} = 5.78 \ \xi^{1/2}. \tag{6.13}$$

The parameter ψ equals $\tan\phi$ and ξ is the distance from the leading edge of the shear plane measured along the shear plane. The shear plane temperature, u, is

$$u = \frac{1}{\sin\phi} \frac{q}{vC_p}(T\sin\phi) = 534 \ (T\sin\phi). \tag{6.14}$$

The results of this calculation are given in table 6.2.

Table 6.2
Temperature distribution on the shear plane
due to shear plane heating by plastic
deformation at $v = 50$ cm s^{-1}

ξ	$y^{1/2}$	$T\sin\phi$	u (°C)
0.01	0.578	0.78	416
0.02	0.817	0.90	481
0.04	1.15	0.96	513
0.06	1.42	0.99	529
0.08	1.64	1.00	534
0.10	–	″	″
0.12	–	″	″
0.14	–	″	″
0.16	–	″	″

2.3.3. Shear plane temperature

Shear plane heating and laser heating contribute additively to determine the temperature distribution on the shear plane during LAM. The larger the power absorbed from the laser beam for a given set of machining conditions, the higher is the average temperature on the shear plane. If a sufficiently high average temperature is attained, then the dynamic shear yield stress and thus the cutting force may be decreased. In the case of a force limited cut, this makes possible an increase in material removal rate. In this section, we will show how the analyses of heat flow due to shear plane heating and laser heating may be employed to predict the amount of absorbed beam power required for a specific increase in material removal rate.

Fig. 6.11 is a plot of static yield stress versus temperature for Inconel 718. The behavior of this alloy is typical of precipitation hardenable alloys that should respond to LAM. The sharp decrease in static yield stress at temperatures greater than 650°C reflects a basic change in the dislocation precipitate interaction. We will assume that the dynamic shear yield stress varies in a similar manner with respect to temperature. For the purpose of this calculation, we will assume that the ratio of the dynamic shear yield stress to the static yield stress is constant for the temperature range of interest.

Let us calculate the absorbed beam power required to effect a 25% increase in material removal rate beyond the current machining limit for Inconel 718. Such an increase can be attained by increasing the feed from

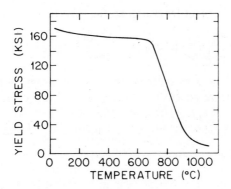

Fig. 6.11. Static yield stress versus temperature for Inconel 718.

0.0250 cm to 0.03125 cm, while keeping the cutting speed and depth of cut the same. Ordinarily, such an increase would result in an increase in cutting force; however, by laser heating material on the shear plane, so as to decrease its average dynamic shear yield stress, it is possible to increase the feed while keeping the cutting force the same.

To calculate the decrease in the average dynamic shear yield stress required to keep the cutting force the same, we consider the equation

$$F_c = \frac{1}{\cos(\phi - \gamma)} (F_s \cos \gamma + F_R \sin \phi). \tag{6.15}$$

The force acting parallel to the shear plane, F_s, is given by the equation

$$F_s = \bar{\tau}_s \frac{a_c a_p}{\sin \phi}, \tag{6.16}$$

where

$$\bar{\tau}_s = \frac{1}{A_s} \int\int \tau_s \, dA. \tag{6.17}$$

The force acting parallel to the rake face of the cutting tool, F_R, is given by the equation

$$F_R = \bar{\tau}_R L a_p, \tag{6.18}$$

where

$$\bar{\tau}_R = \frac{1}{L a_p} \int\int \tau_R \, dA \tag{6.19}$$

and L is the chip contact length along the rake face of the tool. Thus

$$F_c = \frac{\cos\gamma}{\cos(\phi - \gamma)\sin\phi}\bar\tau_s(a_c a_p) + \frac{\sin\phi}{\cos(\phi - \gamma)}\bar\tau_R L a_p. \tag{6.20}$$

In section 2.3.2, it was assumed that F_c was proportional to $a_c a_p$. This behavior follows from eq. (6.20), if we set $L = L_0 a_c$, where L_0 is a constant. Setting $\bar\tau_R L_0 a_c a_p = \bar\tau_R L_0(0.0250)(0.125) = F_R = 539$ N [see eq. (6.7)], we find that $\bar\tau_R L_0 = 1.725 \times 10^5$ N cm^{-2}. Thus if $\phi = 20°$ and $\gamma = 8°$, then

$$F_c = 2.95\bar\tau_s(a_c a_p) + 6.04 \times 10^4(a_c a_p).$$

Setting $F_c = 905$ N, $a_c = 0.0250$ cm and $a_p = 0.1250$ cm, we find that $\bar\tau_s = 7.77 \times 10^4$ N cm$^{-2} = 777$ MPa. Thus without laser heating the ratio of the average dynamic shear yield stress to the average static yield stress is $\bar\tau_s/\sigma_y = 777$ MPa/1069 MPa $= 0.727$.

If we keep $F_c = 905$ N when we increase the feed (a_c) to 0.03125 cm and ϕ, $\bar\tau_R$ and L_0 do not change, then we must decrease $\bar\tau_s$ to 5.8×10^4 N cm$^{-2} = 580$ MPa. This corresponds to a 25% decrease in average dynamic shear yield stress. The average static yield stress is 580 MPa/0.727 = 798 MPa or 115 800 psi. If ϕ decreases due to laser heating, then $\bar\tau_s < 580$ MPa. Although this behavior has not been documented, it would be surprising because decreasing ϕ moves the shear plane closer to the laser beam causing the average shear plane temperature to increase. If L_0 increases due to laser heating, then $\bar\tau_s < 580$ MPa. If $\bar\tau_R$ decreases due to laser heating, then $\bar\tau_s > 580$ MPa. One of the basic features of laser heating is, however, minimal heating of material contacting the rake face of the tool. Thus significant changes in L_0 or $\bar\tau_R$ are not considered to be likely.

In order to attain an average dynamic shear yield stress in the shear plane of 580 MPa, which corresponds to an average static yield stress in the shear plane of 798 MPa (115 800 psi), we heat with the laser. The temperature increment due to laser heating added to that due to shear plane heating must give a temperature distribution such that

$$\sigma_y = \frac{1}{A_s}\int\int\sigma_y\,dA = 115\ 800 \text{ psi} \tag{6.21}$$

where values of σ_y corresponding to a particular temperature are obtained from fig. 6.11.

If the average dynamic shear yield stress is 580 MPa, then $q = \bar\tau_s v_s = 5.8 \times 10^4 \times 0.51 = 2.94 \times 10^4$ W cm^{-2} where $v_s = v\cos\gamma\cos^{-1}(\phi - \gamma)$ is the change in velocity parallel to the shear plane as the metal crosses the plane.

According to eq. (6.14), the temperature increment due to shear plane heating equals $404(T\sin\phi)°C$. The absorbed laser beam power, which must be found by trial and error, must provide a sufficient temperature increment so that eq. (6.21) is satisfied.

Fig. 6.12 shows the shear plane divided into elements for the calculation of temperature increments and the average static yield stress. Fig. 6.13 shows the temperature increment due to laser heating (300 W absorbed power), the temperature increment due to shear plane heating and the resulting shear plane temperature distribution. Fig. 6.14 shows the resulting distribution of static yield stresses obtained from figs. 6.11 and 6.13, which give an average value satisfying eq. (6.21). The predicted laser beam power, must be regarded as a minimum value. If ϕ decreases due to laser heating or

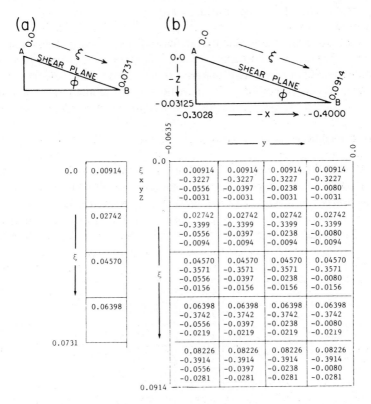

Fig. 6.12. Shear plane divided into elements for calculation of temperature increments and the average static yield stress.

83	223	484	724
84	214	445	651
81	196	390	558
76	172	327	459
69	147	267	366

+

311	311	311	311
380	380	380	380
396	396	396	396
402	402	402	402
404	404	404	404

=

394	534	795	1035
464	594	825	1031
477	592	786	954
478	574	729	861
473	551	671	770

LASER HEATING   SHEAR PLANE HEATING   SUM
INCREMENT   INCREMENT

Fig. 6.13. Temperature distributions on shear plane due to laser heating at 300 W absorbed power and shear plane heating (temperature given in °C).

L_0 increases without a compensating change in $\bar{\tau}_R$, then more power would be required.

An evaluation of the technological feasibility of LAM of Inconel 718, requires experimental verification of the calculations presented here. The economic benefits of increased material removal rate must be compared to the increased costs associated with laser heating. A detailed investigation of the technological feasibility of LAM is in progress (Tipnis et al. 1980).

3. Laser machining

The $CW-CO_2$ laser has been used to shape ceramics by turning and milling. In this section we present results on such shaping of Si_3N_4, SiAlON and SiC workpieces. A straightforward analysis is presented, which predicts the feed and power corresponding to a specific surface roughness grade and effective material removal rate.

156	156	92	12
156	156	75	12
156	156	98	21
156	156	130	58
156	156	154	108

Fig. 6.14. Static yield stress distribution on shear plane after laser heating. The average static yield stress on the shear plane is 116 KSI.

The apparatus for laser turning is identical to that used in LAM except that no mechanical cutting tool is used, see fig. 6.1. After reflection by the cross-slide mirrors the laser beam is focused by a ZnSe lens on the workpiece along a radial direction for turning or along the turning axis for facing.

Laser machining in the milling configuration is obtained by employing a scanner to sweep the focused beam back and forth in a direction perpendicular to the direction of workpiece motion. If the advance of the workpiece during one cycle of the scanner is less than the width of the groove, then a nearly continuous swath of material will be removed. Most scanners deflect the light with a sinusoidal motion. As a result the light dwells longer on the material at the end of each stroke than it does on that in the middle. This leads to a nonuniform rate of material removal and a cut bottom with a cross section shown in fig. 6.15a. By placing metallic masks over the sample at the ends of the scan the possibility of cutting too deeply is eliminated and a flat bottomed cut as shown in fig. 6.15b can be produced.

A description of the materials that have been shaped in our experiments is given in table 6.3.

3.1. Turning results

Fig. 6.16 shows the cross section of the grooves of a laser machined spiral in SiC. The grooves were cut in the facing configuration using the TEM_{00}

Fig. 6.15. Cross section of laser milled grooves in silicon nitride, (a) unmasked, (b) masked. 900 W (CW) CO_2 laser; scanner frequency: 65 Hz, scanner amplitude at focus: (a) 0.3 cm, (b) 0.9 cm; translation speed: (a) 0.41 cm/s, (b) 0.63 cm/s; two passes. Marker represents 0.5 mm.

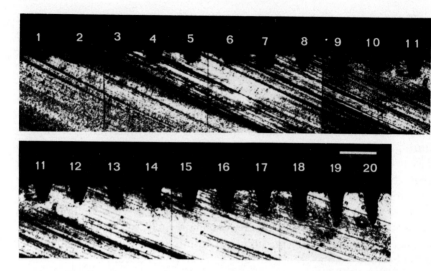

Fig. 6.16. Photomicrographs of laser machined spiral grooves on SiC. Marker represents 0.4 mm.

mode laser; the high numbered grooves were closest to the turning axis and thus correspond to the lowest cutting velocities. It is evident that the groove depth increases with decreasing cutting velocity while the groove width decreases slightly, being approximately equal to the beam diameter. In fig. 6.17 the groove depth is plotted as a function of dwell time given by the equation, $\theta = b/v$, where b is the beam diameter and v is the cutting velocity.

Fig. 6.18 shows the departure of the groove shape from nearly gaussian which occurs when a multimode laser was used and an O_2 gas jet was added

Table 6.3
Materials investigated

Material	Description
SiC	NC 435 (Norton Co.), a sintered SiC containing 20 vol% free Si.
Si_3N_4	NC 132 (Norton Co.), a hot-pressed Si_3N_4.
SiAlON	A hot-pressed SiAlON by General Electric Co.
Alumina	AV30 produced by McDaniel Inc.

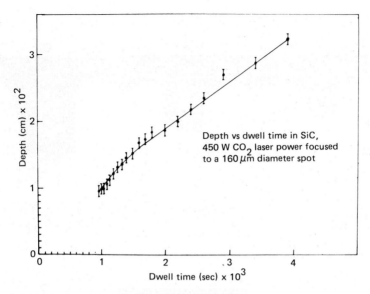

Fig. 6.17. Depth of groove versus dwell time for laser machining of silicon carbide. 450 W (CW) CO_2 laser, beam focused to 160 μm diameter spot; ambient atmosphere.

to supplement the burning process. The asymmetry seen in this figure can become extreme as seen in fig. 6.19 where for a long dwell time the groove is actually curved. This is thought to be caused by the asymmetric spatial distribution and some light guiding into the most deeply cut part of the groove.

When one wishes to obtain faster cutting, the laser-induced burning process can be enhanced by a collinear jet of reactive gas. Fig. 6.20 shows the measured cut depth versus dwell time obtained using N_2 and O_2 gas jet assists to cut Si_3N_4. The depth of cut is deeper than without either gas jet assist but the form of the relationship between depth and dwell time is the same as shown in fig. 6.17.

A major concern in evaluating the potential of shaping with a laser is the quality of the resulting surface. Fig. 6.21 shows the topography of laser machined grooves in SiC. Although no cracks were observed in the grooves of this material or the others examined, deposits suggestive of a condensate or oxide were observed. The mechanical properties of specimens with laser machined surfaces have not yet been investigated.

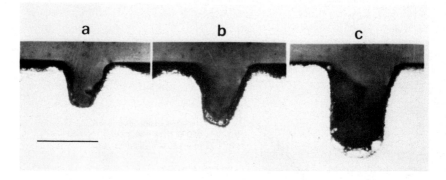

Fig. 6.18. Cross section of laser machined grooves in silicon nitride versus dwell time: (a) 2.9×10^{-3} s, (b) 4.8×10^{-3} s, (c) 14.6×10^{-3} s. 900 W (CW) CO_2 laser, O_2 coaxial gas jet. Marker represents 0.5 mm.

Fig. 6.19. Curved groove in silicon nitride when using high power and long dwell time. 1300 W (CW) CO_2 laser, N_2 gas jet with large orifice; scan speed: 2.7 cm/s. Marker represents 0.5 mm.

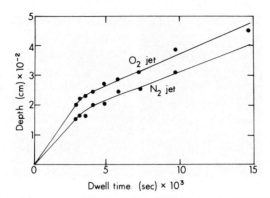

Fig. 6.20. Depth versus dwell time data for silicon nitride. 900 W (CW) CO_2 laser, oxygen and nitrogen gas jet assists.

Fig. 6.22 shows the cross-sections of grooves 2 and 18 at a higher magnification than in fig. 6.16. It appears that their shape is approximately gaussian. The mechanism of material removal is believed to be oxidation resulting in gaseous products.

3.2. Milling results

Fig. 6.23 shows three laser machined milled slots in hot pressed Si_3N_4. These results show that increasing the overlap of a pass with the previous

Fig. 6.21. Surface of a laser machined "O" ring groove in silicon carbide. Illumination incident obliquely from the right. Marker represents 0.4 mm.

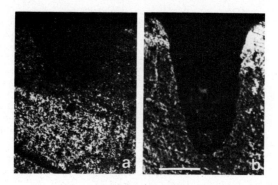

Fig. 6.22. Cross section of two of the grooves in fig. 6.16, (a) ring #2, (b) ring #18, showing the nearly gaussian shape and the independence of the base width of dwell time. Marker represents 0.1 mm.

pass (decreasing the feed of the workpiece) does not necessarily result in a smoother surface. Fig. 6.24 is a series of profilimeter traces of the cross section of the laser machined surfaces for increasing overlap. The profilimeter has scanned perpendicular to the milled grooves (i.e. parallel to the workpiece motion). A ground surface profile is included in fig. 6.24 to facilitate quick evaluation of the laser machined surface. In fig. 6.25 similar data is shown but for a higher laser power. It is clear in this case that the surface roughness increases for too small a feed. Photos of the roughened surfaces are shown in fig. 6.26. The regular pattern may be the result of incident laser light interacting with material that remains hot following prior

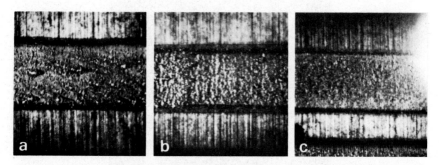

Fig. 6.23. Top views of laser milled surfaces in silicon nitride. 900 W (CW) CO_2 laser; scanner frequency: 65 Hz, scanner amplitude: 0.7 cm unmasked. Translation speed (cm/s) and depth of groove (mm): (a) 0.15 and 0.42, (b) 0.25 and 0.29, (c) 0.63 and 0.19.

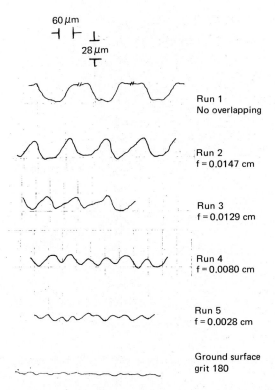

Fig. 6.24. Profilimeter recordings of silicon nitride surfaces after laser machining with different degrees of overlap of successive passes. 340 W (CW) laser, model 300; scan velocity: 37.5 cm/s.

passes. This model requires more careful study and is now being evaluated both experimentally and theoretically.

3.3. Analysis

The shape of the laser machined grooves can be described employing the normal curve

$$D(y) = D_0 \exp(-y^2/a^2). \qquad (6.22)$$

Tracings of grooves 2 and 18 along with normal curves corresponding to the parameters given in table 6.4 are shown in fig. 6.27. The total area under the

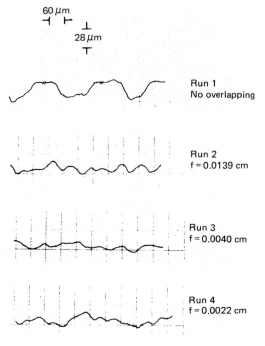

Fig. 6.25. Same as fig. 6.24, except 420 W (CW) laser, model 1003.

normal curve is

$$A = D_0 a \pi^{1/2}. \tag{6.23}$$

The material removal rate is

$$Z = Av \tag{6.24}$$

and the specific cutting energy neglecting power expended by the lathe is

$$\rho = P/Z, \tag{6.25}$$

where P is the incident beam power. Values for A, Z and ρ for grooves 2 and 18 are also given in table 6.4. It is interesting that the cross section of groove 2 is less than that of groove 18 but the material removal rate of groove 2 is the greatest. Since the incident beam power was the same for both grooves, the specific cutting energy the groove 2 must be the least. Thus, the faster the beam scans the surface, the more efficient becomes the laser machining process. Although the origin of this effect has not been

Fig. 6.26. Top view of laser milled surfaces of silicon nitride, showing evidence of interactions between successive passes when the overlap is too great. 630 W (CW), scan velocity: 91 cm/s. (a) Run 1, groove overlap 0.00154 cm, (b) run 2, groove overlap 0.00215 cm.

determined, it may be due to lower conductive losses or possibly better coupling of the laser beam to the workpiece at high speed. The latter may be caused by excessive ejecta at low speeds blocking the incoming light. This ejecta may even become ionized and then form an opaque plasma. Various models exist in which such a plasma reradiates energy to the surface of a metal and thus increase the coupling. As yet there is insufficient evidence to determine the role of such a process in laser machining of ceramics.

Actual turning or facing involves the partial overlapping of laser machined grooves as shown in fig. 6.28. In the following analysis, we shall assume that the shape of the groove cut by the laser in turning is the same

Table 6.4
Parameters describing laser machined grooves

Groove	$D_0(\mu m)$	$a(\mu m)$	$A(10^{-8} m^2)$	$Z(10^{-9} m^3 s^{-1})$	$\rho(GJ m^{-3})$
2	125	100	2.22	3.42	131
18	300	75	3.99	2.19	204

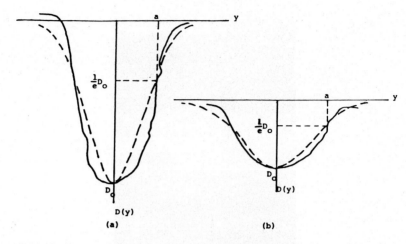

Fig. 6.27. Tracings of the groove shapes in fig. 6.22 and fitted gaussian approximations.

as in fig. 6.16 even though in turning it is cut into the edge of a step. The surface roughness (R), which is equal to the sum of the absolute values of all the areas above and below the mean line divided by the sampling length (Boothroyd 1975) is given by the equation

$$R = 4\overline{abc}/f. \tag{6.26}$$

The effective material removal rate, which is the material removal rate corrected for the overlapping of grooves is given by

$$Z' = vD_0a\pi^{1/2} - v\left(\overline{geh}\right). \tag{6.27}$$

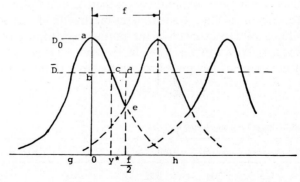

Fig. 6.28. Model of overlapping grooves for evaluating surface roughness and rate of material removal.

The areas $\overline{abc}$, $\overline{cde}$ and $\overline{geh}$ needed to evaluate the surface roughness and the effective material removal rate can be evaluated with the help of tables giving the ordinates of the normal curve and the areas under the normal curve (Kennedy and Neville 1976). Accordingly, the depth of the laser machined groove is given by

$$D(y) = D_0(2\pi)^{1/2}f(z),\tag{6.28}$$

and the area of the laser machined groove is given by

$$\int_0^y D(y)\,\mathrm{d}y = D_0 a\pi^{1/2}F(z)\tag{6.29}$$

where

$$z = y2^{1/2}/a,\tag{6.30a}$$

$$f(z) = \frac{1}{(2\pi)^{1/2}}\exp(-z^2/2).\tag{6.30b}$$

$$F(z) = \int_0^z f(z)\,\mathrm{d}z.\tag{6.30c}$$

The surface roughness can be determined from eq. (6.26) and is given by

$$R = \frac{4}{f}\left[D_0 a\pi^{1/2}F(y^*2^{1/2}/a) - \overline{D}y^*\right],\tag{6.31}$$

where y^* is the abscissa value when $D(y)$ equals the mean depth, see fig. 6.28. The mean depth can be found by equating $\overline{abc}$ [the quantity in brackets in eq. (6.31)] to $\overline{cde}$, and solving for $\overline{D}$, where

$$\overline{cde} = \overline{D}\left(\frac{f}{2} - y^*\right) - \left[D_0 a\pi^{1/2}F(f2^{1/2}/2a) - D_0 a\pi^{1/2}F(y^*2^{1/2}/a)\right].\tag{6.32}$$

Following this procedure, we obtain

$$\overline{D} = \frac{2D_0 a\pi^{1/2}}{f}F(f2^{1/2}/2a).\tag{6.33}$$

The parameter

$$y^* = \frac{a}{2^{1/2}}f^{-1}\left(\frac{\overline{D}}{D_0(2\pi)^{1/2}}\right).\tag{6.34}$$

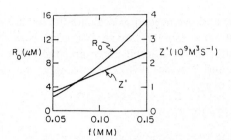

Fig. 6.29. Theoretical plot of roughness and effective material removal rate as functions of feed.

The effective material removal rate is obtained from eq. (6.27) and is given by

$$Z' = 2vD_0 a\pi^{1/2} F\left(f 2^{1/2}/2a \right). \tag{6.35}$$

Eqs. (6.31) and (6.35) are employed in fig. 6.29 to make a plot of surface roughness and effective material removal rate for a groove cut with incident power density of 22.3 GW m^{-3}, beam diameter ($1/e^2$) of 160 μm and dwell time of 1.05×10^{-3} s^{-1} (see groove 2, table 6.4, fig. 6.22). The feeds corresponding to ISO roughness grades N8, N9 and N10 are indicated on table 6.5. A feed of 0.05 mm rev^{-1} is normally available in engine lathes. As a rough guide for comparison purposes, turning operations in metals normally give surface roughness greater than 1 μm (Boothroyd 1975).

Consider turning a large radius piece to the shape shown in fig. 6.30 using the focused laser beam as the cutting tool. The piece of material removed by

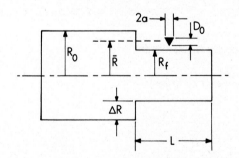

Fig. 6.30. Geometry of workpiece for laser shaping.

the laser beam at any time is indicated by the black triangle and has the shape shown in figs. 6.22 and 27. Its width is $2a$ and its depth is D_0. If this is fed parallel to the axis of rotation by an amount f per turn there will be

$$N_t = L/f \tag{6.36}$$

turns per pass. Since each pass cuts to a depth D_0

$$N_p = \Delta R / D_0 \tag{6.37}$$

passes are required in order to achieve the desired shape. On the average, each turn takes $2\pi \bar{R}/v$ seconds where v is the speed with which the surface rotates under the beam. Thus the time required for cutting is

$$t_c = \frac{\Delta R}{D_0} \frac{L}{f} \frac{2\pi \bar{R}}{v}. \tag{6.38}$$

If after each pass one takes a time T_R to reposition laser beam then the total time taken in this part of the process is

$$t_R = \frac{\Delta R}{D_0} T_R. \tag{6.39}$$

The total time for turning this piece is then

$$T = t_r + t_c = \frac{\Delta R}{D_0} T_R + \frac{\Delta R}{D_0} \frac{L}{f} \frac{2\pi \bar{R}}{v}. \tag{6.40}$$

In fig. 6.17 data is shown for laser cutting SiC which is fitted very well by the following relationships:

$$D_0 = m/v \quad \text{for } v > v_c,$$

$$D_0 = \frac{m'}{v} + b \quad \text{for } v < v_c, \tag{6.41}$$

where $v_c = (m - m')/b$. Inserting this experimental data in the expressions for T gives

$$T = \frac{\Delta R}{m} T_r v + \frac{\Delta R}{m} \frac{L}{f} 2\pi \bar{R} \quad \text{for } v > v_c \tag{6.42}$$

and

$$T = \frac{\Delta R}{m' + vb} T_R v + \frac{\Delta R}{m' + vb} \frac{L}{f} 2\pi \bar{R} \quad \text{for } v < v_c. \tag{6.43}$$

It is clear that the time spent repositioning the laser beam makes a major contribution to the machining time. However, there is no need to reposition the beam because the focused laser beam is a cutting tool that can cut equally well in both directions. Instead of repositioning, consider cutting both ways with some care to ramp the power down at the end of each pass to avoid cutting too deeply while the beam is turned. Thus the machining time is

$$T = \frac{\Delta R}{m} \frac{L}{f} 2\pi \bar{R} \quad \text{for } v > v_c \tag{6.44}$$

and

$$T = \frac{\Delta R}{m' + vb} \frac{L}{f} 2\pi R \quad \text{for } v < v_c. \tag{6.45}$$

The first of these equations shows that for $v > v_c$ the machining time is a constant determined by the material machined (this enters through m), the dimensions of the piece to be machined (ΔR, L and R) and the feed which one selects in order to achieve a desired surface roughness.

Since $m/v > (m'/v) + b$ or $m > (m' + vb)$, the time required when machining at speeds where eq. (6.45) holds is always greater than when machining at higher speeds. This means that for laser machining of ceramics, one can minimize the machining time by taking shallow cuts at high speeds.

Ignoring surface finish requirements, the maximum feed that is permissible is

$$f = 2a.$$

Thus the minimum machining time is

$$T = \frac{\Delta R}{m} \frac{L}{2a} 2\pi \bar{R}.$$

To achieve a desired finish smoothness, the last pass or two could be done with a smaller feed.

The preceding results can be used to calculate the machining time required to obtain desired surface roughness. The data in fig. 6.17 for laser

Table 6.5
Laser machining times

Roughness	f(mm)	T(s)
N8	0.060	105
N9	0.083	76
N10	0.130	48

machining SiC in room air shows that $v > v_c$ if the dwell time is less than ~ 1.3 ms and that $m \approx 0.1$ cm/s. Under these circumstances a piece with $R_0 = 6$ mm, $R_F = 4$ mm, and $L = 10$ mm can be laser machined in the times indicated in table 6.5.

3.4. Applications

Fig. 6.31 contains two macrophotographs of a $\frac{1}{4}'' \times 20''$ thread turned on Si_3N_4 by laser machining. In fig. 6.31 the rectangular rod from which the threaded section was turned can be seen. Fig. 6.31a also shows a region in which we demonstrate the ability to produce concave surfaces.

The negatively curved surface was made by focusing the laser beam so that it struck the center of the rectangular surface at normal incidence. Thus the corners were out of the focal region and irradiated at near grazing incidence where the light-to-material coupling was less. As a result more material was removed near the center than near the corners and a concave surface was established. A convex surface (i.e. the round section which was threaded in fig. 6.31) was made by adjusting the focus and angle of

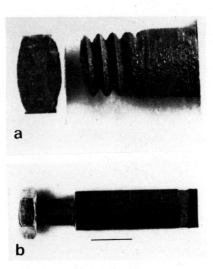

Fig. 6.31. Microphotographs of laser machined Si_3N_4. The rectangular rod was laser turned to a $\frac{1}{4}''$ (6.25 mm) round and then threaded with 20 threads per inch. The groove at the other end demonstrates the ability of laser machining to make concave curved surfaces. Marker represents 1 cm.

incidence so that the corners were near the focus and the light near normal incidence at the corners. This makes the central part of the surface out of the focal region and illuminated at near grazing incidence. Thus more material could be removed near the corners than near the center when the light was so focused.

When laser machining was used to thread Si_3N_4 the focused beam was at normal incidence on the cylindrical surface. However, to make a proper thread the position of the focus with respect to the previous spiral groove must be adjusted. This is shown in fig. 6.32 where the sequence of focal positions are sketched.

Fig. 6.33 shows a laser machined piece of SiAlON which originally had the square cross section remaining in the middle of the piece. The maximum peak to valley nonuniformity of the smooth surface is 7.5 μm which confirms the potential of laser machining SiAlON to desired surface quality

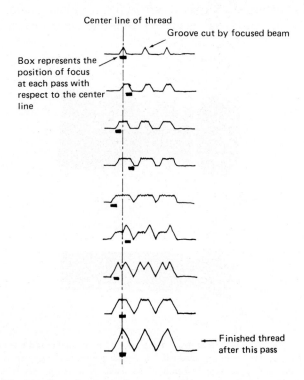

Fig. 6.32. Procedure for laser machining screw threads in ceramics.

Fig. 6.33. Laser machined $\frac{1}{2}''$ (12.5mm) by 13 screw thread in SIAlON. Marker represents 1 cm.

Fig. 6.34. Laser machined "O" ring grooves in alumina. Marker represents 1 cm.

specifications. Fig. 6.33 also shows the $\frac{1}{2}'' \times 13$ screw thread which we laser machined on the other end of the laser turned SiAlON rod.

Fig. 6.34 shows O-ring grooves cut in alumina using the laser machining process. The black material is thought to be the binder used in preparing AV 30.

4. Summary

The concept feasibility of LAM has been demonstrated. The cost of LAM in relation to its benefits has not been established at this time. Important factors such as the fatigue resistance of laser machined components have not been investigated.

Laser machining of ceramics, a form of controlled burning of materials, has also been demonstrated. The strength of laser machined components, however, has not been established. When the capability of this technique is properly applied, fine machining of extremely hard materials may be possible.

References

Boothroyd G., 1975, Fundamentals of Metal Machining and Machine Tools (McGraw-Hill, New York) p. 134.

Esquivel O., Mazumder J., Bass M. and Copley S.M., 1980, Shape and surface relief of continuous laser-melted trails in Udimet 700, in: Rapid Solidification Processing, Principles and Technologies, II, eds., Mehrabian R., Kear B.H. and Cohen M. (Claitor's Publishing Div., Baton Rouge, LA), p. 180.

Gorsler F.W., 1980, Current machining limits for turning aircraft engine materials, in: Annual Technical Report – Advanced Machining Research Program, ed., Flom D.G., DARPA Contract No. F33615-79-C-5119, Ch. 14.

Jones M.G., 1980, Pulse laser experiments, in: Annual Technical Report – Advanced Machining Research Program, ed., D.G. Flom, DARPA Contract No. F33615-79-C-5119, Ch. 19.

Kennedy J.B. and Neville A.M., 1976, Basic Statistical Methods for Engineers and Scientists (Dun-Donnelley Publishing Co., New York).

Plankenhorn D.J., Hill V.L. and Rajagapal S., 1980, Design, construction and operation of the experimental facility for laser-assisted machining, in: Annual Technical Report – Advanced Machining Research Program, ed., Flom D.G., DARPA Contract No. F 33615-79-C-5119, Chp. 17.

Rajagapal S., Hill, V.L. and Plankenhorn D.J., 1980, Laser-assisted machining of Inconel 718, in: Annual Technical Report – Advanced Machining Research Program, ed., Flom D.G., DARPA Contract No. F 33615-79-C-5119, Ch. 18.

Rosenthal D., 1946, Trans. ASME **68** 849.

Tipnis V.A., Mantel S.J. and Ravignani G.L., 1980, Economic modeling, in: Annual Technical Report – Advanced Machining Research Program, ed., Flom D.G., DARPA Contract No. F 33615-79-C-5119, Ch. 20.

Weiner J.H., 1955, Trans. ASME **77** 1331.

LASER PROCESSING OF SEMICONDUCTORS

W.L. BROWN

Bell Laboratories
Murray Hill, New Jersey 07974, USA

Laser Materials Processing, edited by M. Bass
© *North-Holland Publishing Company, 1983*

Contents

1. Introduction

1.1. Current uses and history before 1974

Lasers currently appear in only quite modest ways in semi-conductor processing facilities. Pulsed lasers serve as scribing and lettering devices on semiconductor wafers. In this case, relatively low power, fast repetition rate pulsed lasers focused to small spots locally damage and erode the semiconductor material in patterns determined by a moving stage. The patterns can provide lines for easy cleavage of semiconductor chips from a large wafer and they can provide alphanumeric symbols for chip identification. Pulsed lasers also serve to trim resistors and to break optional conducting paths in logic wiring.

CW lasers are important in interferometers that are essential parts of systems controlling the precise motion of tables for electron beam lithography. At an earlier time a scanned laser beam was a writing tool itself in such lithographic mask making, but the increased demands of spatial resolution and writing speed have made electron beams the current choice for that function.

In none of these applications is the laser used to alter a semiconductor device directly or to process semiconducting material for device fabrication.

In the late 1960s experiments were carried out by Fairfield and Schwuttke (1968) and Harper and Cohen (1970) using pulsed lasers for forming alloy diodes, but these efforts did not reach a level of application viability in their time. The same damaging regime that currently provides scribing has been suggested as a possibility for impurity gettering in semiconductor wafers (Pearce and Zaleckas 1979). In this case a region (or even a whole slice) might be damaged in an array of overlapping spots, the defects so created being sinks for unwanted diffusing impurity species. This approach is in competition with a number of other gettering techniques and has not yet been selected as the process of choice.

1.2. New prospects

Starting in 1974 a group in Kazan in the USSR began to experiment with "laser annealing." They were apparently originally motivated by the desire to find a new material for recording holograms (Shtyrkov et al. 1975a). This interest shifted to finding a new way to remove the damage introduced into semiconductors by the process of ion implantation and to incorporate the ion implanted impurities on electrically active sites in the material (Shtyrkov et al. 1975b). Annealing for these two purposes is conventionally being carried out for silicon in furnaces at ~ 1000°C for times of the order of half an hour. Lasers offer the possibility of effecting the annealing in much shorter times and without subjecting the whole thickness of the wafer to high temperatures at all. The potential benefits of such annealing include reduction in wafer warpage and distortion, reduction in the indiffusion of unwanted impurities and reduction in chemical decomposition of compound semiconductors. The focusing, reflection and absorption characteristics of laser light also afford the possibility of locally annealing selected regions without heat treating other regions at all.

The Kazan group was soon followed by other eastern European groups (Kachurin et al. 1975, Klimenko et al. 1975, Kutukova and Strel'Tsov 1976, Geiler et al. 1977, Krynicki et al. 1977). In 1976 the work at Novosibirsk was discussed at the 5th International Ion Implantation in Semiconductors Conference in Boulder, Colorado (Kachurin et al. 1977). A group in Italy (Foti et al. 1977, Vitali et al. 1977) began experiments shortly afterward and in 1977 work began at Stanford University (Gat and Gibbons 1978; Gat et al. 1978a), Oak Ridge National Laboratory (Young et al. 1977, 1978) and Bell Laboratories (Brown et al. 1978, Celler et al. 1978, Golovchenko and Venkatesen 1978). A large number of laboratories around the world have subsequently become involved. Reports of this work have been presented at symposia at several conferences whose proceedings contain a comprehensive account of this rapidly developing field: the 1978 Materials Research Society Meeting in Boston published in "Laser–Solid Interactions and Laser Processing – 1978" edited by S.D. Ferris, H.J. Leamy and J.M. Poate, AIP Conf. Proc. No. 50, AIP, New York (1978); the 1979 Electrochemical Society Meeting in Los Angeles published in "Laser and Electron Beam Processing of Electronic Materials," edited by C.L. Anderson, G.K. Celler and G.A. Rozgonyi, ECS Proc. Vol. 80–1, The Electrochemical Society, Inc., Princeton, 1980; the 1979 Materials Research Society Meeting in Cambridge, published in "Laser and Electron Beam Processing of Materials," edited by C.W. White and P.S. Peercy, Academic Press, New York, 1980 and the 1980 Materials Research Society Meeting in Boston published in "Laser and Electron Beam Solid Interactions and Materials Processing" edited by T.W. Sigmon, L.D. Hess and J.F. Gibbons, North-Holland, Amsterdam, 1981.

The subject matter of the experimental efforts in the West started, as the work had evolved in Kazan, with laser annealing of ion implanted silicon to remove the damage of electrically activated dopant impurities introduced by ion implantation. These interests have spread to many cases where ion implantation damage is not involved at all, for example to the removal of dislocation networks introduced by high temperature diffusion and to reduction in the misfit defects in epitaxial silicon on sapphire. More will be said about these in section 6. In addition, recent efforts have concentrated on semiconductor processing of quite different kinds, having nothing to do with defects in originally crystalline material at all. Among these are the epitaxial growth of evaporated silicon deposited on a crystalline silicon substrate, the growth of large grain polysilicon from small grain poly formed by chemical vapor deposition processes and the formation of metallic silicides from metal overlayers on silicon. These will be discussed in sections 5 and 6. One of the most intriguing and important areas of recent work is in the growth of crystalline semiconductors on noncrystalline substrates. It will be discussed in section 5. It offers the possibility of inexpensive crystalline material preparation, avoiding the steps of pulling large crystals from a molten pool and then cutting and polishing them into thin wafers for device fabrication. It also offers the unique and intriguing possibility of forming multiple thin layers of crystalline silicon interspersed with insulating layers for possible three-dimensional integrated circuit development. These last efforts have often also been referred to under the title "laser annealing" but here the title is applied in the generic sense to describe the broadening field of laser processing of semiconductors. This broader view is the subject of this chapter.

A quite different area of laser application to integrated circuit processing involves laser photochemistry (Deutsch and Osgood 1979, Ehrlich et al. 1980). This can, for example, provide localized deposition of a metal on a semiconductor surface by selective dissociation of a metal–organic compound at an appropriately chosen ultraviolet wavelength. There are many intriguing scientific as well as technological aspects of this work, but they will not be discussed further in this chapter.

At present there are not known to be any lasers used for semiconductor material modification on actual device fabrication lines. The possibilities represented by the recent laser processing research which is discussed in this chapter have not yet made their way into production.

2. Laser interactions with semiconductors

2.1. Absorption

All semiconductor laser processing depends upon the way in which laser energy is introduced into a material and this invariably starts with excitation of electrons, for it is to electrons that the photons of a laser can couple. The strength of the coupling is heavily dependent on the laser wavelength and the material in question. The absorption coefficient for silicon is illustrated in fig. 7.1. The wavelength of some of the lasers used presently in the semiconductor processing work are also indicated. Notice that the impurity and structural details of the material are important – amorphous

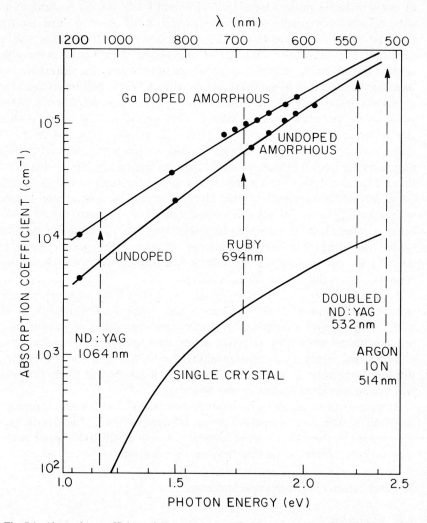

Fig. 7.1. Absorption coefficient of Si as a function of photon energy and wavelength and the wavelength of lasers commonly used in semiconductor laser processing studies.

silicon is quite different than crystalline silicon (Brodsky et al. 1970); heavily impurity doped silicon is quite different from pure silicon (Bean et al. 1979). These differences provide at the same time a complexity and a benefit: a complexity because the laser power densities at which processes can be carried out have to take this into account; a benefit because the selective absorption allows strong coupling to occur in particular regions. For example, if regions of amorphous silicon exist in a sea of crystalline material, Nd:YAG 1.06 μ light will couple to them and much less strongly to the sea. The amorphous regions can thus be selectively processed.

The excitation mechanisms contributing to fig. 7.1 are of two major types: hole–electron pair generation and free carrier absorption. These are illustrated schematically in fig. 7.2 by the processes marked a and b. In crystalline silicon a photon of $\gtrsim 1.1$ eV is capable of breaking an electronic bond, promoting an electron from the valence to the conduction band of

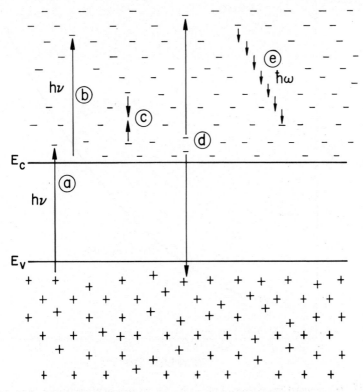

Fig. 7.2. Principal processes in electronic excitation and decay in a semiconductor.

silicon and thus providing a free electron and a corresponding free hole. If the silicon is heavily doped (Bean et al. 1979) or if it is heavily damaged, as in amorphous material (Brodsky et al. 1970), the electronic band structure of the silicon is altered and hole–electron pair generation takes place for somewhat lower energy photons. The band gap is also temperature dependent, narrowing with higher temperatures. Particularly for 1.06 μ Nd:YAG

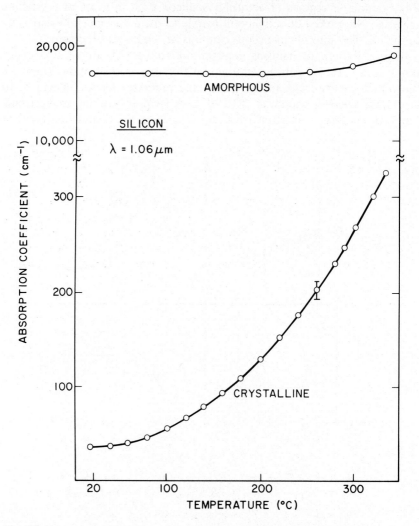

Fig. 7.3. Temperature dependence of the absorption coefficient of crystalline and amorphous Si.

light, whose photons are marginally too low in energy to excite hole–electron pairs in pure silicon at room temperature, the band narrowing has a major effect on the absorption coefficient. This variation is shown in fig. 7.3 (von Allmen et al. 1979).

In the second excitation process (b) of fig. 7.2, photons are absorbed by free electrons (or free holes) and are promoted to higher energy states within the conduction (or valence) bands. The absorption coefficient due to this process depends directly on the density of free carriers present for excitation. More heavily doped material or material at higher temperatures in which the equilibrium hole–electron concentration is higher is more strongly absorbing.

2.2. Reflection

Only a fraction of the light that strikes a surface is available for absorption because of reflection. The reflection coefficient is strongly dependent on the wavelength of the light as shown in fig. 7.4, though the changes are not by orders of magnitude as in fig. 7.1 (Eden 1967). The reflection is also dependent on details of the material. It is governed by the complete complex index of refraction whose imaginary component gives the absorption coefficient. For visible wavelengths (at 633 nm for example, for He/Ne laser light) the reflection coefficient of crystalline silicon is 0.35 and of amorphous silicon is 0.40.

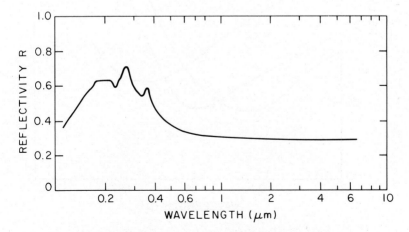

Fig. 7.4. Reflectivity of Si as a function of wavelength.

For wavelengths in the infrared, free carrier excitation processes strongly influence the reflectivity as shown in fig. 7.5 for heavily doped silicon of different doping concentrations (Miyao et al. 1981). In addition to the single particle excitation of free carriers as in fig. 7.2, process b, the free carriers can undergo collective excitations. These plasma excitations are sloshing modes of the electrons with respect to the positively charged donors (as in fig. 7.5) or of electrons with respect to holes when both are present at high concentrations. The plasma excitations are characterized by a plasma

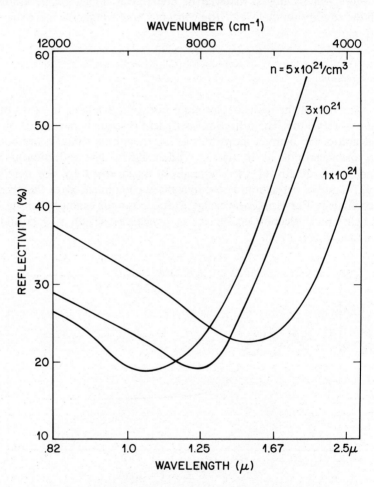

Fig. 7.5. Reflectivity of heavily doped Si in the infrared.

frequency $\omega_p \propto (ne/m)^{1/2}$ where n is the electron density and m the effective mass of the vibrating particles. If the plasma frequency exceeds the optical frequency, the plasma can respond to an incoming electromagnetic wave and the reflection coefficient is high. The shift of the reflectivity curves of fig. 7.5 with increasing doping density are due to an increase in the plasma frequency. Notice that for 5×10^{21} electrons/cm^3 the high reflectivity is still in the infrared. In metals, where the carrier density is $\lesssim 5 \times 10^{22}$/cm^3 and the plasma frequency is correspondingly higher, the reflectivity may be high in the visible.

Phase changes in materials alter their electronic properties and hence their reflectivities. When silicon melts it becomes a metal and its optical reflectivity at 633 nm is ~ 0.65 rather than 0.4 or 0.35 for amorphous or crystalline silicon. The penetration properties of the light into the material are also dramatically changed and the concept of a skin depth becomes important. Light is attenuated in molten silicon with a characteristic depth $\sim 0.01 \mu$ in contrast to the absorption depths (the reciprocal of the ordinate in fig. 7.1) of microns or more for crystalline silicon and tenths of a micron for amorphous material.

The time dependence of the optical interaction of a high power laser beam with a semiconductor is clearly quite complex. The laser may generate free carriers which in turn alter the absorption and reflection coefficients (Lietoila and Gibbons 1979a). The laser also heats the material (more will be said about this in sections 2.3 and 3.1). This narrows the band gap and also promotes phase changes from amorphous to crystalline (to be discussed in section 5) and from solid to liquid (to be discussed in sections 5 and 6) with corresponding changes in the optical parameters.

2.3. Energy transfer to heat

All of the processes discussed in sections 2.1 and 2.2 involve electronic excitation in the material. None of them directly sets the atoms of the material in motion, activates lattice vibrations, makes the material hot. The energy of the laser increases the density of holes and electrons and heats them as well, but at this point the energy still is in the electronic system, not in atomic motion. Within the electronic system there are rapid exchanges in energy illustrated by processes c and d of fig. 7.2. The characteristic times for these processes are shown in fig. 7.6 (Brown 1980, Yoffa 1980b). Process c is electron-electron scattering, one way in which the electrons (and holes) share their energy among themselves. The rate of this process and the rate of electron–plasmon collisions (c' in fig. 7.6) increase in proportion to the electron density. Process d is an Auger event. An electron and hole recombine to give their energy to a third carrier – an electron, as illustrated in fig. 7.2. Since the process involves three carriers it is much faster at high carrier concentration. The characteristic time (d in fig. 7.6) is proportional to $1/n^2$. The inverse of this process is also going on, namely, impact ionization. A hot electron gives up energy to create a hole–electron pair. But all of these processes are still electronic, keeping the energy that is input from the laser circulating among the electrons and holes, keeping them hot and in fast contact with one another. The vibrational state of the lattice of atoms may still be cold. It is only as electrons couple their energy to the lattice by generation of phonons (elementary lattice vibrations) that the material gets hot. Because phonons have quite low energy, about 20 phonons must be generated to transfer 1 eV of energy from the electronic to the atomic system (process e in fig. 7.6). Yoffa (1980b) evaluated the time for

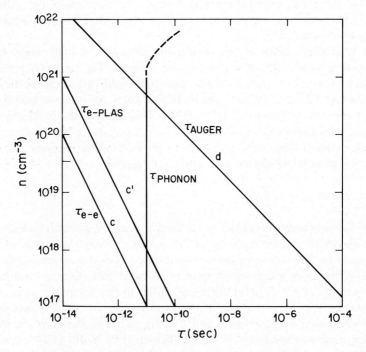

Fig. 7.6. Characteristic times for electronic decay processes in Si as a function of electron density. The tick mark at $n = 3 \times 10^{19}/\text{cm}^3$ corresponds to the equilibrium electron concentration at the crystalline melting point.

this transfer as $\sim 10^{-11}$ s for silicon at carrier concentrations up to $\sim 10^{21}/cm^3$. Above that concentration she estimates that screening effects will increase the time as shown by the dashed curve at the top of the τ_{phonon} line of fig. 7.6. Even at concentrations above $10^{21}/cm^3$ this process is fast. On a time scale of nanoseconds (and perhaps 10^{-10} or 10^{-11} s) the electronic system is in equilibrium with the lattice. Energy flows rapidly from the laser beam to the electrons to the vibrating lattice. Laser processing of semiconductors can thus be viewed as dominantly thermal and that will be the basis for most of the discussion in the remainder of this chapter.

There is a large body of experimental data that is explicable by a thermal model of laser processing as will be discussed in sections to follow. A definitive experimental measurement of the energy transfer time from the electronic to the atomic system has not yet been made at the high energy densities appropriate to laser processing to verify the expectations cited above. Several experiments have given preliminary results. Lo and Compaan (1980) are carrying out an extremely difficult time resolved Raman scattering experiment designed to measure the lattice temperature versus time by comparing the Stokes and anti-Stokes line intensities. They report (Compaan and Lo 1980, 1981) the result that at a time of ~ 30 ns after a 7 ns pulse the lattice temperature is $< 600°C$, implying a long energy transfer time. Time resolved optical transmission and reflectivity measurements are being extended to the picosecond regime by Murakami, Gamo and their associates (Murakami et al. 1980, Gamo et al. 1981). Their preliminary results suggest the energy transfer time is $> 10^{-11}$ s but $< 10^{-8}$ s. This question surely needs to be answered for the fundamental science involved. If the time turns out to be long, $\gtrsim 10^{-8}$ s, a great many puzzling material questions will be raised.

3. Models of the process

3.1. Thermal model

The thermal model of "laser annealing" is based on the rapid transfer of energy from the electronic to the atomic systems as discussed in section 2.3 above. As far as semiconductor processing is concerned, the laser is an intense source of localized heat. There is no question that on the millisecond or submillisecond time scale appropriate to CW lasers the thermal model applies. Such a laser produces a steady state hot spot in the material. The consequences of this hot spot can be in solid phase rearrangement of atoms of a damaged semiconductor or in crystallization at the cool edge of a moving melted spot if the conditions are such as to bring the maximum temperature of the material above its melting point. These material changes will be discussed in more detail in section 5.

For fast laser pulses in the nanosecond and even in the picosecond regime, energy transfer is generally believed to be fast enough so that material processes that are stimulated are dominantly thermal. For nanosecond pulses to be effective they must bring a material to its molten state; solid phase rearrangements of atoms take place too slowly to be significant in ~ 100 ns or less. A region of the material at the surface is quickly melted. Either crystalline or amorphous solid layers are transformed to a liquid and lose their identity, dislocations cease to exist, precipitates dissolve, and impurity atoms rapidly diffuse in the liquid. As the liquid cools by heat conduction into the interior it refreezes, usually in a crystalline state. A high concentration of dopant impurities can be frozen in and as it turns out the impurities are dominantly located on substitutional lattice sites in the crystal. The whole process of melting and refreezing can be complete in 10s or 100s of nanoseconds. The process is an extreme example of splat cooling so widely studied in metals and is closely parallel to "laserglazing" in metallic systems. More detailed consequences of the process will be discussed in section 6.

In addition to the changes in materials that are induced by a Q-switched pulse and which will be discussed in the context of the thermal model in section 6, one of the most persuasive experiments in support of a simple thermal picture is that carried out by Auston et al. (1978, 1979a, b) who measured time resolved reflectivity. A schematic illustration of the experiment and one result are shown in fig. 7.7. A He–Ne CW laser at 633 nm was used to monitor the reflectivity of the center of a spot illuminated by a Nd:glass laser, in this case at 530 nm. The starting material was a ~ 500 Å layer of amorphous Si formed by arsenic ion implantation into crystalline Si. The reflectivity undergoes dramatic changes from its starting value that corresponds to reflection from amorphous Si. The flat topped region of high reflectivity corresponds to reflection from liquid Si. The small dip at the beginning of the plateau is due to liquid Si at temperatures above the melting point. The reflectivity falls finally, in this case after ~ 600 ns to a value corresponding to reflection from crystalline Si. The slightly higher reflectivities R'_a and R'_c just after R_a and before R_c correspond to hot amorphous and hot crystalline Si. In terms of a thermal model this time dependence is easily explained. The heating pulse melts a layer at the surface. When the melting occurs, the subsequent absorption of the heating pulse is confined to the skin depth (~ 100 Å) and heats the layer above the

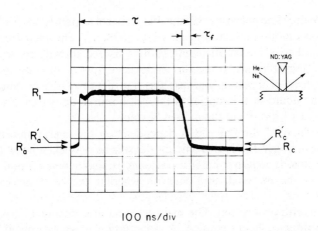

100 ns/div

Fig. 7.7. Time resolved reflectivity from Si during and following a Q-switched Nd:YAG glass laser pulse of 30 ns, 2.75 J/cm² at 532 nm. Reflection measured at 633 mm.

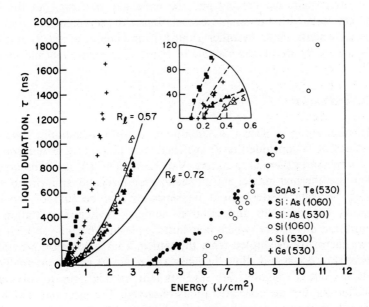

Fig. 7.8. Duration of high reflectivity plateaus (fig. 7.7) as a function of laser energy density for Si, Ge and GaAs.

melting point. The duration of the flat top corresponds to the time the surface stays molten (Auston et al. 1978, 1979a, b). The melt front propagates inward to a maximum depth (not measured directly by this experiment) and then retreats to the surface as heat flow to the interior extracts the latent heat of freezing. Because the skin depth of the probing laser is short in the liquid Si it is not until the last hundred angstroms freezes that the reflectivity drops from the molten Si value.

The duration of the flat top is dependent on the energy density of the laser-pulse – longer for higher energy density as a thicker layer is melted and more time is required for it to refreeze. It also depends upon the laser coupling to the material, and thus on whether the Si is amorphous or crystalline and on the laser wavelength, as discussed in section 2 in connection with absorption. The melt duration also depends, of course, on the semiconductor being studied. A compilation of Auston et al.'s results (1979a, b) is given in fig. 7.8. All of them are consistent with a simple melting picture. In each case there is a threshold below which no high reflectivity state is observed, corresponding to the energy density below which melting does not occur. Accurate calculations of the energy input are difficult because of the rapidly changing interactions of the laser as hole–electron pairs are created and the band gap narrows, but the data agree very well with the simple expectations. Measurements of transient optical reflectivity made by other groups (Murakami et al. 1979) and (Liu and Wang 1979) are also explicable in terms of a simple thermal melting model.

3.2. Plasma model

The Russian group in Kazan (Khaibullin et al. 1978, Khaibullin 1980) has been reluctant to conclude that Q-switched laser processing of semiconductors is dominantly thermal in nature. Van Vechten (1980) has subsequently rejected the thermal model in the nanosecond regime and proposed a so-called "plasma model." In it he supposes that because of the high generation rate of holes and electrons under Q-switched excitation the concentration of carriers (which constitute a plasma) become so high that they are thermally decoupled from the lattice. That is, phonon generation processes slow down and allow the plasma to retain its energy internally for periods of 100s of nanosec. He suggests that the high density plasma is responsible for the material alterations observed. The electrons and holes provide a high density of broken bonds which give the material a great deal of freedom to rearrange, to anneal, to crystallize and for impurities to diffuse. In effect, he describes the material as a fluid, but one that has not undergone a phase transformation to a normal liquid and one that is cool. Only the electronic system is hot.

Van Vechten (1980) attributes the high reflectivity of the experiments of Auston et al. (1978, 1979a,b) to reflection from the plasma, not from liquid Si. As indicated in section 2, fig. 7.5, for the plasma frequency to exceed the He–Ne optical probing frequency requires a plasma density $> 5 \times 10^{21}/cm^3$, in fact $\geq 2 \times 10^{22}/cm^3$. The plasma density will decrease with time, whether by diffusion into the interior (Yoffa 1980a) or by recombination as energy does flow to the lattice. The plasma frequency will correspondingly drop and when it drops below the probe frequency, the reflectivity will fall. For different probe frequencies the fall should occur at different times. Measurements by Nathan et al. (1980) at two different probe frequencies failed to show any difference in the duration of the flat topped pulse. Their results are shown in fig. 7.9. It could be suggested that the plasma density may fall very rapidly as soon as it gets below some critical value where the assumed decoupling from the lattice is again restored. If so, no measurable difference might be expected in the experiment of Nathan et al. (1980).

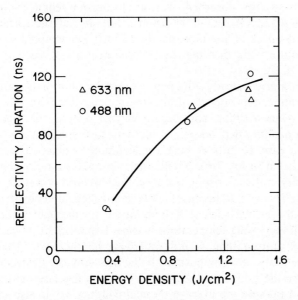

Fig. 7.9. Duration of high reflectivity plateau at two probe wavelengths (633 and 488 nm).

There is no question that holes and electrons are present in the Q-switched laser case and that since they are the route through which energy must reach the lattice they will always be present at somewhat higher density than the equilibrium value corresponding to the lattice temperature. However, theoretical considerations give no support for the idea that the equilibrium time will be $> 10^{-9}$ s, (Yoffa 1980a,b; Dumke 1980) and thus for laser pulses $\gtrsim 10^{-8}$ s, the carrier concentration will never be substantially out of equilibrium with the lattice either during or after the pulse. Note that the equilibrium electron–hole concentration in crystalline Si at the melting point is only $3 \times 10^{19}/\text{cm}^3$. The predictive capability of the plasma model is at this time extremely limited. It remains to be seen what new insights it can give to material processing.

4. Methods of approach

4.1. CW lasers

A CW laser beam with a power of the order of 10 W focused to a circular spot of ~ 40 μ in diameter is a very potent localized source of heat. If the laser light is strongly absorbed (with an absorption length short compared to the spot diameter) heat flows from a hot disc at the surface into the interior. If the back of a semiconductor wafer is well coupled to a heat sink, the temperature of the disc and the material near it will reach a steady state in times ~ 10 μs.

Calculation of the steady state temperature distribution is complicated by the temperature dependence of the thermal conductivity. However, Lax (1978) has given a simple and elegant way of relating the solution in this case of temperature dependent conductivity to the equivalent temperature independent case. Results of such a calculation (Williams et al. 1978, Hill 1981) are shown in fig. 7.10. Notice that the relevant parameter is not the power density in the laser spot, but P/a, the absorbed power per unit radius of the spot. This is true because of the hemispherical geometry of thermal flow from the heated surface disc. Because the thermal conductivity decreases with increasing temperature it takes less incident power to reach a desired temperature than would otherwise be the case. The heat sink temperature is also quite important. If the heat sink is elevated above room temperature, the added temperature increment the laser must supply is reduced and because the average thermal conductivity is also reduced, the added increment is even easier to obtain.

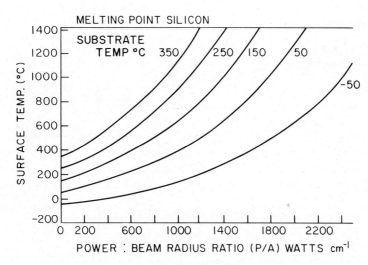

Fig. 7.10. Calculated maximum temperature in Si illuminated with highly absorbed laser light of power P in a spot of radius a with different substrate temperatures.

In CW laser studies of semiconductor processing the laser spot scans over the sample surface, its hot spot with it. At the scanning speeds employed, ~ cm/s, (Gat and Gibbons 1978, Gat et al. 1978a) and with spot sizes of 20–40 μ, the dwell time, $t = 2a/v$, is of the order of milliseconds, much longer than the time to establish the steady state. The spot motion can be produced by moving the beam with a mirror deflection system. This approach was discussed by Gat and Gibbons (1978) and is illustrated in fig. 7.11a. Alternatively, the sample can be moved under the laser beam as in fig. 7.11b. The latter approach is an easier way to scan large areas because of the expense of telecentric lenses required to preserve the laser focus if the laser beam is deflected over long distances. Simple $x - y$ raster scanning has been used in both scanning modes, usually with substantial overlap of the hot region from one scanned line to the next, to more nearly expose all the material to the same heat treatment. Scanning has primarily been done with two CW lasers: Nd:YAG and argon ion lasers. It has been done primarily in air though it clearly is feasible to control the surrounding atmosphere or to work in vacuum.

The time to process a whole 3 in diameter wafer in this way is long,

$$t_{\text{CW}} = \frac{A}{\pi a^2 f} \frac{2a}{v} = \frac{2A}{\pi a f v},$$

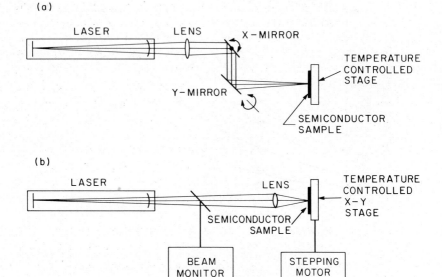

Fig. 7.11. Two methods of CW laser scanning: (a) mirror deflection of the optical beam, (b) motion of the target stage.

where A is the area to be scanned, a is the beam spot radius, v is the scan velocity and $1 - f$ is the fractional scan overlap. For $a = 20$ μ, $v = 3$ cm/s, $f = 0.5$ and $A = 44$ cm^2, $t = 9 \times 10^3$ s. The process can be carried out more expediently using a line focused beam scanned perpendicular to its long dimension. The beam power requirements increase in this geometry, but not in proportion to the area of the line because heat flow is now semicylindrical rather than hemispherical.

4.2. Pulsed lasers

Pulsed lasers of many different wavelengths, pulse lengths, peak powers and repetition rates have been used in studies of laser processing. Pulse lengths as short as 30 ps from mode locked lasers and as long as μs from free running lasers have been explored, but the vast majority of experiments have been done in the Q-switched region between ~ 10 and ~ 100 ns. A few experiments have been done with excimer lasers in the UV and with

Nd:YAG lasers quadrupled at 265 nm and with CO_2 lasers at 10 μ, but the most extensive work has been done either with ruby lasers at 682 nm or with Nd:YAG or Nd:glass lasers at 1.06 μ or frequency doubled at 532 nm. These wavelengths are shown in fig. 7.1 in relation to the absorption properties of Si. Doubled YAG and ruby wavelengths have quite strong absorption even for pure crystalline Si. The fundamental of YAG at 1.06 μ is marginally beyond the absorption edge for crystalline Si and its absorption coefficient is a strong function of temperature, doping density and defect concentration as discussed in section 2.

The phenomena of pulsed laser processing to be discussed in section 6 occur for Q-switched pulses in the $\sim$ 0.2–10 J/cm^2 range. In some cases, pulsed energies of a joule or more have been used to irradiate an area $\sim$ 1 cm^2 in a single pulse. Higher energies yet are certainly available, but at the present exploratory stage, large areas are not very important since existing evaluation techniques can easily be carried out in areas < 0.1 cm^2.

The spatial uniformity of the energy density over the irradiated spot is of great importance. Some of the interesting processes have rather narrow energy windows. In the 1.06 μ case with Si there is positive feedback (von Allmen et al. 1979) in transient heating as a result of the band edge shift with temperature (see section 2.1). If the beam has a hot spot, that hot spot may easily be accentuated to the damage point. In a few cases, spatial filtering has been used to provide mode control and produce a clean gaussian spot or even a spot with a flatter profile. However, an easier solution to the uniformity question that does not involve such a large loss in multimode beam energy is the use of an homogenizer. This idea, due to Cullis and his associates (1979), is illustrated in fig. 7.12a. By forcing the beam to follow the curve and taper of a clear quartz rod by multiple internal reflections, modes and paths are scrambled so that at a particular short distance ($\sim$ mm) from the output end of the rod the energy density is extremely homogeneous and flat topped as shown in fig. 7.12b. The quartz of the homogenizer must be of very good quality and the ends must be kept very clean to avoid damage to the homogenizer itself at the $\sim 10^7$–10^8 W/cm^2 peak powers needed. The efficiency of energy transmission is about 60%.

Rather than using a single high energy laser pulse, another approach is to use a high repetition rate, low energy-per-pulse laser and carry out processing of a desired area with multiple overlapping spots. In this case the beam can be carefully shaped to avoid hot spots and the spot size chosen to bring the available energy per pulse into the appropriate energy density range. The scanning can be accomplished by beam deflection or by target motion

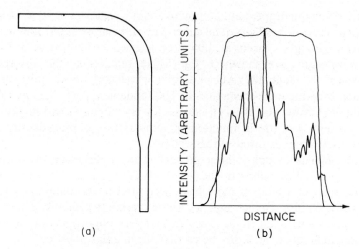

(a)   (b)

Fig. 7.12. Optical homogenizer for producing a uniform laser beam (a) schematic of the quartz rod geometry (b) the measured beam uniformity in a target spot with and without the homogenizer.

as in the case of CW beams discussed in section 4.1. The pulsed processing time, t_p, for a wafer can be much shorter than for the CW case.

$$t_P = \frac{A}{\pi a^2 f v} = \frac{A\epsilon'}{\bar{P}f},$$

where A is the area to be scanned, $a =$ the spot radius, $v =$ the pulse repetition frequency and $f =$ the area overlap factor, $\bar{P}$ the average laser power and ϵ' the energy per unit area needed for processing. With a 1 W average power, $\epsilon' = 1$ J/cm^2 and $f = 0.1$, the time to scan a complete 3 in wafer is $\sim 4 \times 10^2$ s, an order of magnitude less than in the CW case. With more available average power and a larger overlap factor (less overlap) this time can be reduced to ~ 1 min. The fundamental mechanisms of processing in the pulsed laser case (see section 6) require less than a microsecond in contrast to $\sim$ millisecond for annealing mechanisms active in the scanned CW laser case (see section 5). The difference in wafer processing time stems, of course, directly from this difference. So far only Nd:YAG lasers have been used in the scanned pulsed mode since they are commercially available with relatively high average powers even at the 532 nm doubled wavelength (Kaplan et al. 1980).

One attraction of the overlapping spot approach is its potential for processing any pattern of selected areas with dimensions large compared to the spot size. Different regions could even be processed at different energy densities by use of an electro-optic attenuator. It has been found relatively important to have a large amount of overlap between adjacent pulsed spots on a line and between adjacent lines to provide uniformity in processing. Every point of the processed region is thus exposed several times (in fact at $f = 0.1$, ten times) by different parts of the laser spot. The edge of the scanned pattern cannot be made uniform in this way and the processed material has special properties in those regions (section 6.2.1).

4.3. Electron beams

Semiconductor processing has been carried out with electron beams as well as with lasers. The same fundamental processing mechanisms are to be expected since for both types of radiation the energy coupling is to the electrons of the systems (von Allmen 1980) and the same sequence of energy interchanges must go on as discussed in section 2.3. Electron beams are different in two principal and closely related respects. First, the energy deposition in a material by an electron beam has quite a different depth profile than for light because of the penetration characteristics of the electrons which typically have energies of 10s of keV. The depth scale can be increased by increasing the electron energy. Second, the energy deposition is independent of the electronic state of the material. It does not matter whether the electron beam is striking amorphous or crystalline or even molten Si, the electron reflection and energy absorption characteristics are essentially the same. If the electrons were of low energy the details of the electronic structure of the material would matter, but at energies ~ 10 keV such a wide range of excitations are available the result is insensitive to the material state. For the same reason, differences from one material to another are relatively small (von Allmen 1980). This insensitivity can either be an asset or a liability as discussed in section 2.1 in connection with the 1.06 μ laser absorption in Si which is extremely sensitive to material state.

Both pulsed and scanned electron beams have been used. In all cases, of course, processing must be in vacuum. Spire Corporation has pioneered the use of high power nanosecond pulsed electron beams of large area (Minnucci et al. 1980). Beam uniformity depends on subtleties of the multiple field emission cathode and the electric and magnetic fields that exist in the very high current densities of the beam, but Spire has these adequately under control to process 3″ diameter Si solar cells in a single pulse. The scanned electron beams used so far have usually been either electron beam welders (Gibbons et al. 1979a) or modified electron microscopes (Ratnakumar et al. 1979) with their apertures removed to increase the available current. New machines with high power line-focused electron beams are becoming available for scanned beam applications (Knapp and Picraux 1981). As in the laser case, a scanning line offers big advantages in throughput.

Electrons might appear to offer advantages for very fine scale processing since they can be so finely focused. However, the processing time for such serial processing becomes prohibitively long as the size of the process region decreases unless the fundamental processing mechanism can be scaled up in rate correspondingly. In addition, electron beams with high enough power to be effective must be at 10s of keV in energy, where the penetration is the order of a micron. Lateral scattering of an electron beam is comparable to its penetration depth so the region heated and thus processed will be the order of a micron in diameter or more.

Still a different electron beam approach has been pursued by Ahmed and his collaborators at Cambridge University (Shah et al. 1981). They use a fast rastered electron beam as a broad area heater. The rastering is so fast that a steady state is not established under the moving spot and it is only after 1000s of sequential frames that the material reaches a steady state at high temperature in response to the average power input. While this machine may look like the world's most expensive electric toaster, it permits clean, uniform and fast (~ seconds) processing of large areas. Such electron beams are also very energy efficient.

4.4. Ion beams

Even pulsed ion beams are being considered for semiconductor processing. One motivation for this interest is the possibility that dopant atoms might be introduced and annealed in a single pulse or perhaps with multiple pulses of a single ion species, since desired implantation doses usually exceed the fluence needed for an anneal. Ion beams can in principle input energy directly into atomic motion, bypassing the electronic excitation step, if the ions used and their energy are such as to maximize the so-called "nuclear stopping" component of energy loss and minimize the "electronic stopping" component (Gibbons et al. 1975). Relatively high mass ions, bismuth for example, in the 100 keV range satisfy this criterion. Published results (Hodgson et al. 1980) have so far been for ~ 150 keV proton pulses for which electronic stopping is overwhelmingly dominant. It is found, as expected, that the processing results they produce are much the same as for pulsed laser or electron beams. A recent report of a heavy ion experiment

(Baglin et al. 1981) seems to give the same results, adding evidence that the energy transfer from electron to atomic motion is rapid and the processing is thermal.

Focused ion beams have not yet been developed which provide enough localized power to do processing in a scanning mode.

5. Processing with CW beams

5.1. *Solid phase processes*

Annealing of semiconductor device structures to remove defects and to recrystallize amorphous layers is conventionally carried out in furnaces of ~ 1000°C for times the order of half an hour. At these elevated temperatures there is enough atomic motion to permit defects to be healed. It is not practical to shorten the processing time to less than a few minutes because of the thermal time constants involved in bringing whole wafers to a uniform and well-defined temperature. On the other hand, with a focused CW laser beam a localized spot at the surface of a semiconductor can be raised to a steady state temperature in a few microseconds. At an absorbed power per unit radius $P/a = 1.8 \times 10^3$ W/cm the central surface temperature for silicon will be 1000°C if the substrate is at a heat sink temperature of 20°C (Williams et al. 1978). For a spot 20 μ in radius this requires an absorbed power of 3.6 W (or an incident power of ~ 7 W with a reflection coefficient of ~ 0.5). The hot region will extend inward in the steady state heat flow to a depth comparable to a, hence to a small fraction of the typical 250 μ thickness of a Si wafer (Lax 1979). It is possible in this way to stimulate solid phase processes locally. Two examples of this kind are discussed below.

5.1.1. *Laser annealing by solid phase epitaxy*
Mayer and his associates (Csepregi et al. 1976, 1977, 1978; Kennedy et al. 1977) investigated the epitaxial crystallization of amorphous Si layers formed by ion implantation damage in crystalline Si. Their annealing was carried out in a furnace in the temperature range between 450 and 600°C. They found that at a single temperature the amorphous Si regrew from the crystalline substrate outward to the surface at an approximately constant velocity. The velocity is temperature dependent with an activation energy of ~ 2.3 eV. The velocity at a given temperature is also dependent on the type and concentration of impurities present in the material which can either

enhance or retard the atomic rearrangement process required for crystallization.

The earliest CW laser processing experiments turned out to be extensions of these studies at higher temperatures. Fig. 7.13 shows an extrapolation of the furnace data for heavily arsenic doped silicon to temperatures near the melting point (Williams et al. 1978). The extrapolation indicates that regrowth of a 500 Å amorphous layer should occur in times of a few milliseconds at temperatures of ~ 900°C. Based on calculations such as those of fig. 7.10 a temperature of 900°C is produced by an absorbed power per unit radius of ~ 1.7×10^4 W/cm or 0.17 W/μ. It is in just this range of

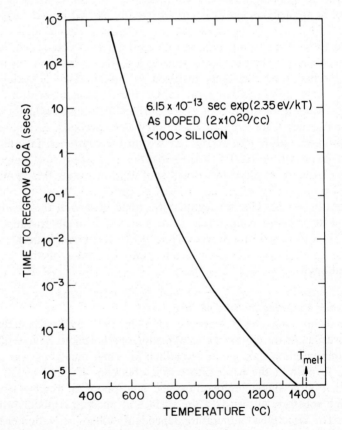

Fig. 7.13. Time for solid phase epitaxial regrowth of 500 Å amorphous Si layers extrapolated from low temperature measurements to the high temperature regime of laser annealing.

powers and dwell times that the experiments of Williams et al. (1978) were carried out. Helium ion channeling and backscattering were used to reveal the crystallization directly, as shown in fig. 7.14. By increasing the scan rate of a CW beam, and thus reducing the hot spot dwell time, partial regrowth was observed; the crystal–amorphous interface had moved only partway to the surface.

Gat et al. (1978a) had previously determined from secondary ion mass spectrometry measurements that in scanned CW annealing there was no measurable change in the depth profile of an implanted impurity as the impurities were electrically activated by the annealing. The channeling measurements confirmed this and identified the process as one of solid phase epitaxy. For the millisecond processing times involved, normal donor and acceptor impurities can be expected to diffuse in the solid phase only a few angstroms. This is in contrast to the situation for half hour furnace anneals at comparable temperatures. Gibbons et al. (1979a) pointed out the simplicity this affords the device designer who can arrange the implanted impurities in a desired depth profile and be confident the profile will be unaffected by laser annealing.

Recent studies by Olsen et al. (1981) have refined these solid phase regrowth measurements beautifully by using time resolved laser interferometry to indicate the thickness of the amorphous layer as it thins during regrowth. They have successfully followed the regrowth velocity to ≥ 1000°C and to regrowth times of 10s of μs. They find the same type of regrowth velocity dependence on impurities and on crystal orientation as found at much lower velocities and temperatures (Csepregi et al. 1976, 1977, 1978, Kennedy et al. 1977).

The topography of amorphous layers regrown by solid state epitaxy is extremely simple. If a single spot is illuminated, a flat and depressed but otherwise featureless disc of crystalline material is produced in a sea of the remaining amorphous layer. The regions are optically distinct because of the difference in the reflection coefficient of amorphous and crystalline material. The disc is depressed because the density of crystalline silicon is greater than that of amorphous material. Such a disc is illustrated in fig. 7.15a. The same featureless surface is found in a single scanned line or in a scanned area if successive lines have sufficient overlap to complete the regrowth. By sharply focusing the laser beam it has been possible to regrow lines < 2 μ in width (Williams 1978). The edges of the regrown regions have sharp transitions to amorphous surrounding material; there is no polycrystalline border. Of course, at the edges of the spot or scanned line or area where the laser light intensity has been lower, the temperature lower, and hence the

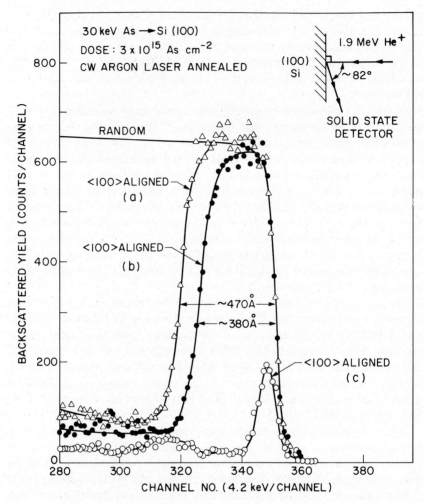

Fig. 7.14. Channeling and Rutherford backscattering measurements of the regrowth of an initially 470 Å layer made amorphous by As ion implantation: (a) before laser stimulated regrowth, (b) after partial regrowth, and (c) after complete regrowth.

regrowth velocity lower, the thickness of the regrowth tapers from complete to none.

It has been found possible to grow epitaxial Si by CW scanning of Si vacuum deposited on a Si substrate (Bean et al. 1979, Hess et al. 1979). This case, however, requires careful cleaning of the interface in ultrahigh vacuum and deposition of Si under the same conditions in order for the epitaxy to proceed. The case of ion implantation amorphized Si has, by its nature, a clean buried interface from which to grow. It has not been possible to grow epitaxial Si from chemical vapor deposited amorphous Si, presumably because of the clean interface requirements.

The epitaxial regions formed by CW laser scanning are not, however, defect free. There are residual dislocation loops and point defect clusters visible in transmission electron microscopy in the regrown layer (Lietoila et al. 1979b, Takai et al. 1980), though fewer than in furnace annealing. There are also electrically active defects found by deep level transient spectroscopy (DLTS) in the region below the originally amorphous material (Johnson et al. 1979). The details of the defect story are complex and still incomplete. The large thermal gradients involved in CW laser processing are one factor contributing to the defects. The temperature drops by 500°C in a few microns in the typical case. The stress associated with the corresponding thermal expansion is also large. This is dramatically evident if a laser spot is allowed to dwell for much longer than necessary to regrow an amorphous layer. A Nomarski interference contrast micrograph of such a case is shown in fig. 7.15b. The hot region has developed slip lines as it plastically deformed. This does not happen instantly because time is required for nucleation of the slip dislocations (Rozgonyi et al. 1981). Typically it occurs in the order of 1 s.

5.1.2 Silicide formation

Silicides are important metallic compounds that are formed between silicon and a wide range of transition metals. They serve in integrated circuits as electrical contacts and junctions to Si and as conducting paths. Normally they are formed by furnace heat treatment of Si coated by evaporation or sputtering with a thin metallic film. The formation temperatures vary widely from Pt_2Si at ~ 200°C to $NbSi_2$ at ~ 800°C. They do not in general grow well epitaxially on Si because of large lattice mismatches, with the exception of $CoSO_2$ and $NiSi_2$ (Ishiwara 1979, Chiu et al. 1980, 1981). They do, however, have flat interfaces. The compound forming reactions proceed by thermally activated solid phase diffusion; the reaction front marches cleanly through the film as revealed in Rutherford backscattering measurements.

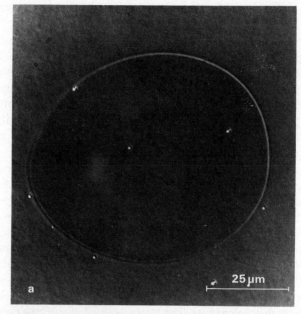

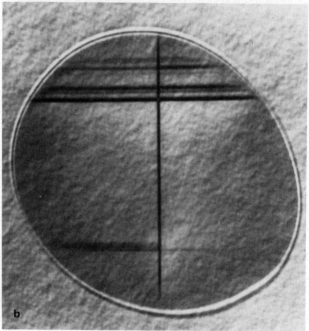

Fig. 7.15. Nomarsky micrographs of a CW laser annealed amorphous Si spot: (a) exposed for 100 ms, (b) exposed for 2 s, showing developed slip.

Since CW lasers are localized sources of heat, it is perhaps not surprizing that they can be used to drive silicide reactions. A complication exists because of the large difference in optical reflectivity of a typical metal and a silicide. To introduce enough power into the system at the beginning in the presence of high metallic reflection, a high incident P/a is required. However, when the silicide forms, the reflectivity drops, the power coupled into the system rises, and the system overheats, melts and is damaged. This problem was circumvented by Shibata et al. (1980) who deposited a thin overlayer of Si on top of the metal film so the reflectivity change (now from Si to silicide) was small. They successfully grew silicides of Pd, Pt, Nb, W and Nb in this way. Controlling the temperature of the scanning spot they could selectively form either the low temperature Pd_2Si or the higher temperature $PdSi_2$. Thermal activation energies can also be obtained for the formation process. These results are quite consistent with the solid phase reactions previously studied by furnace processing. Two exceptions arise in the case of Pt, which shows a Pt_2Si_3 phase and for Pd which shows a PdSi phase, neither of which are normally obtainable in furnace processing. These may result from the high temperature of the moving hot spot which may serve to nucleate metastable phases near to the lowest eutectic temperatures of the Pd/Si and Pt/Si systems (Shibata et al. 1980).

5.2. Liquid phase processes

5.2.1. Grain growth

Chemical vapor deposition of Si produces a very fine grain polycrystalline material important for gate electrodes and conducting paths on MOS integrated circuits. The conductivity of the material (which is normally heavily impurity doped) is limited by electron scattering at the grain boundaries. Gat et al. (1978b) have found it possible to increase the grain size markedly by CW laser scanning. In this case the phenomenon is not one of epitaxial growth of the polysilicon onto a crystalline Si substrate, but the formation of large grain polysilicon on an amorphous SiO_2 or Si_3N_4 insulating layer which in turn overlies a Si substrate.

The results of a single CW scan line are shown in fig. 7.16. Large, needle-shaped grains are formed, a few microns wide and as much as 25 μ long, the long dimension being in the direction of the laser scan. This result is inexplicable on a solid phase basis and indeed is taking place in the liquid phase. Because of the presence of the SiO_2 or Si_3N_4 insulating layer whose thermal conductivity is much less than that of Si, the melting point of Si is reached at lower P/a than would be indicated by fig. 7.10. A molten spot is formed and pulled along with the scanning CW beam. As the spot moves, freezing takes place at its cooler trailing edge. This trailing edge is in contact with various polycrystalline grains along the scanning track which serve as potential nucleation sites (or seeds) for grain growth. It is also in contact with the amorphous underlying SiO_2 or Si_3N_4 whose imperfections may serve as nucleation sites. In this situation it is difficult to control the nucleation and so irregular grains are formed, though very much larger than the grains of the starting material. The liquid phase character of this process is confirmed by observation of major redistribution of dopant impurities throughout the thickness of the polysilicon film, in contrast to the unchanging profiles observed in the solid phase as discussed in section 5.1.1. The larger grain poly has substantially greater electrical conductivity than the starting material. In spite of the remaining grain boundaries, it has been possible to form active MOS devices in this material.

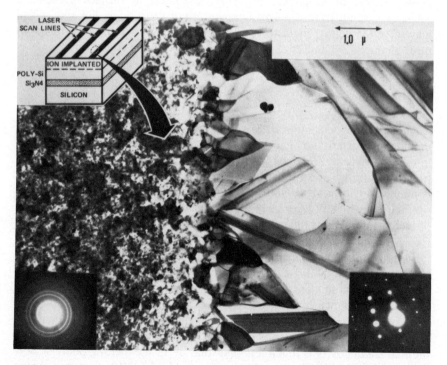

Fig. 7.16. Growth of large grain polysilicon from small grain CVD poly by scanning CW laser beam.

5.2.2. Crystalline semiconductors on amorphous substrates

One of the most unique and important areas that laser processing has contributed to is that of single crystal growth on amorphous substrates. The grain growth studies discussed in section 5.2.1 were its beginning. If it is feasible to produce high quality thin film single crystal semiconductors without the requirement of a crystalline substrate, these techniques will be of immense practical importance. The continued interest in epitaxial Si-on-sapphire material (SOS) attests to the need for insulating substrates even if, as in the SOS case, the substrate has to be single crystal and the quality of the Si as grown on it is significantly poorer than material cut from single crystalline Si or grown by epitaxy on Si substrates. In addition, crystal growth on amorphous substrates allows new possibilities for solar cell materials and the intriguing possibility of three-dimensional integrated circuits using alternate films of crystalline Si and amorphous SiO_2.

Extension of the results of large grain but irregularly shaped polycrystalline material formed by pulling a liquid zone through a polyfilm has concentrated on controlling nucleation. Only partial success has been achieved so far. An early result was that of Gibbons et al. (1979b) who crystallized small isolated islands of chemical vapor deposited polysilicon that had been delineated by lithographic techniques. If the islands were 2 μ wide, comparable to the width of the individual grains of poly formed in a continuous CVD film, they crystallized in individual single crystals. In some cases, grains seem to have a preferred orientation for their surface normal, but are random in orientation in the surface plane. The most common result, however, is that the individual grains are randomly oriented with respect to one another and show no textural preference even for a particular surface normal.

The first isolated islands were rectangular in shape as shown in fig. 7.17a. Other shapes have subsequently been investigated, frequently, the boat shape of fig. 7.17b (Fastow et al. 1981). A laser beam is scanned over the boat, melting first the prow and finally the stern. The prow will be the first to cool and the idea is that it will solidify to form a seed on which the remainder of the boat will crystallize as it cools. It is clear that in this scheme the width of the boat must be less than the width of the scanning beam so that there are no cold polyedges to serve as nuclei. Control of heat flow and hence temperature remains an important problem. The edges of the boat would be expected to be cooler than the middle because of the geometry of cooling to the amorphous substrate, fig. 7.17c. Different approaches have been considered to reduce the edge cooling (Bigelsen et al. 1981). One is to separate the CVD poly islands by only very narrow gaps so that heat flow remains nearly one-dimensional as in fig. 7.17d (assuming the

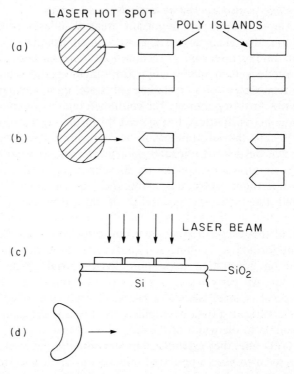

Fig. 7.17. Geometry of islands for single crystal growth on amorphous substrates: (a) rectangular islands, (b) boat-shaped islands, (c) heat flow in cross section from a single heated island, (d) close-spaced islands to control heat flow, and (e) a crescent-shaped scanning beam.

laser beam is large enough to heat neighboring islands to similar temperatures). Bigelsen et al. (1981) have discussed shaping the laser beam to give the desired result: putting more power into the edges of an island than into its center. They have also considered making the beam crescent shaped as in fig. 7.17e (with a convex leading edge). The middle of the trailing edge will then be the first to cool so that seeding from the boat tip will be heavily favored with respect to nucleation of grains from along the edge. The role of the substrate in the nucleation is not yet understood. It may be necessary to control the heat flow so that the interface to the substrate is the last to freeze, not an easy condition to arrange.

Results reported at the Materials Research Society meeting in Boston in November 1980 indicate important improvements in achievable single crystal island size to islands 20 μ wide. These islands are already individually large enough to allow device fabrication to be carried out in them. The material is suitable in quality for metal–oxide–semiconductor (MOS) structures.

The control of crystal orientation is highly desirable. It is particularly important to have the film normal be crystallographically consistent in order to have uniform subsequent processing (oxidation for example) in the grown islands. It would be desirable to control the orientation in the plane as well. There is the hope that these can be achieved by optimizing the shape of the prow of the boat or the way it freezes or the orientationally dependent interfacial energy to the amorphous substrate. Another possibility is to bury or surround the poly islands with oxide or nitride to constrain the growth. Still another approach is to orient the islands of regrowth by epitaxial contact with crystalline Si underlying the SiO_2. Holes in the oxide allow such seeding to take place (Fastow et al. 1981). So far this method has produced epitaxial orientation but only close to the holes.

5.2.3. Explosive crystallization

This title refers to the self-propagating crystallization of amorphous thin films. In such systems, when conditions are insufficient for self-propagation it is possible to control the crystallization by a scanned CW laser beam. Explosive crystallization is included in this section because of its relationship to section 5.2.2 above and because, although it has in the past been thought to be a solid phase process, new information indicates that a liquid phase actually is critically involved.

Fig. 7.18 is a schematic illustration of explosive crystallization. In this case an amorphous film of Ge evaporated on a thick SiO_2 substrate, is

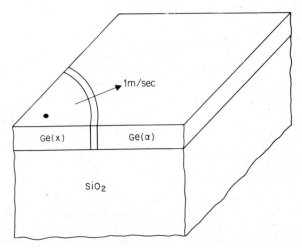

Fig. 7.18. Explosive crystallization of a Ge film on a SiO_2 substrate.

stimulated by a laser at the marked point. Crystallization proceeds outward explosively from the point and results in crystallization of the whole film. This phenomena has been studied for decades and quite intensively in the early 1970s (Mineo et al. 1973, Takamori et al. 1972, Kikuchi et al. 1974). The propagation velocity was measured by Mineo et al. (1973) for Ge and found to be ~1 m/s. The driving force for the crystallization is the energy stored in the amorphous material. The triggering laser heats a small region to a temperature where crystallization can occur. Release of the energy of crystallization heats the adjacent material which in turn crystallizes, and so on. The conditions for self supported crystallization are that the heat released be sufficient to raise the neighboring temperature enough to continue the process. If too much heat escapes to the substrate or if the film is too cool originally for the heat released to bring it to a critical temperature, crystallization may propagate a short distance from the triggering point and stop.

A speed of 1 m/s is too fast for a solid phase process and Gilmer and Leamy (1980) and Gold et al. (1980) postulated that a thin liquid layer exists at the crystallization interface. Fig. 7.19 is a free energy diagram for a semiconductor due to Bagley and Chen (1980). The difference between the free energy for the amorphous and crystalline state is an indication of the

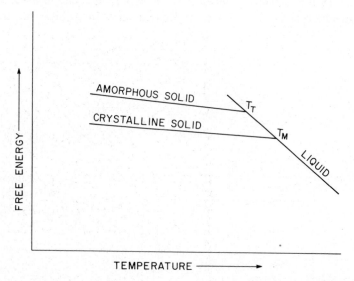

Fig. 7.19. Free energy diagram of an amorphous and crystalline semiconductor.

energy stored in the disorder of the amorphous material. The crystal is the preferred state, but the amorphous state is metastable with an enormously long relaxation time at room temperature. That metastability can relax in the solid phase at an appropriate temperature, as in the epitaxial growth of amorphous semiconductors on a crystalline silicon substrate (section 5.1.1). Notice, however, that the melting point of amorphous material is lower than the normal melting point of the crystal. For Ge this difference is about 200°C. If the amorphous material can be taken rapidly to a temperature T such that $T_m > T > T'$, it will melt and be supercooled with respect to the crystalline solid. Such supercooling can lead to very rapid crystal growth. This is the regime Gilmer and Leamy (1980) theorized was active in explosive crystallization: a thin interface was molten, but at a temperature below T_m.

A recent experiment (Leamy et al. 1981) used impurity markers to look for the passage of the postulated liquid zone. The crystallization of an amorphous Ge film on an amorphous quartz substrate was carried out at a substrate temperature of 300°C using a laser trigger. Both antimony and lead that had been implanted in the shallow layer at the surface of the amorphous Ge film were found broadened into the depth when the film was

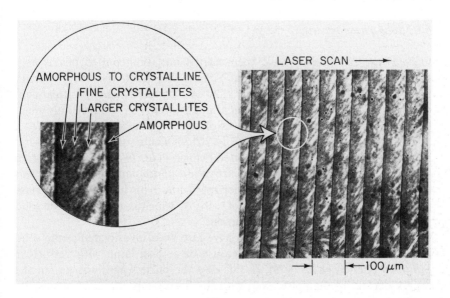

Fig. 7.20. Micrograph of a Ge film crystallized by a scanned laser beam under nonexplosive conditions.

crystallized. The extent of the broadening was consistent with a molten layer whose thickness was ~1% of the film thickness.

If the substrate temperature is lower, or the film is thinner (so a larger fraction of the heat of crystallization is lost to the substrate by conduction) Ge does not explosively crystallize. Fan and his associates (Fan et al. 1980) at Lincoln Laboratory have been studying this subcritical region in detail. Their objective is to make large grain polycrystalline Ge (or ideally, single crystal Ge) on cheap (maybe even metallic) substrates. Such Ge could serve as a template for vapor phase epitaxial growth of GaAs for solar cells – inexpensively. Using a scanned CW beam they observed crystallization patterns such as that of fig. 7.20. The crystallization runs rapidly ahead of the scanning CW beam but stops when not enough energy is being released to raise the temperature high enough to keep it going. It then waits until the heat added by the scanning laser catches up with it, brings the system to the triggering point again, and again it spurts ahead. The control of nucleation, grain size and orientation is important in this case, as it was in section 5.2.2. There are many problems yet to be solved in this fascinating type of crystal growth.

6. Processing with pulsed lasers

6.1. Solid phase processes

Only a few experiments have been carried out using pulsed beams at a sufficiently low energy density that Si does not reach the melting point. Solid phase processes are relatively slow so that a transient heat treatment of hundreds of nanoseconds or even a microsecond can produce only small effects. Using the extrapolated curves of fig. 7.13, the regrowth time for a 500 Å amorphous layer at 1200°C is 20–30 μs. In one experiment, 100 successive Q-switched pulses were used (Miyao et al. 1978). Regrowth of an amorphous layer accompanied by activation of implanted impurities was observed without distortion of the dopant profile. This implies a solid phase process with a regrowth per pulse of ~10 Å, which is reasonable for solid phase epitaxy under the given conditions.

Other pulsed heating approaches have also been investigated in the solid phase regime. Among these was the use of a flash lamp with very close geometry (Cohen et al. 1978). In this case the pulse length was μs but the energy density in the pulse, even with the flash lamp operating at near its explosion limit, was too low to achieve annealing without multiple flashes and an elevated starting temperature. An appropriate fast time scale for solid phase processing is between 0.1 and 10 ms and pulsed optical techniques are not well matched to that range.

6.2. Liquid phase processes

6.2.1. Melting and regrowth

A number of calculations have been published describing the thermal transients associated with pulsed laser radiation of semiconductors (Wang et al. 1978, Baeri et al. 1979, Wood et al. 1980). Fig. 7.21 is from Baeri et al. (1979) on the temperature transient at different depths in Si during and after a Q-switched ruby laser pulse. Near the surface the temperature rapidly reaches the melting point and rises above it. Deeper inside, the temperature maximum occurs somewhat later, delayed by the thermal

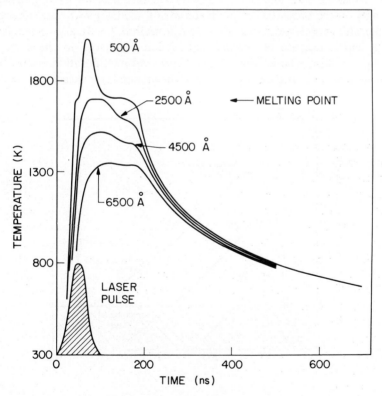

Fig. 7.21. Calculated temperature versus time and depth in Si for a 50 ns ruby laser pulse.

diffusion time from the heated surface. The near surface region melts to a depth of ~ 2500 Å. Deeper than that the temperature maxima do not reach the melting point. The surface cools by thermal diffusion into the bulk of the material. The cooling by radiation or convection at the surface is negligible. The inner edge of the melted region refreezes first, the outermost surface layer, last. A different view of the transient is presented by a plot of the position of the melt front with time, fig. 7.22. The melting layer grows inward rapidly, to different depths for different laser pulse energy densities. The rate of return of the melt front to the surface is controlled by the rate of transport of the latent heat of melting from the surface to deeper inside. Typical speeds for the front as it reaches the surface are a few m/s. The surface layer can remain molten for as much as several hundred nanoseconds as shown.

A threshold for surface melting exists. For ruby lasers with pulse lengths of ~ 20 ns, the threshold is ~ 0.2 J/cm^2. The threshold for melting depends strongly on the properties of the material and the laser wavelength because of the absorption variations discussed in section 2.1. It also depends on pulse length because of thermal conduction of energy to the substrate during the pulse. The maximum melt depth and melt duration are set by a damage threshold. If the temperature of the molten layer exceeds the boiling

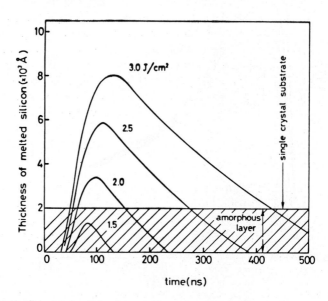

Fig. 7.22. Calculated melt front penetration versus time and laser pulse energy density.

point of the material, the surface will be disrupted and material will be blown off. The energy density at which this occurs depends on the details of the laser pulse shape. After the surface is melted, subsequent input energy is strongly absorbed (for any of the lasers used) in the shallow skin depth of the metallic liquid and continues to heat that layer (see section 2.1). If the laser pulse is long, there is more time for inward diffusion of heat from the surface and the maximum surface temperature is lower for a given energy density.

In fig. 7.22 an amorphous surface layer is indicated with a thickness of 2000 Å. The properties of the regrown material are different depending on whether or not the melt penetrates completely through this layer. If only part of the amorphous layer is melted, when solidification occurs it takes place on a still-amorphous substrate and polycrystalline material is produced. On the other hand, if the maximum melt depth exceeds the thickness of the original amorphous layer, the underlying substrate is single crystal and serves as a template for epitaxial growth: the regrown layer is then single crystal. The features of an energy density threshold for epitaxial crystallization are observed in Rutherford backscattering measurements by the appearance of a crystalline layer with the orientation of the substrate. The polycrystalline state that exists at energy densities below the epitaxial threshold are observed by TEM.

In any laser pulse there is a variation in energy density across the laser spot. Material illuminated by the lower density portions of the spot will not melt as deeply as that exposed at the center. As a result, in liquid phase pulsed processing of amorphous layers, a polycrystalline border is always found around the single crystal central region. This is illustrated schematically in Fig. 7.23. A transmission electron micrograph of the border region is shown Fig. 7.24 where the material is amorphous on one side, single

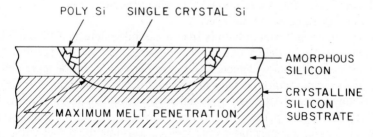

Fig. 7.23. Schematic of the formation of polysilicon and single crystal Si from an amorphous Si layer on a single crystal substrate by pulsed laser melting.

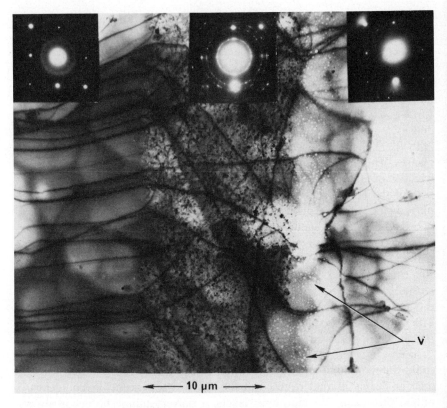

Fig. 7.24. Transmission electron micrograph of the edge of a pulse laser melted spot in amorphous Si, showing the transition from single crystal to polycrystal to amorphous material as in fig. 7.23.

crystal on the other and polycrystalline in between. This is in contrast to the solid phase epitaxial case (section 5.1.1) where the border is not polycrystalline but a tapered amorphous layer.

Liquid phase epitaxial growth is free of extended defects if the melt layer has penetrated into defect free underlying single crystal. If the underlying crystal is defective, those defects result in imperfections in the regrown layer also. In ion implantation, defects are created beyond the limit of the region that is amorphized. Although epitaxial single crystal material is regrown if the melt front just penetrates the amorphous layer, defect free epitaxial material requires a somewhat deeper melt to consume the defect tail as well.

Regrowth speeds in the liquid phase processing region are high. They are totally determined by the heat flow properties of the system. The speed can be changed from ~1 m/s to ~10 m/s by manipulating heat flow parameters. Temperature variations of the substrate alter the thermal conductivity. Higher growth speeds occur at lower temperatures where the thermal conductivity is higher and extraction of the latent heat of solidification is correspondingly faster. Changing the pulse length and pulse energy density also influences the regrowth velocity (Baeri et al. (1980). The effect of these parameters is shown in fig. 7.25. The qualitative features are straightforward. Because heat flow from the surface into the interior is taking place throughout the duration of the pulse, as well as afterward, a long pulse leaves a shallower temperature gradient while producing a given melt depth than does a short one. The shallower gradient reduces the subsequent rate of heat extraction and the velocity of final regrowth is lower. In the same way, to refreeze a thickly melted layer (produced by a higher pulse energy density) requires more total heat flow to the interior than to refreeze a

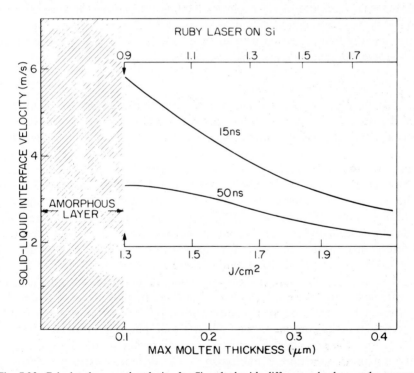

Fig. 7.25. Calculated regrowth velocity for Si melted with different ruby laser pulse energy densities and two different pulse ranges. The velocity is the average for regrowth of the last 1000 Å layer (assumed to be originally amorphous).

thinner one. The heat extraction time increases, the thermal gradients decrease and once again the final velocity is lower.

Throughout the range of parameters explored with nanosecond pulsed lasers, single crystalline material is formed in regrowth. Crystal growth speeds of 1–10 m/s are enormous compared to the speeds of some mm/h which are typical of normal semiconductor crystal growth. It might seem surprizing that single crystal organization of atoms can occur so quickly: a monolayer of atoms must be added in $\sim 10^{-10}$ s. Such times are not short, however, compared with the rate at which atoms of a liquid are striking the liquid–solid interface as it grows. These times are $\sim 10^{-12}$ s. Thus, Spaepen and Turnbull (1979) expect crystallization up to speeds of $\sim 10^2$ m/s. A few experiments have been done with picosecond Nd:YAG 0.53 μ pulses by Liu et al. (1980) on initially single crystal Si. They find evidence for formation of an amorphous Si ring around a crystalline center. By the same reasoning used above in discussing the faster growth speeds for lower energy density pulses that have produced shallower melting, the border of a picosecond pulse-melted spot should solidity more rapidly than the center. Liu et al. estimate the freezing speed in their case at $\sim 10^2$ m/s, in the range where amorphous Si might be expected.

6.2.2. Impurity incorporation

This topical area is one of the most intriguing that has come out of studies with pulsed lasers on semiconductors. It is equally relevant to the study of metals, but because impurities play such a central role in the electrical properties of semiconductors, it has received a great deal of attention in the last few years. Fig. 7.26 shows the depth distribution of arsenic in Si, originally introduced by ion implantation in a shallow layer (~ 500 Å) near the surface and subsequently annealed with different energy density Nd:YAG pulses (Williams et al. 1981). At energy densities above a threshold, the arsenic redistributes as a result of the laser processing. That threshold corresponds to the threshold for melting. As the laser energy density increases, the arsenic distribution continuously broadens. This is a consequence of liquid phase diffusion of arsenic in molten Si with a diffusion coefficient $\sim 10^{-4}$ cm^2/s. In molten layer times of 100s of nanoseconds this diffusion broadens the arsenic distribution by 100s of angstroms, as observed. This is in contrast to the CW laser solid phase epitaxial case (section 5.1.1) for which no alteration in the impurity profile is found, even though the hot processing time is milliseconds. The difference is in the diffusion coefficient which changes by 8 orders of magnitude from the solid to the liquid phase.

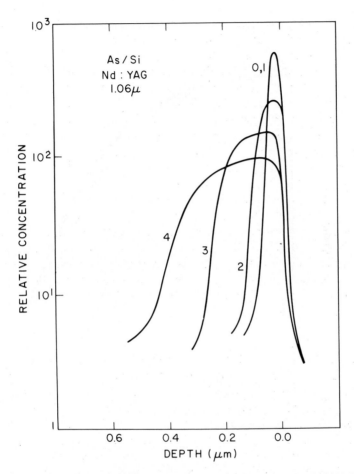

Fig. 7.26. Arsenic depth profiles before and following laser melting with 75 ns Nd:YAG pulses at 1.06 μ and different power densities: 0, before laser melting; 1, 43 MW/cm^2; 2, 62 MW/cm^2; 3, 84 MW/cm^2; 4, 105 MW/cm^2.

Fig. 7.27 is the case for boron (White et al. 1978) whose distribution was measured by secondary ion mass spectroscopy after different numbers of Q-switched ruby laser pulses. The profile broadens with the first pulse and broadens still more with the second and third pulses. This is as expected for diffusing species given more time to diffuse. It is important to recognize that the diffusion time is not the same for all parts of the molten layer because

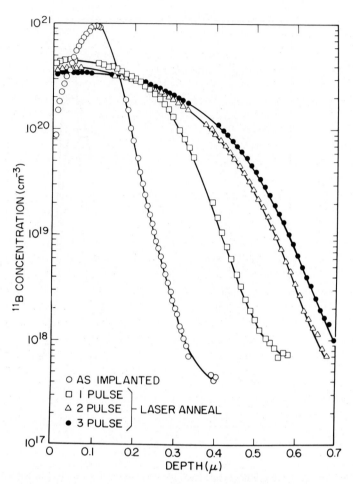

Fig. 7.27. Boron depth distributions before and after ruby laser melting of Si ion implanted with boron. The broadening of the boron distribution for successive pulses is shown.

the deeper regions melt last and refreeze first and so are molten for a shorter total time.

The impurity incorporation in fast liquid phase epitaxy is remarkable in two particular respects: first, the impurities are incorporated with very high efficiency on substitutional lattice sites and second, the concentration of impurities that can be incorporated in this way exceeds the maximum solid solubility of the impurities in Si, in some cases by order of magnitude. The process is not an equilibrium process, of course, and thus it affords the opportunity to form metastable systems.

Fig. 7.28 is a schematic illustration of a solid growing from a liquid. As a solid grows, impurity atoms as well as host atoms are continually striking the liquid–solid interface. The incorporation of the impurities is expressed by a segregation coefficient $k = C_s/C_1$. In the case illustrated, k is quite low, the liquid has a much higher concentration of impurities in it than does the solid. In fast crystal growth k is quite different than at growth near equilibrium. It is nevertheless feasible to extract a value, k', by fitting the experimental depth distribution of impurities found after laser melting. This procedure has been carried out by several groups, but the two cases shown in fig. 7.29 are from Oak Ridge for arsenic, a normally quite soluble impurity and for indium, a relatively insoluble impurity. Arsenic has a normal distribution coefficient of 0.3. However, its depth distribution after laser melting is well fitted, as shown in the curve of fig. 7.29a, by a value of $k' = 1.0$. For indium, as shown in fig. 7.29b, a substantial fraction of the initially implanted impurity segregates to the surface. Indium has an equilibrium k of 4×10^{-4}. The distribution that would have been expected on the basis of this equilibrium k is shown in the figure. The incorporated concentration is nearly 2 orders of magnitude higher than this prediction. The measured distribution in depth and the segregated impurity fraction are well accounted for by a $k' = 0.15$. By varying the amount of impurity initially present in the material, a maximum concentration which can be incorporated in pulsed laser growth is found. The results of this procedure for several impurities are shown in table 7.1 due to White et al. (1980b). The maximum solubilities are a factor of four to 500 above the equilibrium solid solubility.

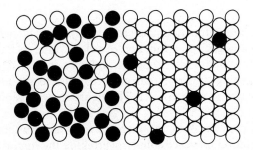

Fig. 7.28. Schematic of a liquid/solid interface. The normal distribution coefficient $k = C_S/C_L$ is small in this case.

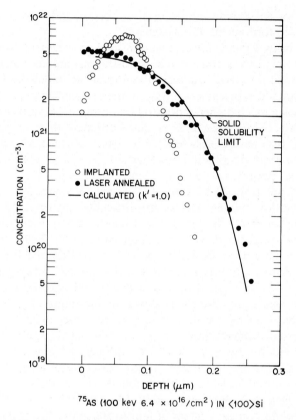

Fig. 7.29a. Depth distributions of As as-implanted and following laser melting. The fitted solid curve is calculated for the nonequilibrium k' value indicated.

The value of k' is a function of regrowth velocity. The indium data point from fig. 7.29b and data as a function of velocity due to Baeri et al. (1980) are shown in fig. 7.30. The regrowth velocity in the latter results was changed by changing substrate temperature, pulse length and pulse energy density as discussed in section 6.2.1. The figure shows a rapid increase in k' at ~ 3 m/s.

It is attractive to think of the increase in k as resulting from the fast moving solidification interface simply overtaking the impurity before it has time to diffuse away in the liquid. However, this picture is too simple. The diffusivities of different impurities in the liquid state vary only a little from

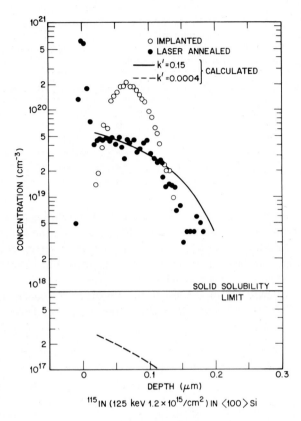

^{115}IN (125 keV 1.2×10^{15}/cm^2) IN $\langle 100 \rangle$ Si

Fig. 7.29b. Depth distribution of In as-implanted and following laser melting. The fitted solid curve is calculated for the non-equilibrium k' value indicated.

one to another, yet the k's span nearly an order of magnitude at high regrowth velocity. The process of non-equilibrium segregation often referred to as solute trapping, has been discussed by Wood et al. (1980), Cahn et al. (1980) and Jackson et al. (1980). It is valuable to examine the question by considering the temperature of the growing interface. Fig. 7.31 due to Jackson is an estimate of the growth of Si as a function of temperature. If $T > T_m$ the solid melts; growth is negative. At $T = T_m$ no growth occurs. For $T < T_m$ the growth rate increases with the degree of undercooling. The curve of fig. 7.31 indicates that for a 5 m/s growth rate the undercooling must be ~ 60°C. In refreezing after laser melting the solidification rate is almost

Table 7.1

Comparison of maximum substitutional dopant concentration C_S^{MAX} and segregation coefficient K' for pulsed laser melting with their equilibrium values C_S^0 and K_0, respectively, from White et al. (1980b)

Dopant	C_S^0 (cm^{-3})	C_S^{MAX} (cm^{-3})	C_S^{MAX}/C_S^0	K_0	K'	K'/K_0
As	1.5×10^{21}	6.0×10^{21}	4	0.3	1.0	3.3
Sb	7×10^{19}	1.3×10^{21}	18	0.023	0.7	30
Bi	8×10^{17}	4×10^{20}	500	0.0007	0.4	571
Ga	4.5×10^{19}	4.5×10^{20}	10	0.008	0.2	25
In	8×10^{17}	1.5×10^{20}	188	0.0004	0.15	375

entirely determined by heat flow considerations (the liquid is going to freeze at that rate, no matter what crystal properties result) but the temperature of the interface at which freezing occurs is a parameter set by the material itself. A cooler interface increases the sticking probability for impurities striking it and thus k increases as velocity increases.

The very presence of impurities at the interface influence the growth temperature. Solutes in general depress the freezing point. The zero growth

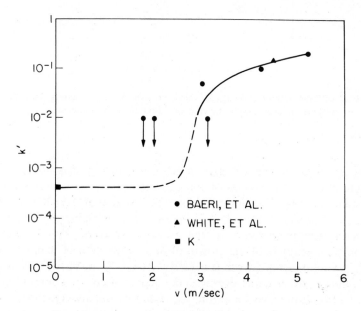

Fig. 7.30. Regrowth velocity dependence of k' for In in Si.

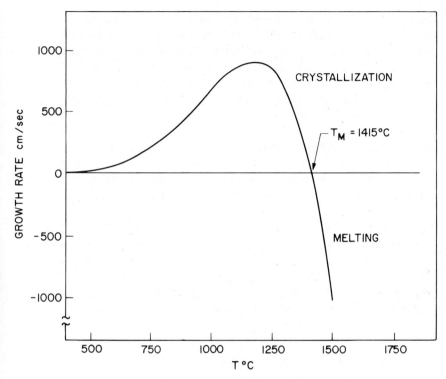

Fig. 7.31. Approximate crystal growth rate versus temperature for Si.

crossover on the curve of fig. 7.31 will be shifted to lower temperatures and no doubt the shape of the curve of growth versus undercooling will change as well. Unfortunately there is still very little known about details of this kind for the fast growth rates that are interesting in the pulsed laser processing case.

Velocity dependent measurements of bismuth incorporation in Si have recently been made which show a crystal orientation dependence to k' (Baeri et al. 1981). The results are given in fig. 7.32. At all velocities, k' is greater for (111) than for (100) growth. This result seems explicable on the same argument of interface temperature discussed above. The (111) orientation of Si normally grows more slowly than the (100) because larger collections of atoms have to be organized to grow a (111) layer than a (100) layer. This is true at low growth velocities (even in the solid phase) and must be true at high velocities as well. But since the growth rate is set by heat flow variables, in order for the (111) orientation to grow at the same rate as the (100) the interface must be cooler. A cooler interface will promote sticking of bismuth impurities and more will be incorporated in the solid. Thus, k' will be larger for (111) than for (100). It would be extremely valuable if it were possible to measure the interface temperature directly to test these arguments experimentally.

6.2.3. Constitutional supercooling
The melting and regrowth of surface layers by Q-switched laser pulses are primarily planar processes, lacking in planarity only to the extent that the laser intensity or its reflection or absorption coefficients are non-uniform.

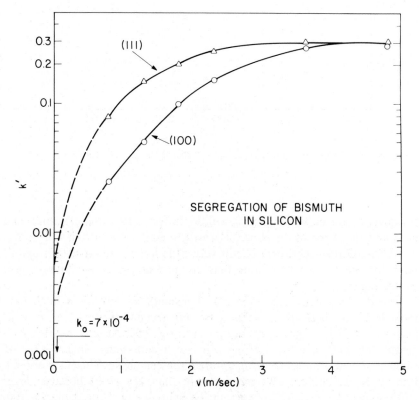

Fig. 7.32. Nonequilibrium k' for Bi in Si with regrowth in 2 different crystallographic directions.

This is not always the case, however. In some situations, impurities at the liquid–solid interface cause major perturbations in the crystal growth and result in large lateral non-uniformity. The first reported case of this kind was for platinum in Si. Regrowth velocity dependence of k' for Pt was observed by Cullis et al. (1980), whose depth distribution data are shown in fig. 7.33. At the lowest regrowth velocity (with the substrate at 620 K to reduce the thermal conductivity) the Pt was dominantly segregated to the surface. At the highest velocity (regrowth at 77 K), however, a large fraction of the Pt was retained in the solid at concentrations far above the solid solubility of Pt. As observed by channeling and Rutherford backscattering only about half of the Pt is on substitutional lattice sites. TEM measure-

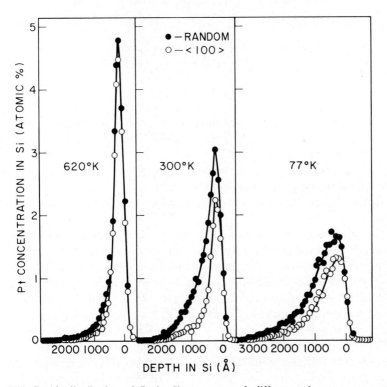

Fig. 7.33. Depth distribution of Pt in Si regrown at 3 different substrate temperatures producing growth velocities of 1 m/s for 620 K, 2 m/s for 300 K and 3 m/s for 77 K. Channeled and random distributions show that at high velocity about half of the platinum is on substitutional lattice sites.

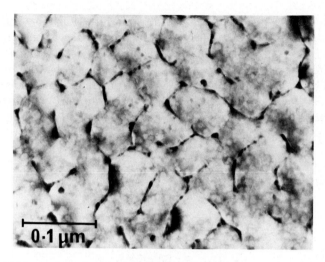

Fig. 7.34. TEM of microstructure of platinum in Si illustrating the formation of cells by constitutional supercooling.

ments shown in fig. 7.34 reveal the reason for this. About half of the Pt is dissolved in the Si (on lattice sites) and the remainder is in the walls of a cellular structure whose typical scale is ~ 1000 Å. The columns of Si (heavily doped with Pt) in the center of the cells are single crystal material epitaxial with the underlying crystalline silicon substrate. The cell walls are a highly Pt-rich alloy which is polycrystalline and nonepitaxial with the Si lattice.

This phenomena was identified as due to constitutional supercooling and has been observed subsequently in a number of other impurity systems. A schematic illustration of the process is shown in fig. 7.35. As a solid regrows from the Pt–Si liquid, if the local density of Pt is high anywhere along the interface, the temperature of freezing at that point is reduced. Solidification is correspondingly delayed at that point. Pt from neighboring regions of the interface, which are incorporating much more Pt than they would like, diffuses laterally to the still-liquid local spot and adds more Pt to it, reducing the freezing point still further. The process is clearly unstable. The interface breaks up into cells whose walls are the Pt-rich material. The cell size is found to be dependent on the interface velocity as expected for constitutional supercooling. At higher velocity there is less time for lateral motion and the cells are smaller. This systematic variation with velocity has

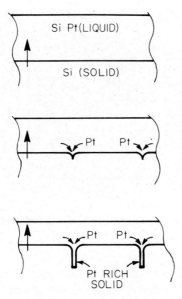

Fig. 7.35. Schematic of the constitutional supercooling process.

been observed for the case of In in Si as shown in fig. 7.36 (Baeri et al. 1981).

The cellular structures of constitutional supercooling have now been observed in a number of impurity systems at high impurity concentrations: Pt (Cullis et al. 1980), In (Baeri et al. 1981), Ga and Fe (Narayan and White 1981), for example. When the concentration exceeds the maximum that can be incorporated substitutionally in the solid, the excess appears in these periodic separated phases. From the standpoint of device material, irregularity of this kind seems undesirable. On the other hand, periodic structures with such a fine scale may find special applications in their own right.

6.2.4. Dislocations and precipitates
The liquid phase of Q-switched laser melting offers important possibilities for improving the quality of thin layers of material that some previous processing step has left filled with precipitates or with dislocation networks. High temperature diffusion can produce defective layers of this type. When such a region is melted, dislocations cannot exist since there is no ordered lattice of atoms in a random liquid. Precipitates also can dissolve. They require a time for dissolution, but the time can be very short. Narayan and

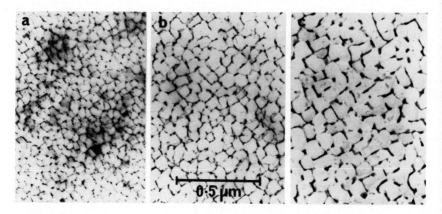

Fig. 7.36. Cell size dependence on growth velocity for In in Si.

White (1980) have investigated these processes using TEM. In fact, they have used the depth dependent disappearance of small dislocation loops and precipitates as indications of the penetration depth of the melt front.

Rapid solidification of the melted region can occur without reappearance of the defects. The incorporation of concentrations of impurities well above the solid solubility limit has been discussed in section 6.2.2. Phosphorus precipitates, for example formed by high temperature diffusion, are dissolved in the liquid and regrown as substitutional impurities in the solid. High concentration systems of this kind are metastable at temperatures well above room temperature but it has been shown that precipitation does occur at sufficiently high temperatures and long times. Examples of this kind have been studied for antimony by Revesz et al. (1979) and for arsenic by Chu et al. (1980).

Dislocation-free regrowth is the common mode for fast liquid phase epitaxy. This is remarkable because of the enormous strain that may be accommodated in the solid. In Si regrown layers containing boron at $> 10^{21}/cm^3$, the lattice constant can be smaller than for Si by more than 1% and still the region is dislocation free. X-ray, channeling and TEM studies of this case (Larson et al. 1978) have shown that the lattice of the thin regrown layers is held in tension by the thick crystalline substrate and has the lattice constant of the substrate. In the dimension perpendicular to the surface, the lattice is not clamped and it contracts. Although the system would have lower energy if it were dislocated, plastic flow does not occur at room temperature in Si. Furthermore, as studies by Hofker et al. (1979)

have shown, even at high temperatures dislocations are not readily formed, presumably because of an absence of dislocation nucleation sites.

Dislocations can be formed during rapid regrowth if there are dislocations at the maximum penetration depth of the liquid phase to nucleate them. A study of such a case has been carried out by Narayan and Young (1979), who investigated the crystallography of the dislocations that are produced.

6.2.5 Point defects

While fast liquid phase epitaxy produces semiconductor layers free of extended defects, they are not free of point defects. In retrospect this does not seem surprizing. In the very strong temperature gradients ($\sim 10^7 - 10^8$ °C/cm) and fast temperature transients ($10^9 - 10^{10}$ °C/s) encountered in the process, defects can form and even diffuse into the unmelted region. These defects are too small to be observed in TEM and their existence and identity are determined from measurements such as deep level transient spectroscopy (DLTS) which probes the electronic energy levels in the band gap of a semiconductor, especially those levels far from the band edge (hence deep in the band).

A DLTS spectra for a laser melted Si layer can be seen in fig. 7.37. Two distinct peaks – observed at temperatures of ~ 100 and ~ 190 K in a temperature scan – correspond to defects with energy levels 0.19 and 0.33 eV below the conduction band edge of Si (Benton et al. 1980a,b). The $\sim 10^{14}$/cc defect concentration, while small as an atomic fraction ($< 10^{-8}$), is not small from the point of view of the electrical properties of the material. Deep levels of this kind are effective and unwanted traps and recombination centers for holes and electrons in a semiconductor. Fig. 7.37 also shows various attempts to anneal the defects out by low temperature furnace annealing. In fig. 7.37c, a DLTS spectrum after annealing was done in a hydrogen plasma at 200°C, no defects can be detected. The defects themselves have apparently not been removed in the process, but primarily compensated or passivated by protons diffusing into the Si from the hydrogen plasma. Note that in fig. 7.37d, following a heat treatment in vacuum following the plasma treatment, defects are again revealed.

The understanding of complex defect questions such as these is emerging slowly. These defects are very important in bipolar semiconductor devices where carrier lifetime is important. The defect question in the laser annealing of compound semiconductors is even more complex. In this case, not only can there be missing lattice atoms (vacancies) and complexes of these with the intentionally present dopant impurities or other undesired impuri-

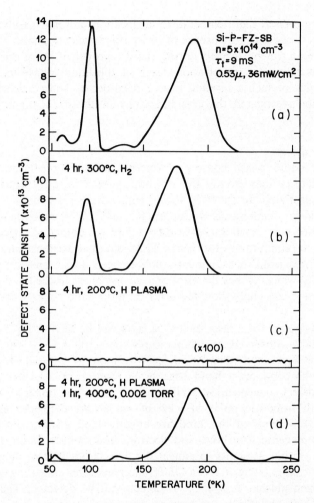

Fig. 7.37. Deep level transient spectroscopy measurements of the defect state density: (a) after laser annealing, (b) after subsequent heat treatment in H_2, (c) after heat treatment in H plasma, and (d) after subsequent anneal in vacuum.

ties in the material, but there can be anti-site defects as well: an atom A on a B site, for example. For making heavily doped contacts on compound semiconductors, laser melting may be a viable process, but for electro-optic devices (light emitting diodes and solid state lasers, for example) defects are nonradiative recombination paths for carriers that can drastically reduce the optical efficiency. One study of this kind by Parsons et al. (1980) has involved GaAlAs.

6.2.6. Surfaces

Pulsed laser melting of semiconductors provides new provides new possibilities for the preparation of clean, ordered and doped surfaces. The idea was first discussed by Bedair and Smith (1969) and is now being studied in several research laboratories (Zehner et al. 1980, Cowan and Golovchenko 1980). By melting a surface in ultrahigh vacuum, contaminant impurities on the surface can be evaporated. Alternatively, they may be dissolved in the first micron or so of the material to reduce their atomic concentration by several orders of magnitude. Auger spectroscopy measurements of surface contamination show the efficiency of the surface cleaning to be at least as good as the more standard repeated sputter and anneal prescriptions which, of necessity, heat the whole sample (and its attachments as well). Ordered surfaces are also produced from the liquid phase as observed by low energy electron diffraction. Since impurity redistribution in a molten layer is known to occur as a result of laser melting, it should be feasible to ion implant dopant atoms of any chosen kind into the semiconductor and subsequently bring them to the surface as selected surface constituents. Work in this area is at a very early stage. Whether it will have a direct bearing on production in semiconductor materials is not at all clear at this point, but it offers a fascinating new tool for the preparation of clean, ordered and doped crystalline surfaces.

7. Application prospects

A great many possible applications of lasers (or other directed energy beams) to the processing of semiconductor devices have been suggested since the surge of interest in this field starting in the mid-1970s. These suggestions range from ways of adjusting the turn-on voltage of Zener diodes by the pulsed annealing of individual diodes at power densities selected for each diode, to single pulse annealing of large areas of ion implanted amorphized silicon for solar cells. Several reviews have been published discussing these in considerable detail (Gibbons 1980, Gibbons et al. 1980, Tokuyama 1980, Miller 1980). This section will not attempt a comprehensive survey of these ideas, but will instead choose a few representative (and some of the most promising) examples from the pulsed and CW laser processing regimes.

7.1. Self-aligned processing

The term "self-alignment" refers to a situation in which some existing structure defines the region of processing in a processing step. Self-aligned ion implantation of the source and drain in MOS devices, as shown in fig. 7.38a, is very important. In this case, the gate electrode itself is a self-align-

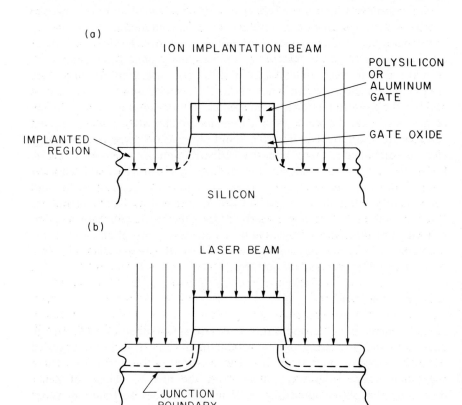

Fig. 7.38. Self-aligned processing: (a) self-aligned ion implantation, (b) self-aligned laser melting.

ing mask to prohibit implantation of the gate region without the need for a separate and painstakingly applied lithographic masking step. This same idea has been investigated in pulsed laser processing for annealing the damage of ion implantation and for electrically activating the ion implanted impurity atoms, as in fig. 7.38b. In comparison with furnace annealing, the laser processing offers the advantage of minimal extension of the source and drain under the gate because there is no prolonged high temperature processing which allows extended impurity diffusion. The difference in effective channel length and gate-to-source and -drain capacitance are quite important in short channel MOS devices. The annealing also only takes place where it is needed; the rest of the structure is not subjected to high temperature at all.

It is attractive to consider using an oxide layer as a mask, but this has complications as pointed out by Hill (1979). The masking property of an oxide depends on the enhanced reflectivity it provides. It works as a reflective (in contrast to antireflective) coating when its thickness is correct. At the edges of a window in the oxide, however, the oxide thickness goes from full to zero and there is a thickness at which the film acts as an antireflective coating, increasing the light coupled into the material. The result is a spike in the melt depth at the edge of the window and a corresponding variation in the junction depth and impurity profile because of that spike, both quite undesirable. This situation is illustrated in fig. 7.39a. Hill (1980) has shown that by using aluminum over the oxide for reflection (depending now on reflection from the material property not from the film thickness) the junction spike can be eliminated, fig. 7.39b. Similar results have been obtained by Koyanagi et al. (1979) and Miyao et al. (1980) using a polysilicon layer as a mask over the oxide.

There is another interesting feature of the self-aligned ion implantation and laser processing by liquid phase epitaxy. As noted in section 6.2.1, around a melted and epitaxially regrown region in a broad amorphous layer there is a border of polysilicon material. However, when only the region exposed through the window is damaged by ion implantation (the ions not reaching the Si outside the window) the laser melting which follows will consume all of the amorphous material and the crystalline template for epitaxy will exist on the sides as well as below the melted region. No polysilicon is formed; the material is free of extended defects.

It is generally undesirable to melt the SiO_2–Si interface over long distances. The thermal expansion of SiO_2 and Si are so different that when melting occurs, there is relative motion of the SiO_2 on the surface of the molten Si which results in motion of oxide features relative to one another

(a)

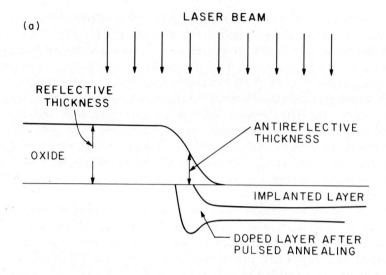

(b)

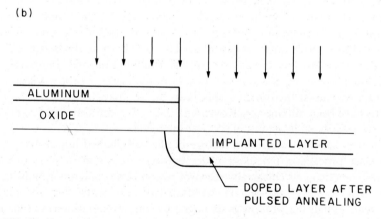

Fig. 7.39. Interface effects at: (a) an oxide window edge, (b) when a sharp edge is formed and protected with Al.

and formation of structural damage at the interface when refreezing occurs (Hill 1980, Stephen et al. 1980). Under appropriate masking conditions this long-range melting need not occur, but not all the secondary problems have yet been solved.

7.2. New materials and structures

The replacement of furnace processing steps by laser processing steps with equivalent goals is one approach to the application of lasers in semiconductor device manufacture. Gains may be realized this way as in the control of channel length in short channel MOS devices discussed in section 7.1. Another approach, however, is through the creation of new materials and structures that other processing methods simply cannot provide. CW laser scanning for solid phase annealing without impurity redistribution is marginally in this category and so is localized silicide formation by control of a scanning CW beam. On the other hand, separately adjusting impurity profiles in the base and the emitter regions of bipolar transistors is simply not feasible in furnace processing. With laser melting it can be achieved by lateral control of melt depth. The very high dopant density, no extended defect material produced by lasers in Q-switched melting and regrowth are unique. Whole new systems of alloys are accessible by this process. Doping of materials often can be realized very inexpensively by pulsed melting. A desired constituent can be sputtered or evaporated or laser photo-chemically deposited (Deutsch and Osgood 1979, Erhlich et al. 1980) on a surface and melted into it. It is even possible to use a metal–organic liquid layer over the surface as a source of metal atoms to be incorporated during laser melting of the surface region (Stuck et al. 1981).

The formation of crystalline materials on amorphous substrates is another special result of the laser (ore-beam) melt regime. It shares uniqueness with grapho-epitaxy (Geis et al. 1979) and more recently with special very high temperature short time heat treatments (Geis 1980). This regime is simply not available in normal furnace processing because a liquid layer cannot be controlled over large areas and long times. The achievements, problems and hopes in this area were discussed in section 5.2.5 for CW laser melting. The process has also been tried with some success in the pulse melted regime (Miyao et al. 1980, Tamura et al. 1980) in bridging structures as shown in fig. 7.40. Polysilicon, put down by chemical vapor deposition, covers an SiO_2 layer and reaches down to underlying Si at windows in the oxide. When a Q-switched pulse melts the polysilicon, it would be expected to refreeze first where it is in contact with the crystalline Si because there the thermal conduction to the substrate is greatest. This serves as a seed for

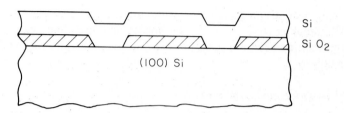

Fig. 7.40. Bridging structure between holes in an oxide layer. The overlying bridge is epitaxial with the silicon in the oxide holes.

regrowth of the layer on top of the oxide. The crystallization front in this case moves at a very high speed since the desired crystallization is proceeding along the surface while cooling is taking place both along the surface to the holes and through the oxide. Nucleation at the oxide surface undoubtedly limits the distance over which single crystal material can be formed. The morphology of the regrown layer is interesting. Since the last part to freeze is the center of the bridge, and since molten Si has a higher density than crystalline Si, the final freezing tends to produce a ridge of thicker material in the center of the bridge.

8. Summary and outlook

The laser processing of semiconductors is still quite early in its development stages. As of the end of 1980 there were not yet any cases known to the author where it was actually a part of a production line. Much of the basic science is qualitatively and in some cases even quantitatively understood, but scientific study of new phenomena and exploration of their technological potential are continuing hand in hand.

Lasers offer the broad processing advantages envisioned in section 1.2: localization of thermal treatment in 3 dimensions and, thus, freedom from wafer degradation due to homogeneous, lengthy, high temperature processing. Lasers allow processing to be carried out in either the solid or the liquid phase by choice of the laser power, wavelength, time scales and geometry and the material properties, geometry and starting temperature; a very wide range of parameters available to the semiconductor process designer. From the point of view of the author, however, the most exciting prospects for this new tool are in the unique materials and structures it provides ways of fabricating and these seem to arise primarily from the ability to reach and control the liquid phase. The liquid phase is controllably accessible with either CW or pulsed beams. With CW lasers it is controlled by confinement of the high temperature regime by the geometry of heat flow. With pulsed beams it is controlled by the fast temperature transient. The liquid phase enables the following: local adjustment of impurity distribution, high concentration impurity incorporation, elimination of extended defects, the formation of small scale cellular alloy structures, the smoothing of sharp corners on semiconductor or insulating layers, the growth of epitaxial single crystal layers, the growth of large grain polycrystalline thin film materials on amorphous substrates, and the prospect for the growth of single crystals on noncrystalline substrates. These phenomena are challenging opportunities for materials research. The unique materials that lasers enable will surely make laser processing part of the semiconductor technology of the 80s.

References

Auston D.H., Surko C.M., Venkatesan T.N.C., Slusher R.E. and Golovchenko J.A., 1978, Appl. Phys. Lett. **33** 437.

Auston D.H., Golovchenko J.A., Simons A.L., Slusher R.E., Smith P.R., Surko C.M. and Venkatesan T.N.C., 1979a, in: Laser–Solid Interactions and Laser Processing, 1978, eds., Ferris S.D., Leamy H.J. and Poate J.M., AIP Conf. Proc. No. 50 (American Institute of Physics, New York) p. 11.

Auston D.H., Golovchenko J.A., Simons A.L., Slusher R.E., Smith P.R., Surko C.M. and Venkatesan T.N.C., 1979b, App. Phys. Lett. **34** 635.

Baeri P., Campiasano S.U., Foti G. and Rimini E., 1979, J. Appl. Phys. **50** 788.

Baeri P., Poate J.M., Campisano S.U., Foti G., Rimini E. and Cullis A.G., 1980, Appl. Phys. Lett. **37**.

Baeri P., Foti G., Poate J.M., Campisano S.U., Rimini E. and Cullis A.G., 1981, in: Laser and Electron Beam Solid Interactions and Materials Processing, eds., Gibbons J.F., Hess L.D. and Sigmon T.W. (North-Holland, Amsterdam) p. 67.

Bagley B.G. and Chen H.S., 1978, Laser–Solid Interactions and Laser Processing, eds., Ferris S.D., Leamy H.J. and Poate J.M. (American Institute of Physics, New York, 1979) p. 97.

Baglin J.E.E., Hodgson R.T., Chu W.K., Neri J., Hammer D. and Chen L.J., 1981, Nucl. Instr. and Meth. **191** 169.

Bean J.C., Leamy H.J., Poate J.M., Rozgonyi G.A., Van der Ziel J.P., Williams J.S. and Celler G.K., 1979, J. Appl. Phys. **50** 881.

Bedair S.M. and Smith H.P., 1969, J. Appl. Phys. **40** 4776.

Benton J.L., Doherty C.J., Ferris S.D., Kimerling L.C., Leamy H.J. and Celler G.K., 1980a, in: Laser and Electron Beam Processing of Materials, eds. White C.W. and Peercy P.S. (Academic, New York) p. 430.

Benton J.L., Doherty C.J., Ferris S.D., Flamm D.L., Kimerling L.C. and Leamy H.J., 1980b, Appl. Phys. Lett. **36** 670.

Bigelsen D.K., Johnson N.M., Bartelink D.J. and Moyer M.D., 1981, in: Laser and Electron

Beam Solid Interactions and Materials Processing, eds., Gibbons J.F., Hess L.D. and Sigmon T.W. (North-Holland, Amsterdam) p. 487.

Brodsky M.H., Title R.S., Weiser K. and Pettit G.D., 1970, Phys. Rev. **B1** 2632.

Brown W.L., 1980, in: Laser and Electron Beam Processing of Materials, eds., White C.W. and Peercy P.S. (Academic Press New York) p. 20.

Brown W.L., Golovchenko J.A., Jackson K.A., Kimerling L.C., Rozgonyi G.A., Leamy H.J., Miller G.L., Poate J.M., Rodgers J.W., Sheng T.T., Celler G.K. and Venkatesan T.N.C., 1978, in Rapid Solidification Processes, Claitors Publ. Div., Baton Rouge.

Cahn J.W., Coriell S.R. and Boettinger W.J., 1980, in: Laser and Electron Beam Processing of Materials, eds., White C.W. and Peercy P.S. (Academic, New York) p. 89.

Celler G.K., Poate J.M. and Kimerling L.C., 1978, Appl. Phys. Lett. **32** 464.

Chiu K.C.R., Poate J.M., Feldman L.C., and Doherty C.J., 1980, Appl. Phys. Lett. **36** 544.

Chiu K.C.R., Poate J.M., Rowe J.E., Sheng T.T. and Cullis A.G., 1981, App. Phys. Lett. **38** 988.

Chu W.K., Mader S.R. and Rimini E., 1980, in: Laser and Electron Beam Processing of Materials, eds., White C.W. and Peercy P.S. (Academic, New York) p. 253.

Cohen R.L., Williams J.S., Feldman L.C. and West K.W., 1978, Appl. Phys. Lett. **33** 751.

Compaan A. and Lo H.W., 1980, Proc. Conf. on Defects in Semiconductors, Oiso, Japan.

Compaan A., Lo H.W., Aydinli A. and Lee, M.C., 1981, in: Laser and Electron Beam Solid Interactions and Materials Processing, eds., Gibbons J.F., Hess L.D. and Sigmon T.W. (North-Holland, Amsterdam) p. 15.

Cowan P.L. and Golovchenko J.A., 1980, J. Vac. Sci. Tech. **17** 1197.

Csepregi L., Mayer J.W. and Sigmon T.W., 1976, Appl. Phys. Lett. **29** 92.

Csepregi L., Kennedy E.F., Gallagher T.J., Mayer J.W. and Sigmon T.W., 1977, J. Appl. Phys. **48** 4234.

Csepregi L., Kennedy E.F., Mayer J.W. and Sigmon T.W., 1978, J. Appl. Phys. **49** 3906.

Cullis A.G., Webber H.C. and Bailey P., 1979, J. Phys. E: Sci. Instr. **12** 688.

Cullis A.G., Webber H.C., Poate J.M. and Simons A.L., 1980, Appl. Phys. Lett. **36** 320.

Deutsch T.F. and Osgood R.M., 1979, Appl. Phys. Lett. **35** 175.

Dumke W.P.,1980, Phys. Lett. **78A** 477.

Eden R.C., 1967, Photoemission Studies of the Electronic Band Structure of GaAs, Ga P and Si (Stanford El. Labs Tech. Rep. No. 5221-1).

Ehrlich D.J., Osgood R.M., Jr. and Deutsch T.F., 1980, Appl. Phys. Lett. **36** 916.

Fairfield J.M. and Schwuttke G.H., 1968, Sol. St. Electron. **11** 1175.

Fan J.C.C., Zeiger H.J., Gale R.P. and Chapman R.C., 1980, Appl. Phys. Lett. **36** 158.

Fastow R., Leamy H.J., Celler G.K., Wong Y.H. and Doherty C.J., 1981, in: Laser and Electron Beam Solid Interactions and Materials Processing, eds., Gibbons J.F., Hess L.D. and Sigmon T.W. (North-Holland, Amsterdam) p. 495.

Foti G., Rimini E, Vitali G. and Bertolotti M., 1977, Appl. Phys. **14** 189.

Gamo K., Murakami K., Kawabe M., Namba S. and Aoyagi Y., 1981, in: Laser and Electron Beam Solid Interactions and Materials Processing, eds., Gibbons J.F., Hess L.D. and Sigmon T.W. (North-Holland, Amsterdam) p. 97.

Gat A. and Gibbons J.F., 1978, Appl. Phys. Lett. **32** 142.

Gat A., Gibbons J.F., Magee T.J., Peng J., Deline V.R., Williams P. and Evans C.A., 1978a, Appl. Phys. Lett. **32** 276.

Gat A., Gerzberg L., Gibbons J.F., Magee T.J., Peng J. and Hong J.D., 1978b, Appl. Phys. Lett. **33** 775.

Geiler H.D., Goetz G. and Klinge K.D., 1977, Phys. Stat. Sol. **A41** K171.

Geis M.W., Flanders D.C., Smith H.I. and Antoniadis D.A., 1979, Appl. Phys. Lett. **35** 71.

Geis M.W., 1980, private communication.

Gibbons J.F., 1980, in: Laser and Electron Beam Processing of Electronic Materials, eds., Anderson C.L., Celler G.K. and Rozgonyi G.A. (Electrochemical Society, Princeton) p. 1.

Gibbons J.F., Johnson W.S. and Mylroie S.W., 1975, Projected Range Statistics (Dowden, Hutchinson and Ross, Stroudsberg, Pa.).

Gibbons J.F., Gat A., Gerzberg L., Lietoila A., Regolini J.L., Sigmon T.W., Pease R.F.W., Magee T.J., Peng J., Hong J., Deline V., Katz W., Williams P. and Evans C.A., Jr., 1979a, in: Laser–Solid Interactions and Laser Processing, 1978, eds., Ferris S.D., Leamy H.J. and Poate J.M., AIP Conf. Proc. No. 50 (American Institute of Physics, New York) p. 365.

Gibbons J.F., Lee K.F., Magee T.J., Peng J. and Ormond R., 1979b, Appl. Phys. Lett. **34** 831.

Gibbons J.F., Lietoila A., Nissin Y.I. and Wu F.C., 1980, in: Laser and Electron Beam Processing of Materials, eds., White C.W. and Peercy P.S. (Academic, New York) p. 593.

Gilmer G.H. and Leamy H.J., 1980, in: Laser and Electron Beam Processing of Materials, eds., White C.W. and Peercy P.S. (Academic, New York) p. 227.

Gold R.B., Gibbons J.F., Magee T.J., Peng J., Ormond R., Deline V.R. and Evans C.A., Jr., 1980, in: Laser and Electron Beam Processing of Materials, eds., White C.W. and Peercy P.S., (Academic, New York) p. 221.

Golovchenko J.A. and Venkatesan T.N.C., 1978, Appl. Phys. Lett. **32** 147.

Harper F.E. and Cohen M.I., 1970, Sol. St. Electron. **13** 1103.

Hess L.D., Roth J.A., Anderson C.L. and Dunlap H.L., 1979, in: Laser and Electron Beam Processing of Electronic Materials, eds., Anderson C.L., Celler C.K. and Rozgonyi G.A. (Electrochemical Society, Inc., Princeton) p. 496.

Hill C., 1979, in: Laser–Solid Interactions and Laser Processing, 1978, eds., Ferris S.D., Leamy H.J. and Poate J.M., AIP Conf. Proc. No. 50 (American Institute of Physics, New York) p. 419.

Hill C., 1980, in: Laser and Electron Beam Processing of Electronic Materials, eds., Anderson C.L., Celler G.K. and Rozgonyi G.A. (Electrochemical Society, Princeton) p. 267.

Hill C., 1981, in: Laser and Electron Beam Solid Interactions and Materials Processing, eds., Sigmon T.W., Hess L.D. and Gibbons J.F. (North-Holland, Amsterdam).

Hodgson R.T., Baglin J.E.E., Pal R., Neri J. and Hammer 1980, Appl. Phys. Lett. **37** 187.

Hofker W.K., Oosthoek D.P., Eggermont G.E.J., Tamminga Y. and Stacy W.T., 1979, Appl. Phys. Lett. **34** 690.

Ishiwara H., Nagatomo M. and Furukawa S., 1978, Nucl. Instr. and Meth. **149** 417.

Jackson K.A., Gilmer G.H. and Leamy H.J., 1980, in: Laser and Electron Beam Processing of Materials, eds., White C.W. and Peercy P.S. (Academic, New York) p. 104.

Johnson N.M., Gold R.B. and Gibbons J.F., 1979, Appl. Phys. Lett. **34** 704.

Kachurin G.A., Pridachin N.B. and Smirnov L.S., 1975, Sov. Phys. Semicond. **9** 946.

Kachurin G.A., Bogatyrev V.A. and Romanov S.I., 1977, Fifth Int. Conf. on Ion Implantation (Plenum) p. 445.

Kaplan R.A., Cohen M.G. and Kin K.C., 1980, Electronics **53** 137.

Kennedy E.F., Csepregi L., Mayer J.W. and Sigmon T.W., 1977, J. Appl. Phys. **48** 4241.

Khaibullin I.B., Shtyrkov E.I., Zaripov M.M., Bayazitov R.M. and Galjautdinov M.F., 1978, Rad. Eff. **36** 225.

Khaibullin I.B., 1980, Proc. 15th Int. Conf. Physics of Semiconductors, Kyoto, 1980, J. Phys. Soc. Japan **49** Suppl. A.

Kikuchi M., Matsuda A., Kurosu T., Mineo A. and Callanan M.J., 1974, Sol. St. Commun. **14** 731.

Klimenko A.G., Klimenko E.A. and Donin V.I., 1975, Sov. J. Quantum Electron. **5**(10) 1289.

Knapp J.A. and Picraux S.T., 1981, Appl. Phys. Lett. **38** 873.

Koyanagi M., Tamura H., Miyao M., Hashimoto N. and Tokuyama T., 1979, Appl. Phys. Lett. **35** 621.

Krynicki J., Suski J., Ugniewski S., Groetzschel R., Klabes R., Kreissig U. and Ruediger, J., 1977, Phys. Lett. **A61**(3) 61.

Kutukova O.G. and Strel'Tsov L.N., 1976, Sov. Phys. Semicond. **10**(3) 265.

Larson B.C., White C.W. and Appleton B.R., 1978, Appl. Phys. Lett. **32** 801.

Lax M., 1979, Appl. Phys. Lett. **33** 786.

Leamy H.J., Brown W.L., Celler G.K., Foti G., Gilmer G.H. and Fan J.C.C., 1981, Appl. Phys. Lett. **38** 137.

Lietoila A. and Gibbons J.F., 1979a, Appl. Phys. Lett. **34** 332.

Lietoila A., Gibbons J.F., Magee T.J., Peng J. and Hong J.D., 1979b, Appl. Phys. Lett. **35** 532.

Liu P.L., Yen R., Bloembergen N. and Hodgson R.T., 1980, in: Laser and Electron Beam Processing of Materials, eds., White C.W. and Peercy P.S. (Academic, New York) p. 156.

Liu Y.S. and Wang K.L., 1979, Appl. Phys. Lett. **34** 363.

Lo H.W. and Compaan A., 1980, Phys. Rev. Lett. **44** 1604.

Miller G.L., 1980, in: Laser and Electron Beam Processing of Electronic Materials, eds., Anderson C.L., Celler G.K. and Rozgonyi G.A. (Electrochemical Society, Princeton) p. 83.

Mineo A., Matsuda A., Kurosu T. and Kikuchi M., 1973, Sol. St. Commun. **13** 329.

Minnucci J.A., Matthei K.W., Greenwald A.C., Little R.G. and Kirkpatrick A.R., 1980, in: Laser and Electron Beam Processing of Materials, eds., White C.W. and Peercy P.S. (Academic, New York) p. 658.

Miyao M., Tamura M. and Tokuyama T., 1978, Appl. Phys. Lett. **33** 828.

Miyao M., Koyanagi M., Tamura H., Hashimoto N. and Tokuyama T., 1980, Jap. J. Appl. Phys. **19** Supp. 19-1, 129.

Miyao M., Motooka R., Natsuaki N. and Tokuyama T., 1981, Laser and Electron Beam Solid Interactions and Materials Processing, eds., Gibbons J.F., Hess L.D. and Sigmon T.W. (North-Holland, Amsterdam) p. 163.

Murakami K., Kawabe M., Gamo K., Namba S. and Aoyagi Y., 1979, Phys. Lett. **70A** 332.

Murakami K., Kawabe M., Gamo K., Namba S. and Aoyagi Y., 1980, Proc. 15th Inst. Conf. on the Physics of Semiconductors, Kyoto, 1980, J. Phys. Soc. Japan **49** Suppl. A.

Narayan J. and Young F.W., Jr., 1979, Appl. Phys. Lett. **35** 330.

Narayan J. and White C.W., 1980, in: Laser and Electron Beam Processing of Materials, eds., White C.W. and Peercy P.S. (Academic, New York) p. 65.

Narayan J. and White C.W., 1981, Phil. Mag. **43A** 1515.

Nathan M.I., Hodgson R.T. and Yoffa E.J., 1980, Appl. Phys. Lett. **36** 512.

Olsen G.L., Kokorowski S.A., Roth J.A. and Hess L.D., 1981, in: Laser and Electron Beam Solid Interactions and Materials Processing, eds., Gibbons J.F., Hess L.D. and Sigmon T.W. (North-Holland, Amsterdam) p. 125.

Parsons R.R., Rostworowski J.A., Springthorpe A.J. and Dymont J.C., 1980, in: Laser and Electron Beam Processing of Materials, eds., White C.W. and Peercy P.S. (Academic Press, New York) p. 373.

Pearce C.W. and Zaleckas V.J., 1979, J. Electrochem Soc. **126** 1436.

Ratnakumar K.N., Pease R.F.W., Bartelink D.J., Johnson N.M. and Meindl J.D., 1979, Appl. Phys. Lett. **35** 463.

Revesz P., Farkas G.Y. and Gyulai J., 1979, Proc. Int. Conf. on Ion Beam Modification of Material, eds., Gyulai J., Lohner T. and Paszter E., Budapest, Hungary, p. 871.

Rozgonyi G.A., Baumgart H. and Phillipp F., 1981, in: Laser and Electron Beam Solid Interactions and Materials Processing, eds., Gibbons J.F., Hess L.D. and Sigmon T.W. (North-Holland, Amsterdam) p. 193.

Shah N.J., McMahon R.A., Williams J.G.S. and Ahmed H., 1981, in: Laser and Electron Beam Solid Interactions and Materials Processing, eds., Gibbons J.F., Hess L.D. and Sigmon T.W. (North-Holland, Amsterdam) p. 201.

Shibata T., Sigmon T.W. and Gibbons J.F., 1980, in: Laser and Electron Beam Processing of Materials, eds., White C.W. and Peercy P.S. (Academic, New York) p. 530.

Shtyrkov E.I., Khaibullin I.B., Galjautdinov M.F. and Zaripov M.M., 1975a, Opt. Spectrosc. **38** 595.

Shtyrkov E.I., Khaibullin I.B., Zaripov M.M., Galjautdinov M.F. and Bayazitov R.M., 1975b, Sov. Phys. Semicond. **9** 1309.

Spaepen F. and Turnbull D., 1979, in: Laser-Solid Interactions and Laser Processing, 1978, eds., Ferris, S.D. Leamy H.J. and Poate J.M., AIP Conf. Proc. No. 50, (American Institute of Physics, New York) p. 73.

Stephen J., Smith B.J. and Blamires N.G., 1980, in: Laser and Electron Beam Processing of Materials, eds., White C.W. and Peercy P.S. (Academic, New York) p. 639.

Stuck R., Fogarassy E., Muller J.C., Hodeau M., Wattiaux A. and Siffert P., 1981, Appl. Phys. Lett. **38** 715.

Takai M., Tsien P.H., Tsou S.C., Röschenthaler O., Ramin M., Ryssel H. and Ruge I., 1980, Appl. Phys. **22** 129.

Takamori T., Messier R. and Roy R., 1972, Appl. Phys. Lett. **20** 201.

Tamura M., Tamura H. and Tokuyama T., 1980, Jap. J. Appl. Phys. **19** L23.

Tokuyama T., 1980, in: Laser and Electron Beam Processing of Materials, eds., White C.W. and Peercy P.S. (Academic, New York) p. 608.

Van Vechten J.A., 1980, in: Laser and Electron Beam Processing of Materials, eds., White C.W. and Peercy P.S. (Academic, New York) p. 53.

Vitali G., Bertolotti M., Foti G. and Rimini E., 1977, Phys. Lett. **A63** 351.

von Allmen M., Lùthy W., Siregar M.T., Affolter K. and Nicolet M.A., 1979, Laser-Solid Interactions and Laser Processing – 1978, eds., Ferris S.D., Leamy H.J. and Poate J.M., AIP Conf. Proc. No. 50, AIP (New York) p. 43.

von Allmen M., 1980, in: Laser and Electron Beam Processing of Materials, eds., White C.W. and Peercy P.S. (Academic, New York) p. 6.

Wang J.C., Wood R.F. and Pronko P.P., 1978, Appl. Phys. Lett. **33**(5) 455.

White C.W., Christie W.H., Eby R.E., Wang J.C., Young R.T., Clark G.J. and Wood R.F., 1978, in: Semiconductor Characterization Techniques, eds., Barnes P.A. and Rozgonyi G.A., ECS Proc. Vol. 78-3 (The Electrochemical Society, Inc., Princeton) p. 481.

White C.W., Wilson S.R., Appleton B.R. and Narayan J., 1980a, in: Laser and Electron Beam Processing of Materials, eds., White C.W. and Peercy P.S. (Academic, New York) p. 124.

White C.W., Wilson S.R., Appleton B.R. and Young Jr., F.W., 1980b, J. Appl. Phys. **51** 738.

Williams J.S., Brown W.L., Leamy H.J., Poate J.M., Rodgers J.W., Rousseau D., Rozgonyi G.A., Shelnutt J.A. and Sheng T.T., 1978, Appl. Phys. Lett. **33** 542.

Williams J.S., Brown W.L., Celler G.K., Leamy H.J., Poate J.M., Rozgonyi G.A. and Sheng T.T., 1981, J. Appl. Phys., **52** 1038.

Wood R.F., Wang J.C., Giles G.E. and Kirkpatric J.R., 1980, in: Laser and Electron Beam Processing of Materials, eds., White C.W. and Peercy P.S. (Academic, New York) p. 37.

Yoffa E.J., 1980a, in: Laser and Electron Beam Processing of Materials, eds., White C.W. and Peercy P.S. (Academic, New York) p. 59.

Yoffa, E.J., 1980b, Phys. Rev. **B21** 2415.

Young R.T., White C.W. and Clark G.J., 1977, Proc. Int. Conf. Photovoltaic Solar Cells (Reidel, Luxembourg) p. 861.

Young R.T., White C.W. and Clark G.J., 1978, Appl. Phys. Lett. **32**(3) 139.

Zehner D.M., White C.W. and Ownby G.W., 1980, Appl. Phys. Lett. **36** 56.

Zehner D.M., White C.W. and Ownby G.W., 1980, Surf. Sci. **92** L67.

CHAPTER 8

Nd:YAG LASER APPLICATIONS SURVEY

STEPHEN R. BOLIN

Raytheon Laser Center
Burlington, Massachusetts 01803, USA

Laser Materials Processing, edited by M. Bass
© *North-Holland Publishing Company, 1983*

Contents

1. Introduction

The neodymium yttrium aluminum garnet (or Nd:YAG) laser has achieved the status as one of the two most widely utilized materials processing lasers. Use of Nd:YAG as a laser was first demonstrated by Geusic (1964). There are several primary advantages of Nd:YAG over other solid state laser materials as a materials processing tool. First, the material exhibits a great deal of flexibility in the manner in which it may be utilized. Nd:YAG exhibits laser action as a continuous wave source, a long-pulse-flash-pumped source, and in various high intensity–short pulse fashions (Q-switch, mode lock, cavity dump). Second, the efficiency of conversion of flashlamp input to laser output is relatively high for the YAG laser.* Third, operation is normally possible without cryogenic cooling of the crystal. Fourth, Nd:YAG can routinely be operated at high average powers and high energy per pulse without damage. Industrial equipment which delivers up to 500 W average power from a single rod and up to 1000 W average power from multiple rod oscillator amplifier configurations are available. No other solid state laser material comes even close to this capability.

The Nd:YAG laser also has several other desirable features for industrial applications:

(1) Nd:YAG lasers are generally quite compact in comparison to gas lasers of equivalent power (the laser head of a typical 400 W unit will measure less than 1 m in length, including beam handling optics. (A typical 400 W unit is shown in fig. 8.1 complete with power supply and cooling system.)

(2) The output wavelength at 1.06 μm allows the use of conventional transmission optics made of fused silica or other glass.

(3) The relatively good absorption characteristics exhibited by most metals at 1.06 μm negates the necessity of using special absorptive coatings (as is the case with CO_2 lasers in some circumstances, such as the heat treating of high reflectivity materials).

---

*Efficiencies of 3–3.5% are observed for Nd:YAG lasers excited by krypton filled flashlamps. This is strongly dependent on neodymium concentration in the crystal. By comparison, ruby laser efficiency is about $\frac{1}{2}$%.

S.R. Bolin

Fig. 8.1. Typical 400 W pulsed Nd:YAG laser system.

The effect of wavelength is especially worth noting with regard to the processing of high reflectivity metals such as aluminum, copper, and their alloys. Whereas the Nd:YAG lasers routinely weld, drill, and cut these metals, only very high power CO_2 lasers are effective on aluminum; CO_2 lasers do not exhibit acceptable results on copper.

The principal materials processing applications of Nd:YAG lasers are summarized in table 8.1 by laser type and average power. Many of these applications will be covered in detail in separate sections.

2. Pulsed Nd:YAG applications

2.1. Laser welding

When America's Apollo lunar probes lifted off in. the late 1960s, they carried with them thermocouple gauges that had been laser welded (Moorehead and Turner 1970).

Today, laser welds are used not only to produce high reliability thermocouples, but also to seal batteries that power digital watches, weld trim rings for electric ranges, and ink cartridges for fountain pens. Yet the capabilities of the laser still are unknown to most production engineers.

In the case of solid state lasers, this may be in great part due to the lack of capabilities present in the early types of equipment. Early applications literature was primarily concerned with joining small diameter wires with single laser pulses delivered at pulse rates of no more than one per second.

It is only within the last six years that this situation has changed to one in which Nd:YAG lasers having significant power outputs are readily available in industrially designed packages. With the availability of units capable of performing "real" applications, the recent growth in Nd:YAG laser usage for materials processing has been astounding.

Table 8.1
Principal application areas of Nd:YAG lasers

Laser type		Principal uses
Pulsed YAG	under 50 W	spot welding, small hole drilling
	100–400 W	seam welding, high speed spot welding, cutting, and large hole drilling
CW YAG (*Q*-switched)	100 W and under	ceramic scribing, resistor trimming, diamond sawing, non-contact marking
CW YAG	200–1000 W	welding, heat treating

Although theories have evolved which relate weld penetration to thermal diffusivity, average power, pulse length, reflectivity, and other factors, actual energies and pulse lengths necessary to accomplish specific welding tasks are generally determined empirically for each material based upon the penetration and weld nugget diameter desired.

The welding ability of a pulsed laser is determined by both its energy per pulse capacity and its average power output capability. In general, a high energy-per-pulse capacity is necessary for the production of welds of a substantial size (penetration and nugget diameter). Desired throughput of product (spot welds per second or seam weld rate) determines the average capability required. The welding capabilities of a CW laser are primarily a function of the average power of the laser and the optical quality of the laser beam (mode structure – purity and stability).

It should be stressed at the onset that Nd:YAG equipment is best suited to "precision" welding applications – "precision" being defined as welds of less than 2 mm penetration. In applications requiring deeper penetration, larger carbon dioxide lasers are being used to some degree by industry for metal welding. Also note that the welding of materials other than metals, such as glass and plastics, is not practical with the Nd:YAG laser system. Because the output frequency, or laser wavelength, is very close to the visible spectrum, light generated by the YAG laser will pass through most transparent materials without generating enough heat to perform welds.

2.2. Spot welding

In principle, it should be possible to produce spot welds at rates up to the maximum pulse rate of the laser welder. Laser welding equipment is now available which produces individual pulses 1 ms long at rates in excess of 200/s. In fact, two limitations prevent this. First, for most applications it is not possible to position parts under the laser beam at these rates. Second, the fact that the laser pulse takes a finite amount of time results in a "blurring" of the laser spot when parts move at high rates "on the fly" during the welding pulse. For example, at 200 pulses/s, if laser spots were to be made 7.5 mm apart, the parts would be passing the laser beam at a rate of 150 cm/s. For a 1 ms pulse length, this would produce a blur of 1.5 mm at the point to be welded. This amount of blur would diminish penetration and create a larger weld spot than acceptable. Although scanning techniques have been suggested in order to eliminate such blurring, to date none have been reduced to practice. At the present time, several successful spot welding applications are known where welding takes place at 20 spots/s, using computer controlled systems with parts moving under a fixed laser beam in a continuous manner. Firing of the laser is controlled by the computer, which detects the position of the table moving the part.

Initially as pulsed Nd:YAG lasers came onto the production floor, such considerations were not meaningful. The equipment available prior to 1975 was primarily limited to pulse rates of under 10 Hz and average powers of less than 50 W.

Among the first applications reported, which was moved onto the production floor, was an automated system for welding wires to terminals on a shunt plate used in telephone switching facilities (Bolin and Maloney 1974). The system employed two 25 cm × 25 cm precision $X-Y$ tables. One was unloaded and reloaded, while the other was directed through its program by a minicomputer. In this manner, up to 130 connections were welded at rates of up to 6 to 8 connections per second. Fig. 8.2 shows a typical relay plate with welded connections.

Fig. 8.2. Typical relay plate with welded connections.

Use of the laser in this application resulted in a great improvement in throughput over the previous hand soldering technique and eliminated the necessity to strip the insulation from wires prior to joining as the insulation was removed by the laser pulse.

Because the Nd:YAG laser beam passes freely through glass, it was found to be a useful tool in repairing expensive vacuum tubes in which a missed weld was discovered after the sealing of the tube (Bolin and Maloney 1975).

Welding of flexure assemblies for use in process controllers was also reported by Bolin and Maloney (1975). Here, 52 spot welds were utilized to create the "tuning fork" shaped device shown in fig. 8.3. Because of the inherently low total heat input of the pulsed laser spot welding process, welds could be made in the thin metal flexure hinges in extreme proximity to the edge of the larger structural components. Making such welds by conventional techniques resulted in unequal cooling rates around the weld, which reduced strength in the weld joint and surrounding heat affected zone.

Fig. 8.3. Nd:YAG spot welded assembly.

A number of spot welding applications were reported (Bolin 1976a) in the following areas:

(1) joining microwave components of tungsten and molybdenum (these are usually joined with the use of a small amount of platinum filler metal);

(2) welding lamp terminal feedthroughs (these consist of two or more pieces of butt welded wire, in which one wire is a material such as dumet, which can be sealed to glass);

(3) gyroscope bearing assemblies (a weld is used to lock a small nut to a shaft);

(4) welding high reliability thermocouple junctions.

Beall (1977) reported on the use of the pulsed Nd:YAG laser for a variety of spot welding tasks in the assembly and repair of aircraft engine parts. Included were the following:

(1) tack welding of vane assemblies of 0.77 mm thick Hastelloy-X prior to brazing;

(2) spot welding in 18 places on a baffle of 0.5 mm thick Inconel 625;

(3) spot welding a double acting valve assembly as the final step in locating the valve closure in its seat (in this application, it was reported that movement due to weld shrinkage was allowed to be no greater than 0.000625 mm; prior to the introduction of laser welding, rejection was 50%; with laser welding, rejection dropped to essentially zero);

(4) repairing cracks in Inconel 718 castings by the addition of filler metal.

In order to most efficiently utilize the high pulse rate capabilities of the Nd:YAG laser as a spot welder, it is desirable to be able to produce welds without stopping part motion, or to weld "on the fly." The usual and most obvious approach to doing this is to use a computer motion control device to coordinate table motion and laser pulsing (Bolin and Maloney 1974). Typically, the controllers necessary for accomplishing this task are both expensive and complex.

Palmquist and Muniz (1977) described a unique approach to this problem, which made use of a relatively simple opto-mechanical arrangement to accomplish the task. As shown in fig. 8.4, a low power helium–neon laser is directed along the same optical path to be taken by the pulsed Nd:YAG welding laser. In their original work, the weld area was scanned by means of a mirror mounted onto a galvanometer arm. During scanning, the helium–neon laser beam was reflected when it encountered specific locations to be welded. These reflections were received by a photodetector and were used to trigger the pulsing of the Nd:YAG laser welder. Since electronic response of under one millisecond was easily achieved, the part did not exhibit any appreciable movement between the time it has reflected the helium–neon beam and the time it was welded by the Nd:YAG laser.

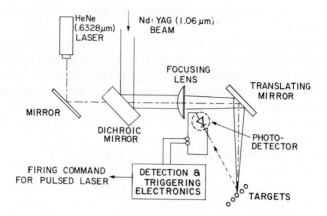

Fig. 8.4. Optical system for automated laser spot welding.

This technique could, of course, be applied to the scanning of larger areas in which the part itself is moved on an $X-Y$ table under a fixed helium–neon beam. This would remove one of the main limitations sited in the original work, which was that processing was limited to one-dimensional targets, i.e., parts lying along a straight line.

The other difficulty sited was that it could only be utilized in the welding of parts which exhibit high reflectivity in the area to be welded and little or no reflectivity in surrounding areas. In the specific application studied, (welding the lead of a telephone switching component), this difference in reflectivity was achieved by having the reflective weld area surrounded by a low reflectivity insulating material.

In addition to simplifying the process control system necessary to perform the task of welding on the fly, this system offers two other unique advantages:

(1) it is self programming, in that the number and spacing of the weld areas may be changed without changing the equipment setup;

(2) since the Nd:YAG laser is triggered by reflection off the part itself, in the absence of parts, the laser will simply not fire.

Welding of flexible circuitry connections offers a particularly difficult challenge since such connections are usually made of highly reflective materials such as copper or aluminum. They are, at the same time, quite thin and, thus, subject to weld burn-through if the weld energy is not precisely controlled. Bauer (1978) reported on the use of the Nd:YAG laser to weld aluminum crimp sleeve connectors which were 0.254 mm thick to aluminum foil conductors which were only 0.051 mm thick. It was noted in this work that, due to the extreme thinness of the foil to be welded, sensitivity was exhibited in several factors, including:

(1) the presence of oxides (which necessitated the lowering of the laser pulse energy to compensate for lowered reflectivity;

(2) crimping together of the connector and the conductor to produce intimate contact and equalize the material thicknesses as much as possible;

(3) attacking the weld area with a laser beam positioned at an angle to the workpiece (30° was found to be the best in this study);

(4) defocusing the beam either above or below the workpiece (defocusing above the workpiece was felt to give better results with regard to having fewer blow holes in the material).

Bosna and Emmel (1980) reported on the welding of flexible printed wiring to multi-pin connectors. A chemical solution called "Ebanol" was utilized in preparing the copper surfaces prior to welding. The prepared surface exhibited a uniform condition of oxidation and required less energy

Table 8.2
Laser weldability of binary metal combinations
(E = excellent, G = good, F = fair, P = poor, – = no data available)

	W	Ta	Mo	Cr	Co	Ti	Be	Fe	Pt	Ni	Pd	Cu	Au	Ag	Mg	Al	Zn	Cd	Pb	Sn
W																				
Ta	E																			
Mo	E	E																		
Cr	E	P	E																	
Co	F	P	F	G																
Ti	F	E	E	G	F															
Be	P	P	P	P	F	P														
Fe	F	F	G	E	E	F	F													
Pt	G	F	G	G	E	F	P	G												
Ni	F	G	F	G	E	F	F	G	E											
Pd	F	G	G	G	E	F	F	G	E	E										
Cu	P	P	P	P	F	F	F	F	E	E	E									
Au	–	–	P	F	P	F	F	F	E	E	E	E								
Ag	P	P	P	P	P	F	P	P	F	P	E	F	E							
Mg	P	–	P	P	P	P	P	P	P	P	P	F	F	F						
Al	P	P	P	P	F	F	P	F	P	F	P	F	F	F	F					
Zn	P	–	P	P	F	P	P	F	P	F	F	G	F	G	P	F				
Cd	–	–	–	P	P	P	–	P	F	F	F	P	F	G	E	P	P			
Pb	P	–	P	P	P	P	–	P	P	P	P	P	P	P	P	P	P	P		
Sn	P	P	P	P	P	P	P	P	F	P	F	P	F	F	F	P	P	P	F	

than an unprepared surface to accomplish welding. Welding results were also more uniform than those achieved on raw surfaces.

Various guidelines have been published for selecting materials suitable for laser welding. Table 8.2 (Gagliano and Zaleckas 1972) serves as a guide for the selection of suitable binary combinations of pure metals. Table 8.3 (Bolin 1980a, b) is based on the observation of pulsed Nd:YAG welding results in a variety of pure metals and alloys.

2.3. Seam welding

Nd:YAG lasers are often employed to produce hermetic and structural seam welds in a wide variety of products. Both continuous wave and pulsed units are employed in these applications; the pulsed Nd:YAG laser produces a continuous seam by the overlapping of spots. With the high pulse rate available in modern pulsed Nd:YAG equipment (typically in excess of 200 Hz), welding rate does not usually suffer. However, it must be kept in mind that seam welding rates for a pulsed laser are inherently limited by the upper limit of the pulse rate and the minimum acceptable weld overlap.

Without satisfactory pulse-to-pulse overlap in pulsed laser seam welding, two difficulties may arise. First, as shown in fig. 8.5, the cross section of the weld will take on a sawtooth shape, producing uneven weld penetration and possibly, insufficient weld strength. Second, if during the course of normal welding a minor contaminant is present, it will often be vaporized as the initial laser pulse strikes the weld area. This vaporization will often produce a tiny defect known as a "blowhole." Such a defect will result in a leak, if the depth of the blowhole extends completely through the weld metal. With sufficient weld overlap, subsequent pulses striking the area of the defect will usually remelt the area and seal the defect. When overlap is not sufficient, blowhole areas will not be remelted; and the resulting leaks lead to part rejection.

Using manufacturers' published data, Marshall (1976) has shown in fig. 8.6 that with regard to weld penetration and weld speed, the pulsed YAG laser exhibits a higher efficiency than does the continuous wave Nd:YAG laser.

Also presented in Marshall's paper was data which demonstrated the feasibility of using the pulsed Nd:YAG laser operating at very high pulse rates to produce a keyhole or "deep" penetration type of weld structure, as shown in fig. 8.7. It was observed that although this structure would be highly desirable for many applications, difficulty was encountered in obtaining welds of uniform quality. It was felt that it would be necessary to have

Table 8.3
Laser welding guidelines

Material	Comments
Aluminum-1100	Pure metal – welds well; no cracking problem
2219	Welds with no cracks – no filler metal required
2024	Requires filler metal of 4047 aluminum alloy to make
5052	hermetic, crack-free welds
6061	
Cu–Zn brasses	Outgassing of zinc prevents good welds
Beryllium copper	Alloys containing higher percentages of alloying agents weld better due to lower reflectivity
Copper	High reflectivity causes uneven welding results – material less than 0.01″ thick may be coated to enhance weldability
Hastelloy-X	Requires high pulse rates to prevent hot-short cracking
Kovar	Presence of phosphorous in electroless nickel plating causes weld cracks
Molybdenum	Usually welds brittle – welds may be acceptable where high strength is not required
Monel	Good ductile welds – good penetration
Nickel	Good ductile welds – good penetration
Steel – carbon	Low carbon – good welds; brittle welds where carbon content is greater than 0.2%
Steel – stainless	300 series – weld well except 303 and 303 SE, which crack; 400 series – welds somewhat brittle – may require postweld anneal
Tantalum	Ductile welds – special precautions against oxidation necessary
Titanium	Ductile welds – special precautions against oxidation necessary
Tungsten	Very brittle welds – requires high energy due to high melting point and thermal conductivity
Zirconium	Ductile welds – special precautions against oxidation necessary

BLOWHOLE SEALED BY FOLLOWING PULSE

GOOD OVERLAP

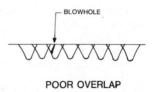

POOR OVERLAP

Fig. 8.5. Effect of insufficient pulse overlap.

equipment capable of much higher pulse rates while maintaining a peak power of several kilowatts.

In general, pulsed Nd:YAG welding, in addition to introducing less heat into the workpiece than continuous wave welding, gives better control of weld penetration, integrity, and appearance. This is particularly true where very thin members are being welded, as in the case of the 0.1 mm thick tantalum diaphragm shown in fig. 8.8.

The Nd:YAG laser has been utilized in seam welding of a wide variety of electronic component device cases. The first high production application for laser sealing of an electronic device was reported by Lockyer (1974). Utilizing a YAG laser pulsing at rates of up to ten times per second welds at rates of 15 cm/min or less were demonstrated. (Today, several relay manufacturers are utilizing equipment with pulses at rates in excess of 100/s and weld at over 150 cm/min.)

Laser welding of relay cans was shown to be an effective way of ensuring that each package is hermetically sealed. In sealing relay cans, a fusion seam weld is accomplished with the high repetition rate pulsed Nd:YAG laser by

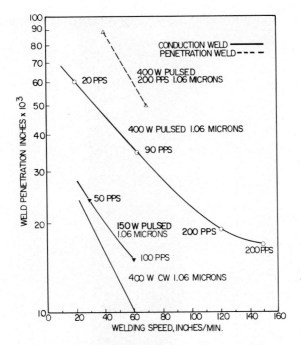

Fig. 8.6. CW and pulsed Nd:YAG weld penetration.

laying down individual spots with 80–90% overlap. In this application the laser is focused onto a 0.75 mm diameter spot so that the energy from the laser beam "couples" into the metal efficiently due to the high "peak power." Thermal effects beyond the actual 0.75 mm spot are almost nonexistent and so there is little chance of damage to the package, the components, or glass-to-metal seals.

Although an atmosphere or vacuum is *not required* for laser welding, the ambient conditions can be used to enhance the weld appearance or to meet other requirements of the package being sealed (such as protection of contents from oxygen or moisture). Cover gas is flowed over the weld area to improve the appearance of the weld. Packages can be welded in a vacuum chamber with a window for the laser beam or in a chamber filled with a gas such as helium or nitrogen.

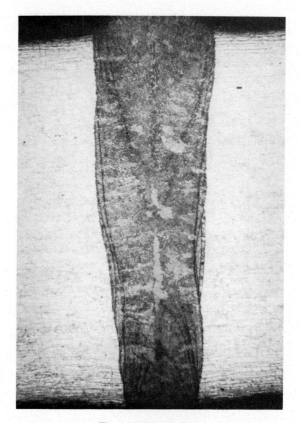

Fig. 8.7. Keyhole weld.

Laser welding has proven to be extremely useful in this application because of its ability to produce welds near heat sensitive glass-to-metal seals. Fig. 8.9 shows a typical laser welded relay case. The case shell is made of a cupronickel alloy, while the header is of kovar or low carbon steel. The inset drawing illustrates how glass-to-metal feedthroughs are located as close as 1.25 mm from the edge of the fusion weld.

A list of proposed advantages and disadvantages for various alternative sealing techniques is given in table 8.4. In practice, the driving considerations are producing a product of acceptable quality most economically. Such factors as product volume, equipment now in house, and other

Fig. 8.8. Laser welded tantalum diaphragm.

Table 8.4
Comparison of alternative sealing methods

Method	Disadvantages	Advantages
Soldering	Long heat-up time – heat conducts to adjacent area, flux clean-up messy and adds chemical into process, which may be undesirable Slow process – high labor content	Low "toolup" costs
Epoxy seal	Epoxy material can get into can undetected; material requires "curing" time Process limited to certain types of leak testing Slow process – high labor content	Low "toolup" costs
Brazing with induction heating	Induction heating slow and tends to heat up large area of can Flux removal problems; magnetic flux around package can damage components; package can warp from general heating	Non-contact process; tooling cost moderate
TIG weld E beam	High heat conducts into adjacent areas; heat sinking required Welding must be done in a vacuum; high initial investment; large floor area required – high heat can transfer to "lead throughs" and components inside; heat-sinking or shielding is usually required High initial investment Vacuum pull down time required	Clean weld – no flux needed
Pulsed YAG laser	High initial investment	Fast, clean welds; no vacuum required; no cleaning required afterward; high reliability welds at high speed; highest yields of all sealing processes; easily automated; compact system size; no distortion; no damage to can, contents, or "lead-throughs;" low tooling costs; lower unit weld costs; caps may be handled immediately after welding

Fig. 8.9. Laser welded relay case.

non-technical issues will also strongly affect the selection of a sealing process.

Pulsed Nd:YAG laser welding has also been applied advantageously to the welding of wet tantalum capacitors (Bolin 1976b). These capacitors consist of a pressed powder tantalum slug submerged in a sulfuric acid electrolyte. This is contained within a pure tantalum case, which must be hermetically sealed. A glass-to-metal seal insulates the central feed-through wire from the cell body. The heat required to weld tantalum by a technique such as inert gas arc welding leads to the cracking of the glass seal and failure of the unit. Using the Nd:YAG laser, the total heat input is normally very low and can be controlled so as to require virtually no heat sinking. The result is welding without damage to either the glass seal or the capacitor contents.

Nd:YAG laser welding has been utilized in the sealing of a variety of lithium batteries. Use was first reported in sealing custom batteries for military applications (Bolin and Maloney 1974) and later reported in the sealing of batteries used in heart pacemakers and consumer applications (Bolin 1980a). Again, the low heat input of the Nd:YAG laser is the key to success in this application. Glass-to-metal seals, as well as reactive battery contents, require use of a welding technique producing negligible temperature rise during the welding process. Another major consideration is that redesign of the battery to place the closure weld in a more remote position is not feasible, as this would waste considerable volumes in batteries designed to be extremely compact power sources.

Another reported application (Bolin 1980b) requiring welded construction of the highest quality is the production of heart pacemaker devices. Attempts to utilize non-welded construction (primarily potting in silicone rubbers) resulted in surgery and expensive recalls for several companies. Today laser welding is the most widely accepted technique for producing hermetic welds in the titanium and stainless steel cases. The total production of the pacemaker industry amounts to some 200 000 units per year, produced by about 35 companies. The four largest companies account for approximately 80% of the total production and utilize their own laser welders.

A particularly difficult microcircuit sealing task was reported using a pulsed Nd:YAG laser. Many circuit packages used in microwave applications are machined from solid blocks of aluminum. Various techniques of producing hermetic seals are successfully utilized, including tungsten inert-gas welding and electron beam welding. However, for many packages, either of these techniques will introduce excessive heat. The small size requirements for packages such as these result in placement of glass-to-metal seals extremely close to the weld area. In the case of aluminum packages, this is particularly critical because such seals must be *soldered* into the case. Excessive heat from the welding process will melt the solder and produce a reject package. In order to apply the laser, it was first necessary to overcome welding difficulties inherent in aluminum – specifically, the tendency to produce centerline weld cracks in aluminum alloys with desirable machining properties. These cracks appear because of stresses arising during the process of cooling from the melting point to room temperature. It was first attempted to relieve these stresses by modification of the part design. Specifically, the area around the weld was machined to produce a thin lip-like structure. This proved to be partially successful since the corners were extremely difficult to machine to the required accuracy, and cracking still occurred in these areas.

The second (and successful) approach taken was to alter the composition of the weld joint by the addition of a filler metal. Specifically, type 4047 aluminum alloy was added to the weld area, as is shown in fig. 8.10. The 12% silicon content of this material prevents the occurrence of centerline cracks by increasing the high temperature strength of the metal. This

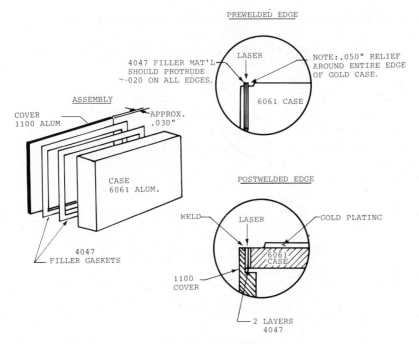

Fig. 8.10. Configuration of aluminum microwave package showing placement of filler metal.

process is now utilized by several microwave components manufacturers making similar products.

The ability to produce a unique welded product makes the Nd:YAG laser useful in performing a variety of sheet metal welding tasks. Two pulsed YAG lasers are used by a manufacturer of electric ranges to produce a continuous hoop of steel used as the trim rings for their electric ranges (Bolin 1980b). Pieces to be welded are sheared from a continuously formed coil of steel. Mating edges are finished so as to produce a tight fitup with a maximum of a 0.005″ gap in metal 0.020″ thick. This is illustrated in fig. 8.11.

The ability of the YAG laser to produce high quality welds even when considerably out of focus makes it unnecessary to fully contour motion to follow the irregular shape of the part. This property (referred to as the depth of focus) greatly simplifies the tooling in the trim ring welding task. The butt welded part made of carbon steel, and chrome plated after laser welding, replaced a part made of stainless steel which used a backup piece

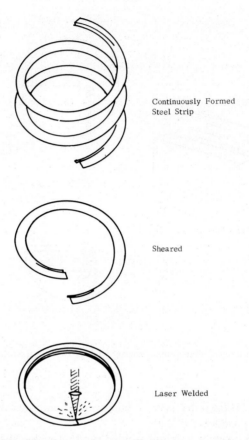

Continuously Formed
Steel Strip

Sheared

Laser Welded

Fig. 8.11. Manufacture of laser welded electric range trim rings.

spot welded to the two sides of the ring. Thus, a better quality part was produced from lower cost material.

2.4. Soldering and brazing

Where the laser is not a suitable welding tool, it may serve as a useful heat source for a new form of precision soldering or brazing (Naugler 1978).

Typically, 1–4 ms pulse lengths have shown promise for reflow soldering tasks. Many of today's electronic components cannot withstand soldering temperature for long periods of time. With the application of a burst of laser pulses at relatively low repeat rates (25–50/s) soldering can take place in less than 1 s. These process parameters ensure precise heat input leading to high product reliability. An automated processing system, including the capability to feed and position components and control laser process variables, can attain high throughput rates.

Brazing of small parts can often be replaced by fusion laser welding. In one application in production, a part slightly over one inch long is assembled from four components – three of stainless steel, and the fourth of molybdenum. The stainless steel components include a tube of stainless steel 0.075 mm in diameter, with walls five mils thick, and another tube of stainless steel with a 0.25 mm wall thickness, both welded into a solid stainless steel body. An end cap of pure TZM molybdenum alloy is attached to the second stainless steel tube in what amounts to a self-brazing process wherein only the steel is melted by the laser. Two distinct advantages of the laser process are freedom from deformation of parts due to the confined nature of delivering heat, and the elimination of the unwanted heat treatment of adjacent material resulting from the precise control of the laser power delivery.

An extensive study of laser brazing was conducted by Witherell and Ramos (1980). Brazes which were silver, nickel, copper, gold, and aluminum based were studied. It was found that laser brazing produced deposited metal which was both finer grained and exhibited a greater hardness than that exhibited from conventional brazing. It was hypothesized that this increase in hardness would be advantageous in improving strength in the ductile alloys (copper, silver, gold, and aluminum) but might lead to brittle joints in the case of alloys which were already quite hard (such as nickel brazing alloys).

3. Percussion drilling

When a high intensity coherent laser beam is focused to a spot of extreme brightness ($> 10^6$ W/cm^2), it will interact with metals and other materials such that the absorbed radiation converted to heat results in controllable material removal.

Some of the advantages of using a laser for drilling are obvious; others are not.

(1) Problems such as drill-bit breakage and wear associated with conventional drilling techniques are nonexistent.

(2) Accurate hole location is simplified because the optics used to focus the laser are also used to align and locate.

(3) Large aspect ratio holes (depth to diameter) can be achieved.

(4) The laser process is easily automated for high-speed hole production.

Early researchers assumed that the material was removed by direct evaporation. Later researchers realized that the laser energy was insufficient to vaporize the amount of material removed and put forward a liquid–vapor model, according to which the material was removed in the liquid as well as the vapor state. However, the mechanism depends upon the laser power density on the workpiece. The liquid–vapor ratio will reduce as the power density increases. Thus, the liquid–vapor ratio can become extremely small, and the limiting case will be a solid-vapor model. This is the case in which Q-switch lasers are used for material removal. The general picture of the power balance diagram given by Chun (1969) is shown in fig. 8.12.

Holes from 0.25 mm to 1.25 mm diameter can be drilled to depths of 0.6 in in most alloys. The drill process employs lasers which generate a maximum beam energy of 50 J per pulse. The laser beam is focused onto the workpiece to melt and expel the material. The number of repetitive pulses required to drill through a workpiece depends on its thickness, the hole angle, and the process parameters. Holes are slightly tapered and have a thin layer of resolidified or "recast" material.

Variable parameters include: aperture limiting diameter, focal length of the lens, depth of focus, beam energy, and number of pulses. Pulse length is approximately 1 ms for most operations. Beam energy is controlled by level of the laser power supply capacitor charge and is monitored by a photodetector placed behind the partially reflective rear mirror of the resonator.

Drill studies have established data, which are guides for selecting parameters for new applications. The results of a study reported by Jollis (1979) is shown in table 8.5. Diameters and depths achievable with the laser compare favorably with other nonconventional drill processes, as shown in fig. 8.13.

The practical range of hole sizes which can be successfully drilled on center with a pulsed laser is a function of a number of factors. In the lower size limit, the limitations are optical; for larger diameter holes, the practical limitation is the energy per pulse available from current industrial laser equipment.

If repeated laser pulses are delivered to the same area, deeper holes can be drilled, as is shown in table 8.5. This suggests that it is more efficient to drill deep holes with multiple laser pulses of low energy than with a single high energy pulse. Holes resulting from multiple pulse drilling are less tapered and better defined than single pulse holes.

Pulsed laser drilling offers cost savings over other techniques in areas which can accept the limitation of the process. Because of the complex

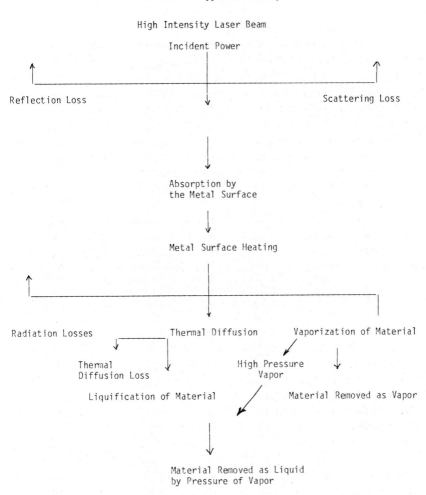

Fig. 8.12. Power balance diagram.

nature of the interaction between the laser beam and the materials to be drilled, a precision hole is not possible in materials more than a few thousandths of an inch thick.

Fig. 8.14 shows how a typical laser drilled hole might appear in cross section in a metal sample. Uncontrollable diameter variations exist, starting with the funnel-shaped entrance to the hole. A thin layer of "recast" clings to the side walls of the hole. This material has been melted and resolidified

Table 8.5
Percussion drilling parameter study, stock thickness (in) = 0.100

Diameter (in)	Lens focal length (in)	Front cavity aperture (in)	Output energy (J)	Depth of focus (in)	Pulses	Taper (in)
0.013	4	0.250	13	−0.085	1	0.0025
0.015	5	0.250	13	−0.030	1	0.0065
0.016	5	0.250	14	−0.033	1	0.0074
	6	0.250	13	−0.098	1	0.0022
0.017	4	0.275	15	−0.033	1	0.0060
	5	0.250	15	−0.034	1	0.0083
0.018	4	0.250	19	0.008	1	0.0114
	5	0.275	13	−0.034	1	0.0053
	6	0.250	14	−0.031	2	0.0065
0.019	4	0.300	15	−0.051	2	0.0044
	5	0.300	12	−0.086	2	0.0010
	6	0.250	15	−0.038	2	0.0072
0.020	4	0.325	14	−0.073	3	0.0017
	5	0.300	13	−0.081	2	0.0024
	6	0.250	16	−0.045	2	0.0079
0.021	4	0.325	15	−0.056	3	0.0034
	5	0.300	14	−0.074	2	0.0038
	6	0.275	14	−0.057	2	0.0045
0.022	4	0.350	14	−0.073	4	0.0008
	5	0.325	11	−0.050	3	0.0004
	6	0.300	12	−0.071	3	0.0008
0.023	4	0.350	14	−0.040	4	0.0017
	5	0.325	12	−0.043	3	0.0017
	6	0.300	14	−0.070	3	0.0020
0.024	4	0.350	15	−0.025	4	0.0030
	5	0.325	12	−0.010	3	0.0024
	6	0.300	14	−0.068	3	0.0031
0.025	4	0.350	16	−0.009	4	0.0041
	5	0.350	13	−0.089	4	0.0004
	6	0.300	16	−0.098	3	0.0039
0.026	4	0.350	16	0.026	4	0.0039
	5	0.350	12	−0.044	5	0.0007
	6	0.325	12	−0.055	4	0.0004

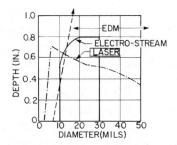

Fig. 8.13. Comparison of achievable hole sizes using various non-conventional drilling techniques.

in such a short time that the frozen metal has a different structure than that of the parent material. Depending on the alloy being drilled, this recast is often of considerable hardness and prone to developing microcracks. However, recast and microcracks are generally limited to 0.05 mm depths. In spite of these handicaps, numerous applications have developed which take advantage of the speed and low cost with which one may drill in very hard and tough materials.

One of the first commercial laser applications was the initial piercing of diamond dies for wire. Each year, thousands of natural diamonds and synthetic composite diamond forms are drilled to precise diameters for the production of dies ranging in diameter from under 0.025 mm to over 1.25 mm and depths to 12.5 mm. This application (believed to be the first industrial use of lasers) was developed by Western Electric Company in 1962 and put into production in 1964. The first system used a ruby rod as the laser medium and operated at pulse rates of 1 or 2 Hz. Current equipment utilizes Nd:YAG and operates at pulse rates of 20 Hz and greater.

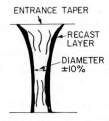

Fig. 8.14. Typical cross section of laser drilled hole.

A feature of early lasers used in this application was a closed-circuit television system connected to the completely shielded work enclosure surrounding the processing operation. With this equipment, the operator was able to monitor an image of the drilling, magnified 60 times, on a TV screen.

The Swiss and Japanese watch industries routinely use flash pumped Nd:YAG systems operating at approximately 20 pps with a pulse width of about 100 μs and an energy of approximately 150 mJ/pulse to drill ruby stones used in timepiece movements. The hole diameter is about 30 microns and requires from 3 to 6 pulses for complete penetration.

A more spectacular application is the drilling of cooling holes in jet turbine components (Heglin 1971). Holes are typically drilled on automated equipment in which the part is mounted on a five-axis CNC machine. The pattern to be drilled in a single piece may include several hundred holes of various diameters and depths at unique angles to the surface. All of this can be accomplished by programming data which has been empirically determined on "set-up" pieces.

Other examples of commercial products produced by percussion laser drilling are fiberglass spinnerettes, bleed holes in hydraulic brakes, and commercial food processing sieves used to debone pieces of meat.

4. Cutting and large hole drilling

By overlapping drilling pulses, a continuous cut may be produced using a high performance pulsed Nd:YAG laser in the same manner that spot welds are overlapped to produce a continuous seam weld. This technique is also employed for producing holes beyond 0.75 mm diameter using beam rotating devices. Quite often a gas jet of oxygen or air is used to assist in removal of the molten metal from the cut zone. Without the gas jet, this molten metal would flow into and refill the area cut by the laser. Fig. 8.15 illustrates a contoured laser cut in progress.

Engel (1978) made a comparison of holes drilled by percussion drilling with those drilled using a gas assist and rotating optics. In general, it was discovered that the use of gas assist and rotating optics produced the following improvements over percussion drilling:

(1) improved hole diameter accuracy and repeatability;

(2) improved hole straightness; and

(3) thinner recast layer on the hole side walls.

Terrell (1980) evaluated the laser process for producing relatively small holes in 2.36 mm thick Hastelloy-X large combustor liners. Traditional hole

Fig. 8.15. Contour laser cutting with 400 W pulsed YAG laser.

production methods were too expensive or too imprecise. Over 9000 holes were generated in each combustor liner ranging in size from 1.17 mm to 12.5 mm diameter. Most of the holes were specified to enter at 25 degrees from the normal of the workpiece.

Pertinent data resulting from this study are:

(1) The diameter control by means of the lens defocus method has a useful range of 0.30 mm through 0.46 mm diameter.

(2) Large holes with a diameter of 0.45 mm (0.018 in) through 2.5 mm are produced more effectively by cutting a circular slug out of the workpiece by trepanning. This method provides a way to control hole diameter tolerances within 25 μm consistently.

(3) Hole sizes over 2.5 mm diameter and other shapes were generated by contour cutting.

Some of the results of this development project were:

(1) the production of functional holes by the least expensive method;

(2) flexibility to change the hole size or location at will; and

(3) inexpensive tooling due to zero tool pressure produced by the laser beam.

It was necessary to find a method to prevent cratering at the hole perimeter from the initial dwell required for laser beam penetration, and to remove resultant burrs consistently and uniformly. This was accomplished in four operations:

(1) To prevent the cratering effect, the laser beam first is moved to the center of the hole and allowed to burn through the part. The beam then is moved, cutting a kerf to the edge of the hole. The lens makes one complete revolution, the slug drops free, the beam moves back to the hole center line and is deactivated.

(2) The burr on the hole exit edge is minimized by applying a liquid laser coat on the bottom surface of the part prior to laser cutting. The laser coat dries rapidly to a powdery chalklike substance that can be removed by water wash.

(3) Burrs are removed flush with the adjacent parent material's top and bottom surfaces by a manual light grinding operation.

(4) A subsequent grit blast operation produces a small uniform break on the hole edges.

Benedict (1980) compared cut surfaces obtained using air, nitrogen, and argon for assist gases. When cutting with air, the cut edge is left with a layer of oxidized, remelted material typically 0.002–0.076 mm deep. This can be removed by conventional means such as hand finishing, reaming, bead blasting, or abrasive flow machining. If the lasercut edge is to be welded in a subsequent processing step, the requirements for recast removal can be eliminated by cutting with argon or nitrogen. Cutting with these gases reduces cutting rates and maximum penetration by 15 to 25%. All lasercut recast layers have residual stresses and may exhibit microcracking, which will affect the fatigue characteristics of the part.

The ability of the laser to cut any material makes it suitable for applications such as cutting diamond machining tools, as well as the tough alloys of aircraft engine components. The programmability of laser cutting, especially in a contouring mode, makes it ideal for short or medium runs (such as prototype or special parts). Typical production applications of laser cutting are limited to metals under 4 mm inch thick, although up to 10 mm thick pieces of Hastelloy-X have been cut with good quality edges.

5. Heat treating

As a precision, high intensity heat source, the YAG laser would seem well suited to performing a wide variety of heat treating tasks. Heat treatment is typically carried out for one of two purposes. One purpose is to soften a metal by raising the temperature to a point where new crystal grains form which do not exhibit the strains present from previous mechanical processing. This is referred to as annealing. The other purpose is to harden a metal by causing the precipitation of a second harder phase (such as iron carbide) within the crystalline matrix. Both processes have been demonstrated using the pulsed YAG laser as a heat source. Tice (1977) utilized 10 and 20 ms pulses to anneal selected areas of CDA-510 phosphor bronze springs. It was found that because the total heat input was more controllable than would be the case for a furnace anneal, the process could be optimized to give a specific, desired amount of anneal. In the traditional furnace anneal, processing is so rapid that only a dead-soft condition can be produced.

Naugler (1978) reported on the use of pulsed YAG in hardening of saw blade teeth of high carbon steel. By controlled pulsing, the hardening was easily confined to the cutting tips of the teeth only, thus preventing hardening of valleys between teeth and preventing cracks associated with such hardening produced by standard techniques.

Although the majority of laser heat treating is performed using CO_2 lasers due to their lower initial cost per watt of output power and their lower operating costs, the YAG laser offers the advantage of high absorbtion on most metals. Metals which are to be heat treated using CO_2 lasers are typically coated with a black material to decrease their reflectivity. Such coatings are unnecessary when the YAG laser is used.

Heat treating rates achievable are limited for pulsed equipment in the same manner as welding and cutting rates are limited by the maximum achievable pulse rate. The use of a high power CW YAG laser overcomes this limitation and such units are used in several commercial applications.

References

Bauer F.R., 1978, Laser welding of electrical interconnections, Topical Report BDX-613-2022 prepared under DOE contract DE-AC04-76-DP00613.

Beall R.W., 1977, Assembly repair of aircraft engine parts using pulsed YAG lasers, in: Program Proceedings, Electro-Optics/Laser 77 Anaheim.

Benedict G.F., 1980, Production laser cutting of gas turbine components, Society of Manufacturing Engineers, Technical Paper, MR 80-851.

Bolin S.R., 1976a, Precision laser welding, in: Program Proc., Electro-Optics/Laser 76, Boston.

Bolin S.R., 1976b, Australian Welding Journal, January–February 23.

Bolin S.R., 1980a, Assembly Engineering, May.

Bolin S.R., 1980b, Assembly Engineering, July.

Bolin S.R. and Maloney E.T., 1974, Limited penetration welding, Society of Manufacturing Engineers, Technical Paper MR 74-956.

Bolin S.R. and Maloney E.T., 1975, Precision pulsed laser welding, Society of Manufacturing Engineers, Technical Paper MR 75-571.

Bosna A.A. and Emmel J.D., 1980, Use of lasers in high speed termination of flexible printed wiring, in: Proceedings of Flexicon, Washington, DC.

Chun M.K., 1969, PhD Thesis, Renselaer Polytechnic Institute, p. 34.

Engel S.L., 1978, Optical Engineering, 17:3 235.

Gagliano F.P. and Zaleckas V.J., 1972, Laser Processing, Lasers in Industry, ed., Charschan S.S. (Van Nostrand-Reinhold, New York) p. 212.

Geusic J.E., 1964, Appl. Phys. Lett. 4 182.

Heglin L. Michael, 1971, Laser drilling applications and equipment, in: Proc. IEEE Machine Tools Industry Conference.

Jollis A.U., 1979, Small hole drilling and inspection with pulsed laser systems, American Institute of Aeronautics and Astronautics AIAA, SAE, ASME 15th Joint Propulsion Conference.

Lockyer J., 1974, A numerically controlled laser for hermetic sealing of relays, in: Proc. of National Relay Conf., Stillwater, Oklahoma.

Marshall H.L., 1976, High-power, high-pulse-rate YAG laser seam welding, Proc. of the Society of Photo-Optical Instrumentation Engineers, 86.

Moorehead A.J. and P.N. Turner 1970, Welding Journal 49 15.

Naugler T.W., 1978, Applications of pulsed, solid state lasers, in: Program Proceedings, Electro-Optics/Laser 78, Boston.

Palmquist J.H. and Muniz J.M., 1977, Enhanced precision welding through optical position detection, in: Program Proceedings, Electro-Optics/Laser 77, Anaheim, California.

Terrell, N.E., 1980, Laser precision small hole drilling, Society of Manufacturing Engineers, Technical Paper MR 80-849

Tice, E.S., 1977, Laser annealing of copper alloy 510, Electro-Optic Lasers, San Diego, CA, USA.

Witherell C.E. and Ramos T.J., 1980, Welding Journal 59 267S.

CONSIDERATIONS FOR LASERS IN MANUFACTURING

SIDNEY S. CHARSCHAN and ROBERT WEBB

Western Electric Company
Princeton, New Jersey 08540, USA

Laser Materials Processing, edited by M. Bass
© *North-Holland Publishing Company, 1983*

1. Introduction

In the last twenty years lasers have become bigger, faster, and more controllable, while material interaction phenomena have become more understandable. Laser processing has evolved from the challenging one-of-a-kind difficult to process applications to products designed to take advantage of the laser's unique ability. New areas for laser applications have developed. It has become obvious that the laser could and should be considered as another machine tool. Fixturing has been improved, numerical control and computer operations introduced, and technology advanced so that the laser is now a critical component of many automated systems. Today's laser processing not only provides a competitive technology with appreciable economic benefits, even for older products, but offers capabilities undreamed of 20 years ago.

Concurrent considerations and evaluations are required prior to making decisions relative to employing a laser in manufacturing. There are two primary interdependent questions: technical feasibility and economic justification.

Assuming technical feasibility to be established by prior history or studies performed in application laboratories on prospective suppliers premises, this chapter will attempt to place in perspective the other considerations. We will provide guidelines for economic justification; give sample specifications for system performance and evaluation; review the preparation required for system installation, operation, and maintenance; make recommendations relative to spare parts and servicing; and last, but not least, clarify the qualification and training requirements of associated personnel.

2. Economic justification

Before committing capital funds there is much background information that needs to be gathered. It should, however, be noted that there are many situations where a detailed study is not warranted. One such example is a semiconductor wafer surface inspection system whose cost is $2000 where the "commercial" system has a unique function well known and accepted in the industry. Another obvious example could be an IC device designed to be laser trimmed to value, where there is no alternative technology to a laser trimmer. Generally, if a system needs to be designed and built for special applications or where the capital acquisition cost is appreciable we recommend that an economic evaluation be made.

2.1. Background information

2.1.1. Quantities
Both annual requirements and quantities should be estimated for the next four or five years – whichever is most acceptable for amortizing purposes and cost determination.

2.1.2. Operating mode
Based on the peak requirements, are 1, 2, or 3 shifts acceptable for production? A decision is required in order to determine both the amortization basis and the available hours for production.

2.1.3. Thru-put
Estimate the cycle-time and downtime for the laser operation. In lieu of more specific information, relative to downtime, an estimate of 1500 h per shift is acceptable for most machine tools. Together with the prior information of the first two items we can now determine how many systems are required.

2.1.4. Costs
Quotes can be obtained (see section 3.3) for the required number of systems, and estimates can be established for the cost of the laser operation.

Estimated capital and operating costs. In many cases the laser is approximately 50% of the system cost. Such items as fixturing, material handling equipment, optical train, computer control, frame support, and covers designed for safe operation comprise the remainder. Reasonable estimates can be made of the laser costs from fig. 9.1 for CO_2 and pulsed Nd:YAG lasers respectively or from trade journals which publish annual Buyers' Guides (Laser Focus, Buyer's Guide; Optical Spectra, Product Table Supplement; EOSD, Vendor Selection Issue).

Operating costs should include: (a) power (which at this time ranges from $0.06/kWh–0.12/kWh); (b) gas, primarily He for CO_2 laser operation at $0.15 ft^3 at STP; (3) water, which we estimate at $0.007/gallon.

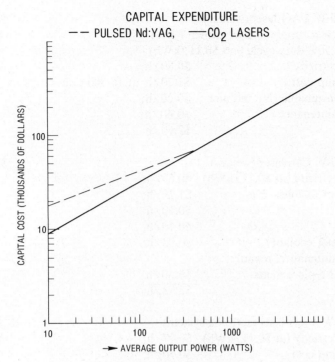

Fig. 9.1. Capital costs of industrial lasers.

The hourly cost could be increased by a factor of 10 if other inert cover gases are required, or O_2 for cutting metals. In addition, a maintenance cost, as well as an estimated cost of repairs per hour should be considered and may be obtained from the vendor.

Estimated operating costs per hour for CO_2 lasers (based on manufacturers catalogs), excluding overheads, maintenance and repair items, are shown in fig. 9.2. Note should be taken that the closed cycle systems – which recycle and minimize gas usage – do have appreciably lower cost per hour per watt operation.

Typical examples. This section illustrates some typical tabulations of estimated costs for several different lasers. Maintenance items are included. One manufacturer, in addition, estimates "Repairs" at 10% of initial capital cost over 5 years; equating to $0.70/h for a 350 W Nd:YAG laser welder.

(a) 150 W YAG laser (welder) $40 000
    (Power supply: 7 kW, chiller 7 kW)
    (at 50% duty cycle and $0.11/kWh)
    Electricity $0.80/h;
    Lamp cost $1.20/h at 16 700 pph;
    Consumable (N_2 and air) $0.20/h;
    Maintenance $0.40/h;
                                   $2.60/h

(b) 500 W CO_2 laser (welder) $80 000
    Electricity (at $0.11/kWh) $0.69/h;
    Consumables: He $1.20/h;
              N_2 $0.10/h;
              CO_2 $0.10/h;
    Water coolant $0.03/h;
    Maintenance, repair
    and replacements $0.80/h;
                                   $2.92/h

(c) 20 W YAG laser (scribing system) $100 000
    Electricity (at $0.11/kWh) $0.77/h;
    Lamp cost $0.50/h;
    Other consumables $0.20/h;
    Maintenance $0.40/h;
                                   $1.87/h

2.2. Economic decision making

There are several bases for deciding to use a laser process – many have been
well illustrated (Charschan 1972).

2.2.1. New products
Which may be more difficult or even impossible to process by conventional
methods.
    Drilling small holes in hard or abrasive materials.
    Welding in mechanically inaccessible areas.
    Trimming extremely small electronic devices.
    Welding electrically dissimilar materials.
    Contouring metals and ceramics.

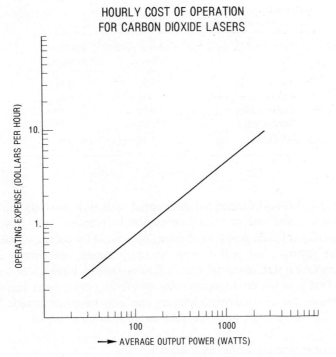

Fig. 9.2. Hourly operating costs for CO_2 lasers.

Welding heat sensitive components.

Selective heat treating of semiconductors and metals – with depth control.

2.2.2. Existing products

Where the newer laser technology may offer significant advantages.

High thru-put capability – adaptable to automation – visible "weld nugget" – small kerf requiring minimal finishing – a sharp tool that does not wear out, need dressing, nor make physical contact.

2.2.3. Savings

It is evident that laser processing offers a competing technology which needs to be evaluated. A simplified comparison of heating costs per semiconductor wafer for a conventional furnace versus a laser surface treatment method is shown in table 9.1 (Kaplan et al. 1979).

Table 9.1
Comparison of heating cost per wafer (Kaplan et al. 1979)

	Conventional furnace	Laser cold processing
Amortization	1.3¢	3.2¢
Utilities at 6¢/kWh	1.8¢	0.4¢
Consumables	0.8¢	0.4¢
Maintenance	1.0¢	0.3¢
TOTAL	4.9¢	4.3¢

More detailed and formalized discounted cash flow analysis can also be used to determine rate of return on capital investment, cash flow rate of return, and breakeven point. Such input data would require estimations for first year savings, five-year average savings, annual operational expense, development capital, discount rate, inflation factors, liquidation factor for capital, first year tax credit, scrap value of existing equipment and tax rate. Comparisons for alternative techniques can also be readily made by such analysis. In many cases, however, such detail is not warranted.

2.3. Example of a successful application

An example of a computer-controlled high-speed laser system is the spot welding of terminals on miniature relays for Western Electric (Muniz and Califano 1978). The system is four times faster than the resistance-welding equipment it replaced.

Over a million of the relays are turned out annually by the company for use on printed-circuit boards in telephone transmission equipment. To properly fit the boards, the relay terminals must be welded to phosphor bronze tabs on adapter plates (see fig. 9.3).

Before the laser system, inspection and mechanical testing for bond strength often ruptured otherwise acceptable spot welds. The tests were necessary because welds created by the earlier resistance-welding system were completely hidden – they could not be inspected visually.

Laser welding produces a highly visible, shiny nugget on the terminals, reducing inspection to a simple visual check. Reliability is additionally increased by making two welds on each tab.

The spot welding system (see fig. 9.4) consists of a 200 W average power pulsed laser and an $X-Y$ positioning system with 0.0001 in (0.00024 cm)

Fig. 9.3. Laser welded relay tabs.

accuracy. Both units operate under the control of a Motorola 6800 micro-processor with four different programs that can be switch-selected to accommodate different configurations of relays. The microprocessor moni-tors the $X-Y$ table position by means of linear encoders and anticipates targets within a 30 ms period, thus permitting welding "on-the-fly" at speeds up to 5 in (12.7 cm) per second, and 20 welds per second. Up to 30 relays are loaded on a fixture that positions each phosphor-bronze tab within 0.002 in. The $X-Y$ table indexes the fixture to preprogrammed positions within the focal plane of the laser beam. Welding pulses lasting for 5 ms and focused to approximately 0.020 in diameter contain about 8 Js of energy per weld at an intensity of 8×10^5 W/cm^2.

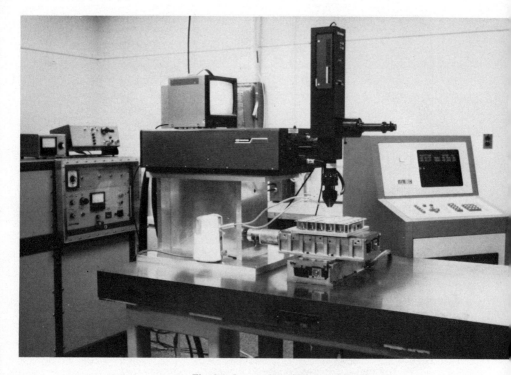

Fig. 9.4. Laser relay welding system.

In this example laser welding was introduced to replace resistance weld-
ing of the relay tabs. Preliminary evaluation of the potential savings used a
discounted cash flow analysis requiring the following "operating assumptions", for an estimated system cost of $76 000:
Laser welding rate four times faster than resistance welding.
Total requirements to be made on one system.
Two shift operation of 1500 h per shift.
The analysis clearly indicated, and experience has borne out, that the
breakeven point could be achieved in appreciably less than one year.
Technical feasibility established that the "operating assumptions" were
reasonable. Detailed consideration was given to developing a system concept involving the laser, work handling, optics and electronics, enclosure
and viewing system, installation area, services such as power and water,
maintenance, and training of personnel. The remainder of the chapter
details the typical steps involved in the implementation of a successful laser
application.

3. Specifications for system

3.1. Technical specifications

3.1.1 Laser type

In most cases the nature of the interaction of the laser beam with the target material will dictate the choice of operating wavelength, which in turn will identify the most appropriate type of laser for the specific application. The two most common types are the optically-pumped Nd:YAG and the electric discharge pumped CO_2 laser operating at 1.064 micron and 10.6 micron wavelengths respectively. Both types may be operated in either a pulsed or continuous manner, with considerable flexibility with regard to duty cycle and pulse duration. As a first step, therefore, it is important to establish what type of laser is most suited to the type of application at hand, whether it be a Nd:YAG or CO_2 laser, or one of the lesser used types such as ruby, argon, excimer, etc.

3.1.2. Output power and energy

The overall average power level capability of the laser should be specified next, and will bear directly upon the overall physical size and cost of the laser system. Margin should be set with regard to derating the actual operating average output power level sufficiently below the maximum power level achievable by the laser, in order to ensure reasonable lifetime of the laser and its expendible components such as lamps, mirrors and optical train. This also holds for the maximum peak power and energy in the case of pulsed lasers. A derating factor of 20% below maximum recommended values for peak and average power is suggested, while a factor of more than 50% probably means that a smaller laser should be considered instead.

3.1.3. Beam mode characteristic

The nature of the material interaction will usually dictate the most desirable optimum beam intensity profile. For the case of symmetry about the laser beam axis, the intensity distribution may be characterized by the Gaussian–Laguerre modes with transverse index notation referring to the spatial power distribution in cylindrical coordinates (Siegman 1971). The lowest order mode has maximum intensity at its center with exponential fall-off as the radius squared, while the next higher-order "mode" is really a superposition of two modes operating out of phase but simultaneously to give a "doughnut" intensity profile. Higher-order Gaussian–Laguerre modes are generally difficult to isolate and maintain in a stable manner, tending to mix together in a "multimode" situation having more or less flat intensity distribution with rapid fall-off at the edge of the beam. While such a multimode admixture may be well suited for laser welding and heat treatment applications, it is important to realize that "multimode" operation cannot be described easily and should not be compared directly between one laser and another without knowing specifically how intensity is in fact distributed in space.

3.1.4. Beam diameter and divergence

Having selected the type, average power and spatial mode distribution for a particular laser system, the next important step is to specify the laser beam diameter and divergence. Together these two parameters define a figure of merit for describing the effective brightness of the laser, since the product of beam divergence and beam diameter at a given point along the beam axis represents an invariant of the system, with the smallest product corresponding to the highest brightness. In measuring either beam diameter or divergence, a criterion must be defined and is generally selected as the $(1/e)^2$ intensity fall-off value whereby 86% of the intensity, for the case of a Gaussian-mode beam, lies within a radius defined as the $(1/e)^2$ beam waist. Alternatively, 50%, 90% or other beam waist criteria may be taken as convenient intensity drop-off points. The important factor is to specify which particular criteria are to be met in a given situation.

3.1.5. Spot size and depth of focus

Once beam diameter and beam divergence have been specified for the laser itself, the final spot diameter at the target must be spelled out. The same criteria used in above section 3.1.4 before focusing should now be used after the beam is transported by an appropriate set of optics to the final target region. The beam delivery optics must be carefully designed in such a way as to achieve this desired final spot size. Additional collimating telescopes and accessories may be required to provide the desired depth of focus with sufficient latitude for fine adjustment above and below the height of the workpiece.

3.1.6. Stability

Fluctuation of laser beam intensity with respect to time at the target zone is another problem for which acceptable limits must be set. For a continuous beam, both long and short-term variations in power level may result from electrical and thermal changes in both the power supply and fluid-cooled mechanical structure of the laser system. An upper bound must be specified for:

(a) long-term, average power drift;
(b) peak-to-peak variations for pulsed output;

(c) jitter with respect to timing synchronization;
(d) spurious pulsations or short term noise;
(e) percentage rms noise level; and
(f) pointing stability with respect to beam angle.

3.1.7. Spectral output

The spectral content of the laser beam may depend upon the temperature and operating power level of the system. In most industrial applications such as welding, drilling, and cutting this is of little significance. However, in holographic, atmospheric scattering and more scientific types of applications, the spectral variations will be of much greater importance. Fluctuations and variations will require additional means of control and should be specified precisely when required.

3.1.8. Lifetime

In optically-pumped solid state lasers (such as Nd:YAG) the mean time during which average output power remains within the stated specification will depend upon the condition of the excitation lamps. While average output may be increased by increasing average lamp current, a limit will be reached after a period of time whereby this is no longer possible. The rate of lamp deterioration usually will increase dramatically as maximum current rating is approached. A corresponding rapid loss of output power will accompany such lamp deterioration, and will set a maximum figure for the useful lifetime of the lamps. This may be specified either as the expected number of hours before lamp replacement, or the total expected number of firings for the case of a pulsed system.

For the case of gaseous discharge lasers (such as CO_2), the end seals of the discharge apparatus may be joined directly to a set of output coupling windows or mirrors. Since the latter are exposed to the discharge, they will generally tend to degrade after a period of time. Restoration to their original condition will require either cleaning or replacement, and again a concomitant decrease in output power will be observed. The mean time between shut down periods for cleaning and maintenance may be stated in the system specifications.

3.2. Legal considerations

3.2.1. Safety regulations

Conformance with both Federal and State safety regulations pertaining to both the manufacture and operation of all laser systems is the first legal consideration which must be addressed. The manufacturer is responsible for providing the following nine safety features on every high power (over class II) laser sold in the USA.

(1) Protective housing which prevents human exposure to laser radiation in accordance with the four stated classes of laser systems detailed in the US Bureau of Radiological Health standards (US Dept. of Health, Education and Welfare, 1979).

(2) Safety interlocks which prevent exposure to any portion of the laser beam in the event protective housing is removed.

(3) Remote controls which prevent all contact with laser emission when not electrically joined.

(4) Key control whereby only authorized personnel may operate the laser system.

(5) An emission indicator which provides a visual or audible signal when the laser is in use.

(6) A beam attenuator which dumps emission exceeding limits of a Class I system into a heat sink collector.

(7) Proper location of electric controls away from any emission area.

(8) Viewing optics safeguarded by proper shutters or attenuators in a failsafe manner.

(9) In cases where the laser beam is scanned, provision for interlocks in the event the scanning system should fail. Details of how such features are to be added to a laser system may be found in Sliney and Wolbarsht (1980).

In addition to the above nine features, the manufacturer is also responsible for proper labeling and attachment of BRH hazard warnings with wording in accordance with the four designated laser classes.

3.2.2. National electric code

Minimum standards of wiring design and installation practice in the USA are established in the National Electric Code. Its rules are written to protect the public from fire and life hazards, and are generally enforced in local municipal ordinances. As a general practice, the manufacturer of laser equipment is not bound by rules pertaining to the installation of electric wiring at a user's facility. However, the manufacturer is expected to provide equipment which is free from electrical fire and accident hazard. The user is expected to ascertain whether additional rules exist within his corporation and local community which apply to equipment connected to existing electric lines.

3.2.3. Electromagnetic interference

After October 1, 1981, the US Federal Communications Commission requires compliance with a new set of electromagnetic interference regulations. By the FCC definition, any industrial device which conducts more than 1000 μV from 450 kHz to 1.6 MHz or 3000 μV from 16 MHz to 30 MHz back into the power lines is prohibited. Furthermore, any device which radiates into the air beyond a 30 m distance is also subject to severe restrictions on the amount of radiofrequency interference permissible. Any products, including all forms and types of lasers, must comply with FCC regulations after October 1, 1981, while all products sold prior to that date have until October 1, 1983 to be retrofitted in such a way as to comply.

3.3. Sample specifications

The following set of laser parameters form a description of a system for marking silicon wafers. Together they represent a sample specification illustrating how requirements for a laser processing system may be spelled out. This sample indicates the type of laser and minimum hourly production thru-put together with information regarding how silicon wafers are to be marked.

Sample specification. This specification is for a silicon wafer serializing system incorporating a continuously-pumped laser which will deliver a train of Q-switched output pulses onto the rear face of silicon wafers. The wafers themselves will be delivered automatically in diameters ranging from 50 to 125 mm, at rates of up to 250 per hour on a continuous two shift per day basis. Serial information will consist chiefly of batch and wafer number, prescribed by a method already adopted.

Therefore, what is sought is a laser serializing system based upon the following set of laser parameters already proven to be effective in achieving desired results.

Laser type – continuously optically-pumped neodymium-doped YAG operating in fundamental Gaussian transverse mode while Q-switched at high repetition rate.

Matrix – serialized mark shall be formed from the continuous overlap of individual focused Q-switched pulses upon the substrate target. (Dot matrix whereby characters are formed from an array of few or several dots is unacceptable.)

Average power – 5.0 to 6.0 W at 1064 nm optical wavelength.

Repetition Rate – adjustable from 4.0 to 6.0 kHz.

Pulse duration – full-width at half-maximum for each Q-switched pulse to be 90 to 100 ns.

Character size – height: 0.115 in minimum, 0.125 in maximum;
width: 0.080 in minimum, 0.100 in maximum.

Line width – 0.0017 in minimum, 0.0020 in maximum.
Line depth – 0.0015 in minimum, 0.0020 in maximum.

Software – system must be capable of sequential marking utilizing the following format: NNNN A to NNNN ZZ, or: MMMMM A to MMMMM ZZ.

Loading/unloading – system shall be equipped to feed silicon wafers, polished face up, by fully automatic means onto a flat-finding station whereby pre-aligned wafers may be serialized automatically without contacting the polished faces in any way. The accompanying figure (fig. 9.5) illustrates the desired spacing and character to be marked onto the bottom unpolished face of each silicon wafer. After laser marking, each wafer is to be automatically reloaded into awaiting empty carriers.

Compatibility – the following types of existing carriers shall be accommodated by the automatic silicon wafer handling system:

model A-82-M for 75 mm diameter;
A-72-39-M 100 mm diameter;
A-72-50-M 125 mm diameter.

All manufactured by "Fluoroware".

Acceptance – final acceptance will be made at vendor's plant facilities before authorization shall be given to deliver system.

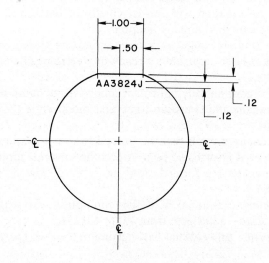

Fig. 9.5. Location of laser identification marking.

4. Manpower, space, and facilities

Three general aspects of a laser application must be considered because they impact on safety and influence the hazard evaluation and application of control measures.
(i) The laser system's capability of injuring personnel.
(ii) The personnel who may use or be exposed to laser radiation.
(iii) The environment in which the laser is used.
A laser classification scheme based on item (i) is recommended by the American National Standards Institute (ANSI) (Z.136.1) and mandated by the federal government (US Dept. of Health, Education and Welfare, 1979) for all manufacturers of laser product. The Federal Laser Products Performance Standard provides for classification labeling and engineering controls appropriate for the hazard. The ANSI standards provide guidance to the user for the various classifications of hazard in the areas of both engineering and administrative controls.

Let us consider the needs associated with the installation of a Class I laser system, with an embedded higher power laser. The normal considerations are power, water, exhaust, space, floor load, and an operator who will only work with a Class I system. If maintenance and/or service is to be performed, then additional training is recommended, and special precautions taken for working with the higher powered Class IIIb or IV laser. This section provides guidance for both cases.

4.1. Manpower

At this time their are no OSHA requirements relative to the safe use of lasers. However, there is a guideline (ANSI Z.136) which is being used by a major portion of the industry both in the USA and abroad, and is being considered in SSRL (Suggested State Regulations for Lasers) for adoption by each state. The standard depends upon first classifying lasers and laser systems according to their relative hazard (the manufacturer's duty) and then specifying appropriate controls for each classification for both the manufacturer and user.

4.1.1. Organization and the LSO
The management (employer) shall establish and maintain an adequate program for the control of laser hazards. Safety and training programs (Smith et al. 1975, Sliney and Wolbarsht, Laser Institute of America (a, b), Amer. Conf. of Gov't Industrial Hygienists) are not required for Class I

lasers and laser systems and for Class IIa laser products certified by the manufacturer in accordance with the Federal Laser Product Performance Standard and operated in the intended use (manufacturer's Class IIa Laser Product).

The program shall include provisions for the following: delegation of authority and responsibility for the supervision of hazard evaluation and control of laser hazards to a trained Laser Safety Officer (LSO). Depending on the extent and number of laser operations, the position of LSO may or may not be a full-time assignment. The Laser Safety Office (LSO) may be designated from among such personnel as the radiation protection officer, industrial hygiene officer, safety officer, laser specialist, laser operator, etc. The LSO may be a user if there are very few laser operations or if the class of lasers and potential hazards are low. Where the number and diversity of laser operations warrants, an associated Safety Committee may be formed and utilized. When there are normally no requirements for an LSO, such as in ownership of Class I lasers and laser systems and manufacturer's Class IIa Laser Products, which contain lasers with a higher classification, the designation of LSO for temporary periods of access for servicing, training, etc., may be the responsibility of the organization requiring access, such as a service organization. However, there shall be a designated LSO for all circumstances of operations of a laser or laser system above Class I and manufacturer's Class IIa Laser Products.

The LSO will have authority to supervise the control of laser hazards. Such authority may include, but is not necessarily limited to, the responsibilities of operation as specified below:

(A) provide consultative services on laser hazard evaluation and controls and on personnel training programs.

(B) Establish and maintain adequate regulations for the control of laser hazards.

(C) Have the authority to suspend, restrict, or terminate the operation of a laser or laser system if he or she deems that laser hazard controls are inadequate.

(D) Ensure that the necessary records required by applicable government regulations are maintained.

(E) Approve before use all protective equipment that will be used for control of laser hazards.

(F) Survey by inspection, as considered necessary, all areas where laser equipment is used. The LSO will also accompany regulatory agency inspectors for laser equipment, such as OSHA, FDA/BRH, state agencies, etc., and document any discrepancies noted. The LSO will ensure that corrective action is taken where required.

(G) Review new installations and modifications to installations that may increase the hazard to personnel or increase the classification of lasers, to ensure that the hazard control measures are adequate.

(H) On notification of a known or suspected accident resulting from operation of a laser, the LSO will investigate the incident and initiate appropriate action. This may include the preparation of reports to applicable agencies.

(I) Approve operation only if satisfied that laser hazard control measures are adequate. These include special operating procedures for maintenance and service operations within enclosed systems, and operating procedures for Class III and IV systems.

(J) Make certain that adequate warning systems and signs are installed in appropriate locations and should approve the location and wording of the signs.

(K) Make sure that adequate training programs are in effect for all employees using lasers.

Education of authorized personnel (operators and others) in the assessment and control of laser hazards. This may be accomplished through training programs.

Application of adequate protective measures for the control of laser hazards, as required.

Management of accident, including reporting accidents and preparing action plans for the future prevention of accidents following a known or suspected incident (manufacturers accident reporting requirements are detailed in the Code of Federal Regulations, 21CFR Subchapter J, Part 1002.20). A guide for the organization of a laser safety program is outlined in Appendix D of the ANS1 Z136.1, 1980 document.

4.1.2. Education

The management (employer) shall provide training on the potential hazards (including bio-effects), control measures, applicable standards, medical surveillance (if applicable), and other pertinent information to the LSO. The training shall be commensurate to at least the highest class of laser under the jurisdiction of the LSO. A safety training programme(s) shall be provided to the users of Class III and IV lasers and laser systems. Users shall include operators, technicians, engineers, maintenance and service personnel, etc., working with lasers. The training shall ensure that the users are knowledgeable of the potential hazards and the control measures on laser equipment they may have occasion to use. A guide for the organization of a training program is outlined in D6 of the ANS1 Z136.1 1980 document.

4.1.3. Implementation
The management (employer) shall provide adequate supervision, personnel training, facilities, equipment, and supplies to control potential hazards of lasers and laser systems.

Operation. The normal laser or laser system in a production mode would most likely be classified as a Class I system. In that operation, the performance of the laser product over its full range of functions is non hazardous, and only minimal instruction of the operator concerning the laser is recommended. Such an operation could simply involve material handling (load or unload) into a chamber whose door is interlocked to prevent inadvertent exposure to laser radiation and press the start button. More complex systems could involve rotary tables, or baffled feed system. However, in all cases, a Class I Certification claims a "safe" system and the operator should treat such a system as any other machine tool. For the operation of Class III or Class IV laser systems the operator should receive training as deemed appropriate by the LSO and the system shall be in a "Laser Controlled Area", see section 4.2.

Maintenance. As distinct from operation and service, maintenance is described as the performance of those adjustments or procedures specified in user information provided by the manufacturer with the laser or laser system which are to be performed by the user to ensure the intended performance of the product. Normally, such procedures may be performed by the operator or set-up man and do not provide access to laser radiation higher than the limit associated with the classification of the system. The adjustment may be as simple as turning up a variac to compensate for a loss in output; or replacement of a lamp, which may be comparable to changing a fuse. Whereas Class I and II laser systems are relatively non hazardous, maintenance of Class III and IV systems shall be only by authorized personnel with appropriate training, and in a Laser Controlled Area.

Service. Service is defined (ANSI) as the performance of those procedures or adjustments described in the manufacturer's service instructions which may affect any aspect of the performance of the laser or laser system. In most cases service personnel could have access to laser radiation in excess of

Class I limits, and therefore require a higher degree of training than either operator or maintenance man. In addition, appropriate control measure compatible with the hazard needs to be provided. Service may be performed by trained personnel from a Service Organization, the supplier of the laser system, or the owner of the laser system. It may be performed at the system or in a separate area set up for testing, alignment, etc. In either case a "Laser Controlled Area" (ANSI) is recommended, see section 3.2.

4.2. Space and facilities

Class I laser products are on the production floor side by side with other drillers, or welders, etc. Maintenance and servicing needs will impact on this decision. Class III and IV laser products need special consideration.

4.2.1. Normal installation

The first industrial application of the laser was to drill small holes in diamond dies (Epperson et al. 1966) used for drawing thin copper wire. As early as 1965 the Western Electric Company performed this operation with a pulsed ruby laser drilling system at 10 J, 0.5 ms pulse length, and focused to a 250 micron spot. Notice the coaxial closed system television (CCTV) in fig. 9.6.

The first industrial application of the laser saved time, and paid for itself many times over. However, the introduction of the laser into a production environment brought with it new and unanswered questions regarding operator safety. Is laser radiation harmful to the human body? Is the effect of laser radiation cumulative? How dangerous are the fumes resulting from the laser action on toxic materials? What minimum safeguards are necessary to protect an operator? The following recommendations were made for the first industrial laser installation.

(1) The laser head and the workstage shall be completely shielded by a light tight enclosure. All access ports to the enclosure shall be fitted with electrical interlocks to prevent an accidental laser discharge while the ports are open.

(2) The laser and its enclosure shall be housed in a light tight room, so that experimental operation can be safely performed with the light shields removed.

(3) The laser room shall be provided with air conditioning and positive ventilation. To relieve the load on air conditioning equipment, power supplies and pumps should be located outside the room.

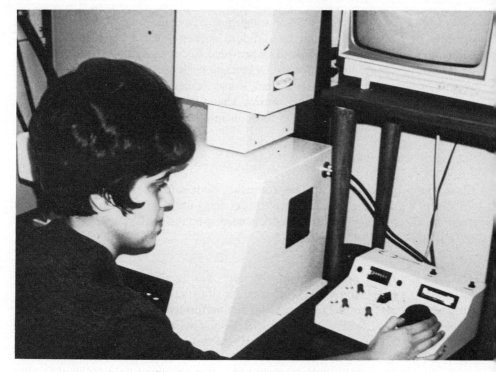

Fig. 9.6. Early pulsed ruby laser drilling system.

(4) The laser enclosure shall be ventilated to draw off any gases or fumes which might be generated by the laser's operation. Baffles shall be used in the exhaust ducts to ensure that no light can be reflected from the enclosure.

(5) Means to visually confirm the well being of the operator in the laser room shall be provided. Voice communication without opening the laser room door shall also be provided.

(6) A high level of illumination shall be provided within the laser room, so that the iris diaphragm of the operators eyes are normally contracted.

(7) The walls of the laser room shall be painted with a light colored very flat paint. Laser light that accidentally hits the walls will be diffused.

To meet the safety requirements, a new light tight, air conditioned, well ventilated, well lighted room was constructed. A second CCTV system and an intercom were installed to provide a means of supervising the well being of the operator who was carefully educated in the safe use of the laser.

Today, a Class I YAG laser drilling system – with a higher repetition rate capability (20/s versus 1/s) has replaced the ruby laser for this application.

Many more industrial production line applications have now evolved. The laser is now used for microwelding, microdrilling, scribing ceramic materials, surface treating, high-speed marking, precision wire stripping, resistor trimming, and integrated circuit manufacture. These types of industrial material processing laser systems now generally make use of a CO_2 laser or a neodymium YAG laser, operating CW or repetitively pulsed. Dependent upon whether the material is to be drilled or welded, a short or longer pulse duration will be employed. The highest CW powers are available from CO_2 lasers; therefore, CO_2 lasers are used for the heavy-duty applications. In any event all of these laser applications present similar hazards. These include viewing the heated material, or exposure to the airborne contaminates produced in this material processing. The amount of toxic material produced by very small units is normally not significant enough to warrant local ventilation. However, in larger units, where significant amounts of material are released, local exhaust ventilation is often needed.

Very high power CO_2 laser systems have entered the commercial market for production welding. These systems most often use dynamic flow CO_2 lasers, and typically have an output power of between 2 kW and 20 kW. This is sufficient for the rapid automated welding of various metal structures including batteries and automotive bodies, and also for cutting aluminum. In most of the production applications, safety policy requires total enclosure of the operation. Both local opaque enclosures of steel or aluminum, and small, room-like enclosures of polymethylmethacrylate (e.g., Lucite or Plexiglas) or similar transparent materials which are opaque to far-infrared laser radiation are found. Plate glass shields would be less suitable since glass would shatter upon beam impact rather than burn.

Ocular and skin hazards associated with laser radiation normally receive the greatest attention in any discussion of laser safety. However, optical hazards in general are not potentially lethal, whereas electrocution is possible under certain circumstances. Although electrical safety precautions have long been standard practice in physics, electronics, and high-power laser laboratories, the large capacitor banks and very high voltage power supplies used in lasers require special attention.

It should be apparent that the normal considerations of floor load and space for operating are tempered by the need to provide for ease of maintenance, and service and the possible safety hazards. In addition some sensitive systems may require vibration isolation or greater electrical (RFI) immunity, see section 3.2.3. There are sufficient applications and documentation (Charschan 1972, Sliney and Wolbarsht 1979, Bellis 1980) to date so that manufacturers should have no problem providing solutions.

4.2.2. Operating and maintenance

Class I or II laser products, which do not permit access to higher levels of laser or collaterial radiation during operating or maintenance, have minimal requirements for floor space. However, access doors may appreciably increase the clearances required for installing the laser product.

Minimal problems with operating or maintenance require assurance that the system suppliers' specification can be met within reasonable bounds in order to produce uniform product with acceptable quality. Normally, meeting voltage and current requirements are not problems, except for fluctuations in the supply where regulation may be required. With the higher power systems an additional cautionary note is required because a reduction in power can prevent ablation of the metal target surface and thereby create hazardous specular reflections.

A similar statement can be made for water coolant, when specified. It is more important to ensure a stable supply; for example: a 59°F specification for a five gallon per minute supply of process cooling water will probably function acceptably at a steady 70°F flow. It is obvious that focusing lenses can be damaged by thermal runaway – with lenses fabricated of semiconductor materials such as Ge, GaAs, ZnSe, and CdTe, as well as by contaminants on the exterior lens surface.

It is most important to ensure that mirror and lens surfaces be carefully and properly cleaned to begin with. A lens tissue dragged through a "puddle" of acetone on the surface does not cause scratches. Next, care should be taken to exclude airborne contaminants, as well as vapor and splatter with particles flying at Mach 2 speeds caused by the laser operation. Air jets perpendicular to the optical axis with sufficient back focal distances for the lens; and/or lens nozzles with concentric jets, at 25 psi (depending on the orifice diameter) have been found to be satisfactory when the air is filtered and water removed from the supply. Nozzles for laser drilling and for cutting are well known in the industry for both standard operation and special needs (Liedtke 1977, Ross 1978).

Class III or IV laser products shall be operated and maintained in a Laser Controlled Area. The requirements for both are listed below.

(A) *Class III laser controlled area should:*
(1) be under the direct supervision of an individual knowledgeable in laser technology and laser safety;
(2) be so located that access to the area by spectators requires approval;
(3) be posted with the appropriate warning sign(s);
(4) have any potentially hazardous beam terminated in a beam stop of an appropriate material; and

(5) have only diffusely reflective materials in or near the beam path, where feasible.

(B) *Laser Controlled Area – Class IV.*
A Class IV laser controlled area shall be designed to fulfill all above items and, in addition, shall incorporate the following safety measures:

(1) safety latches or interlocks shall be used to prevent unexpected entry into laser controlled areas. Such measures shall be designed to allow both rapid egress by the laser personnel at all times ʌd admittance to the laser controlled area under emergency conditions. For such emergency conditions, a "panic button" (control-disconnect switch or equivalent device) shall be available for deactivating the laser.

(2) During tests requiring continuous operation, the person in charge of the controlled area shall be permitted to momentarily override the safety interlocks to allow access to other authorized personnel if it is clearly evident that there is no optical radiation hazard at the point of entry, and if the necessary protective devices are worn by the entering personnel.

(3) All optical paths (for example, windows) from an indoor facility shall be covered or restricted in such a manner as to reduce the transmitted values of the laser radiation to levels at or below the appropriate ocular MPE.

4.2.3. Servicing

It is obvious that poorly trained operators or service personnel could create hazardous situations by improperly cleaning lenses, improper placement of clamps and cooling hoses, misalignment of optical and mechanical parts, and by not setting the proper focus.

With the higher power systems, misalignment of the beam delivery optics can produce the most serious problems. The reduced output power from misalignment can prevent ablation of the metal target surface and thereby create hazardous specular reflections. This should be noted by all developmental and service personnel, as malfunction or reduction in power, in this case, almost uniquely, increases the hazard.

Should removal of panels or protective covers or overriding of interlocks, or both, become necessary (such as for service), and accessible laser radiation thereby exceed the applicable MPE, a temporary laser controlled area should be devised (ANSI). Such an area, which by its nature will not have the built-in protective features as defined for a laser controlled area, shall nevertheless provide for all safety requirements for all personnel, both within and without the temporary laser controlled area.

This may involve the use of protective curtains (Sliney et al. 1981) (or Welder's curtains) to temporarily isolate the area and signs signifying the hazard and recommended action. In some cases, overhead monorails with partitions that can be moved into place have effectively been used to primarily protect personnel in the surrounding areas. It is assumed that the service man is well trained and equipped to deal with the hazard.

Servicing Class IIIb or IV Laser Products which are operated in a Laser Controlled Area, as described earlier is obviously to be performed by a qualified service man in a Laser Controlled Area.

4.3. Examples of a laser installation

A general purpose laser system where heat treating, welding, drilling or scribing can be performed is shown in fig. 9.7. This system has a 500 W CO_2 laser and a 12″ by 12″ moving stage under computer-numeric control. The stage is isolated from the work area by a moving teflon seal to keep the precision mechanical components clean and the work area open. The enclosure itself has a counterbalanced door that easily opens to expose a working area of 50″ by 36″. This makes the system ideal for a wide range of odd shaped parts. A large expanse of plexiglass gives the operator safe viewing of the entire enclosure area. The enclosure is double interlocked for operator safety and BRH compliance.

With microprocessor control of the stages, fast and accurate positioning of the workpiece is available. The microprocessor can also turn the laser power on and off at preset positions or ramp the power for various effects. Additional axes are available under CNC control for rotation of the part, changing the focal length of the lens, control of shuttles or rotating lenses or operating the coaxial gas jet.

4.3.1. Personnel safeguards

In compliance with Federal Regulations (Performance Data for Light Emitting Products) the standard system provides for personnel safeguards. These include (Bellis 1980):

(1) Plasma tube shields. These prevent collateral hazardous emissions from the laser tubes.

(2) Power supply shields. These prevent emission of "soft" X-rays from the high-voltage power supply.

(3) Labels. The BRH requires that a label be affixed to every laser listing its power level and clearly warning of the possible hazard involved.

(4) Emission indicator. A light that goes on near the beam exit when a beam is emitted from the enclosure.

(5) Shutter light. A light provided near the beam exit to show that the shutter is open and a beam is being emitted.

Fig. 9.7. A general purpose laser system. (With permission of Coherent, Inc.)

(6) Time delay. A minimum 10 s time delay is provided in every laser so that if the laser is accidentally turned on, a beam will not be emitted immediately.

(7) Safety beam enclosure. To prevent accidental exposure to the beam while tuning the output optics.

(8) Keylocked controls. To prevent use by unauthorized personnel.

High voltage. The possible danger to personnel is far greater from high voltage than from the laser beam. Appropriate adequate measures have been undertaken to protect the operator or maintenance personnel from high-voltage hazards.

(1) NEMA-style frame and doors. To prevent the machine and operator from external hazards such as water.

(2) Grounding on access panels. This safeguard will automatically ground the laser if the panel is opened.

(3) Total grounding. All system components are electrically grounded.

4.3.2. System safety

(1) High-temperature sensor. To power down the laser when unusual temperatures are encountered.

(2) High-current sensor. To power down the laser when unusual currents are encountered.

The above safeguards apply to the laser itself. Additional safeguards are required when a system is added, to ensure that the operator and maintenance personnel are protected from inadvertent exposure to the laser beam.

(1) Enclosed beam. The laser beam is enclosed by thick-walled tubes from the output of the resonator to the focussing optic.

(2) Work enclosure. An enclosure around the work area ensures that stray beams will not harm passersby.

(3) Interlocks on access panels. The doors into the work enclosure are interlocked to prevent the laser from being on if they are open; a pin prevents the door from being opened when the laser is operating.

(4) Focal optic. The last optical element in a beam delivery is the focusing lens. Beyond the focal point, a beam is diverging and its intensity (power density) reduces as the square of the distance beyond the focal point.

(5) Exhaust system. Many of the materials that are processed give off harmful or toxic fumes when burned. An exhaust system removes these fumes from the operator's environment.

In accordance with BRH requirements the laser supplier provides complete manuals on the laser and system. These manuals detail the operation of the entire system, preventive maintenance schedules, parts lists, recommended spares, and drawings necessary for customer attendance to the system.

4.3.3. Installation requirements

Assistance should be available by the laser supplier for advice before installation as well as at the final hook-up to power and water supplies. We recommend that on all large systems the supplier should be present to turn on the system, check mode and power, and perform general acceptance tests as described in section 3.

Electrical. 208 VAC/3 Phase, 30 A per phase, 12 kW average. In most cases the manufacturer will supply alternatives such as 240 or 440 VAC.

Gas Supply. Pre-mixed gas – for welding: $13\frac{1}{2}\%$ nitrogen, $4\frac{1}{2}\%$ carbon dioxide, 82% helium.

NOTE: this is the recommended mixture only for laser welding. A reputable supplier is required. Striation of the mixed gasses, if stored over a period of time, may be a problem. The lasers are optimized with different percentages of the gases for drilling and cutting, and the manufacturer should supply such information. With needs that vary, mixing valve systems may prove more suitable. Inert gases or oxygen may also be required for special welding or cutting applications.

Water coolant. 59°F temperature, 20 psi, 5 gallons per minute. Such systems have operated satisfactorily with process cooling water provided as high as 75°F provided the outgoing temperature was under 80°F, and the temperature was maintained steady. Higher temperatures caused system shutdown, see section 4.3.2.

Plant process cooling water, especially with an open cooling tower, has been known to provide problems with argon lasers. Sediment and algae buildup cause subsequent insufficient cooling.

There are other coolant requirements, such as lamp coolant for solid state lasers, where an individual "refrigerated, heat exchanger" is needed to circulate chilled DI water at 59°F. In these lasers the temperature should be kept below 70°F. In most cases the laser supplier should specify the needs and provide the necessary facilities.

Floor space. Overall size: 180″ long, 69″ high, 37″ wide. Each side requires 24″ clearance for maintenance and service purposes. In addition the gas bottle or tank storage requirements need be considered. In front at the work area, space for an operator – either sitting or standing – needs to be provided, and material handling or storage needs require consideration.

Further, the probability of servicing the system in situ dictates the need for providing facilities for a temporary or permanent laser controlled area – as described in sections 4.3.2 and 4.2.3.

Miscellaneous. Weight (2800 pounds), sensitivity to vibration, and exhaust system are additional plant engineering considerations that depend on the nature of the intended laser operation.

Common Sense. After the system is installed, users should treat it with the respect that is accorded any other piece of industrial machinery.

5. Spare parts, maintenance and field service

When purchasing a laser system, the user also should consider what will be involved in maintaining it in proper working order. The following sections deal with the questions of keeping a spare parts inventory, routine maintenance during and after the warranty period, and the availability of field service contracts.

5.1. Spare parts inventory

Experience has shown that certain parts of a laser system will generally require replacement either due to wear or failure, while other parts will last indefinitely and will require no attention throughout the life of the system. Knowledge of which parts are most prone to wear or failure may be based upon:

(a) whether or not specific parts are immersed in coolant fluid;
(b) mean time to failure for electrical circuit elements;
(c) whether or not parts must dissipate heat; and
(d) whether or not parts are subject to laser beam itself.

Nearly all those parts which will eventually require replacement will fall into one or more of the above categories. In the first case corrosion and exposure to contaminants in the primary coolant bath is largely responsible for long term degradation of surface area. In the case of optically pumped lasers this may cause a substantial reduction of the reflectance of the water-cooled surfaces which comprise the active pumping chamber. This should be noticeable, however, since reduction in laser output power normally accompanies such surface degradation and therefore warns of its presence. A useful addition to any water-cooled laser system is a pH meter to indicate electrical conductivity and to remind the operator of the need for water change when such conductivity falls beyond an acceptable level.

Failure of electrical circuit elements is generally attributable to three causes, namely:

(a) component ageing;

(b) predictable lifetime; and

(c) premature breakdown.

While the first two causes can be anticipated, the last cannot, and is the most troublesome type. In any case, a spare parts inventory will be required. The chief items to be stocked are those which are unique and obtainable only from the original manufacturer. Such items include printed wiring boards made especially for a particular laser power supply and not available elsewhere. Other items include special capacitors and semiconductor switching devices which have predictable failure rates based upon total number of firings. Also, special hydraulic pumps and thermal sensors which have known service life may be stocked as a precaution against down time. In all, a nominal figure of 10% of original capital expenditure can be expected over the initial five year life of the laser system.

5.2. Field service contracts

Most laser manufacturers offer a one-year warranty on all parts with the exception of expendible items such as flash lamps, water filters, mirrors and additional coated optics. Beyond the first year, some vendors offer annual field service contracts while others offer either limited special service arrangements or none at all. The extent of such availability or lack thereof depends to a great extent upon the size of the vendor's company, and distance from his site location.

For the case where a field service contract is available, it is advisable to purchase such a contract before the first year warranty period has expired. This ensures both parties that the laser is in proper running condition at the onset of the contract period. Otherwise additional assurance that the laser user has not altered or abused the laser may complicate terms of the contract. The contract itself will usually be limited to service of only the laser, and usually will exclude work handling equipment, lenses, mirrors and other interface equipment between the basic laser and the target workpiece. Service contracts also exclude power and water supply, and generally stipulate that the user will perform periodic maintenance, system calibration, cleaning, and lamp and filter replacement.

Those companies which offer limited service contracts are usually restricted by their overall size and employee roster, such that special attention may be given to certain key customers, but overall general service contracts would be financially prohibitive. Furthermore, proximity to their plant facilities will usually determine the extent to which such special attention is available.

Those companies which offer no contracts at all comprise a minority, and the future potential user would be well advised to consider this from the outset when planning extensive future installations. Certainly those users who recognize this situation and have their own in-house laser maintenance and repair facilities may not be adversely affected by it. However, the others less familiar with the nature of lasers may wish to consider such unavailability of field service contracts in their initial vendor selection.

5.3. Performance specifications

How well a laser system accomplishes a given material processing application depends upon how well the process was understood and how well the laser parameters were tailored to achieve desired results. Assuming that an extensive series of feasibility tests was conducted, successful operation of the laser system then depends upon how well design criteria were carried out. Furthermore, it is assumed that the best choice of laser and optimized mechanical feed and parts handling equipment was properly interfaced into the complete system. This takes into account the following.

5.3.1. Duty cycle
How many piece parts per hour and per year must be produced? How many lasers are required to achieve this (allowing for both periodic maintenance, and unscheduled down time)?

5.3.2. Average power or total energy
Based upon computed laser power requirements, what is the smallest laser which can perform the given task? How much better could this same task be performed by multiples of this type as compared with a larger one operating at greater speed and capacity? What are the financial trade-offs versus the down time of one large laser?

5.3.3. Beam delivery
Assuming that piece parts are to be fed to the laser target region at a given rate, the chief question is what movable elements must be incorporated into the system. The following possibilities exist.

(a) Laser remains stationary and parts handling equipment performs all mechanical motion.

(b) Both laser and piece part remain stationary after coarse positioning of multiple gauged cluster of parts, and moving optics delivers beam to individual targets.

(c) Parts are constantly moving, laser is stationary and fires beam "on-the fly" at the command of an external trigger. (Source of this external trigger may be pre-determined, or derived from optical or proximity detector which finds the target.)

(d) Laser is stationary, but combination of moving optics and moving parts stage is programmed to fire at target by a microprocessor which knows both target and beam coordinates.

(e) Laser and target are both stationary, while beam is delivered by means of fiber-optic delivery system.

(f) Entire laser is moving with parts secured to fixed platten.

Certainly, other combinations also exist, but the above list was prepared to suggest the overall flexibility of possible means for directing the beam to the target. Which alternative works best depends upon the overall physical size and complexity of the target and how massive piece parts are compared with the overall size of the laser system. However, one alternative is bound to be simpler than the others for a given application.

5.3.4. Precision

When piece parts are fed to a laser processing system, the degree of accuracy to which the laser beam fires at the correct target position involves the question of precision. The tolerance placed upon hitting a target accurately should be specified with respect to a set of translational, rotational and angular coordinates. This will require specifying:

(a) laser beam alignment and positional stability;

(b) mechanical feed dimensional tolerances; and

(c) beam steering accuracy and repeatability.

Proper choice of design for parts handling equipment must take into account the dimensional tolerances of the piece parts themselves, and the buildup of positional errors due to all parts of the parts handling equipment. During feasibility studies on prototypes, estimates must be made with regard to the precision required when piece parts are ultimately processed at actual production speeds.

5.4. Performance evaluation

Whenever possible, demonstrated performance of a laser system should take place at the vendor's plant facility prior to shipment to the user's location.

This is particularly important if the laser differs in any way from the manufacturer's standard model. Verification of performance and compliance with written specifications prior to final delivery involves a step-by-step procedure which is normally followed by the reputable manufacturer in completing laser system construction, and usually involves nothing more than repetition of this procedure in the presence of the eventual user. In fact, the good will and reputation of the vendor will be largely enhanced by suggesting such a visit by the user prior to shipment simply to reassure him that all matters relating to the performance have been completely tended to.

In addition to system checkout prior to shipment, a vendor's field service engineer should perform final hookup and prove-in of the laser system at the user's facility before final written acceptance takes place. At such time, the field service engineer will verify that the system has been properly installed, that safety regulations have been complied with, and that correct input water supply, electrical lines and operating environment have been installed. Failure to comply with the vendor's recommended operating conditions could void laser system warranty, and again it represents good will in avoiding any subsequent misunderstanding by having both vendor and user agree from the outset that the system was at least initially in proper working order.

6. Conclusion

Today laser processing is a competitive technology offering many advantages. Multifaceted views have been given in this chapter to confirm that large capital expenditures such as may be required for many laser systems can be justified with standard techniques useful for most machine tools.

In addition guidance is provided for obtaining a reliable system as well as how to keep it running. Clarification is also provided on training requirements for laser personnel as well as facilities required for safe operation.

We trust this chapter provides the practical answers required to turn innovative concepts into production reality; and that the application explosion we see now continues to provide new challenges and new solutions.

References

American Conference of Government Industrial Hygienists, A Guide for Control of Laser Hazards (ACGIH, PO Box 1937, Cincinnati, Ohio 45201).
ANSI, American National Standards Institute, 1430 Broadway, New York, NY.

Bellis J., ed., 1980 Lasers-Operation, Equipment, Application, and Design (McGraw-Hill, Book Company, New York).

Charschan S.S., 1972, ed., Lasers in Industry (Van Nostrand Reinhold, New York).

EOSD, Vendor Selection Issue (Milton S. Kiver Publishing, Inc., Chicago, Ill.).

Epperson J.P., Dyer R.W. and Gryzawa J.C., 1966, The laser now a production tool, The Western Electric Engineer, **10** No. 2.

Kaplan R.A. et al., 1979, Electrochemical Society Meeting, Los Angeles.

Laser Focus, Buyers' Guide (Advanced Technology Publications, Inc., Newtown, Mass.).

Laser Institute of America (a), LIA Laser Safety Guide.

Laser Institute of America (b), Laser Safety Training Package (80-slide set with audio-taped cassette on industrial laser safety) (available from Laser Institute of America).

Liedtke H.G., 1977, Laser Drilling Nozzle, Patent #4027137, May 31.

Muniz J.M. and Califano V.E., 1978, High speed microprocesser based pulsed laser welding system, Society of Manufacturing Engineers, Lasers in Modern Industry, 1971, Boston, Mass.

Optical Spectra, Product Table Supplement (The Optical Publishing Co., Inc., Pittsfield, Mass.).

Ross W.A., 1978, Apparatus and Method for Laser Cutting, Patent #4125757, November 14.

Siegman A.E., 1971, An Introduction to Lasers and Masers (McGraw-Hill, New York) p. 330.

Sliney D.H. and Wolbarsht M.L., 1980, Safety with Lasers and Other Optical Sources, A Comprehensive Handbook (Plenum Press, New York).

Sliney D.H. et al., 1981, Semi Transparent Curtains for Control of Optical Radiation Hazards, US Environmental Health Agency, February.

Smith J.F., Murphy J.J. and Eberle W.J., 1975, Industrial Laser Safety Program Management, IBM Corp., System Products Division, Poughkeepsie, NY.

US Department of Health, Education and Welfare, 1979, Code of Federal Regulations, Title 21, Chapter 1, Subchapter J, Part 1040, Performance Standards of Light-Emitting Products.

SUBJECT INDEX